Using MATLAB,
SIMULINK and
Control System Toolbox

A practical approach

Alberto Cavallo
Roberto Setola
Francesco Vasca

Prentice Hall

London   New York   Toronto   Sydney   Tokyo   Singapore
Madrid   Mexico City   Munich

First published in English 1996 by
Prentice Hall Europe
Campus 400, Maylands Avenue
Hemel Hempstead
Hertfordshire, HP2 7EZ
A division of
Simon & Schuster International Group

English language edition arranged through the mediation of
Eulama Literary Agency.

Printed and bound in Great Britain by
Ashford Colour Press Ltd, Gosport, Hampshire

Library of Congress Cataloging-in-Publication Data

Available from the publisher

British Library Cataloguing in Publication Data

A catalogue record for this book is available from
the British Library

ISBN 0–13–261058–2

2 3 4 5   00 99

Contents

1 Introduction   1

I MATLAB   5

2 Fundamentals   7
  2.1 Keyboard input   7
  2.2 Input from external files   9
  2.3 Utility commands   13
  2.4 Exercises   14

3 Vector and matrix manipulation   15
  3.1 Elements of a matrix   15
  3.2 Interval representation   17
  3.3 Matrix manipulation operations   18
  3.4 Special matrices   19
  3.5 Character strings   20
  3.6 Advanced matrix manipulation techniques   21
  3.7 Exercises   21

4 Scalar operations   25
  4.1 Arithmetic operations   26
  4.2 Elementary and transcendental functions   26
  4.3 Examples   30
  4.4 Logical and relational operators   31
  4.5 Exercises   32

5 Matrix operations   **35**
  5.1 Transpose   35
  5.2 Algebraic operations   36
  5.3 Matrix functions   37
  5.4 Logic operations   40
  5.5 Complements on matrices   42
  5.6 More special matrices   50
  5.7 Exercises   51

6 Polynomials   **58**
  6.1 Basic operations   58
  6.2 Interpolation   61
  6.3 Exercises   63

7 Graphics   **66**
  7.1 2D graphics   66
  7.2 Multiple plots   70
  7.3 Axis scaling   73
  7.4 Complex data   73
  7.5 More 2D graphs   74
  7.6 Plotting more graphs on the same window   75
  7.7 3D graphics   76
  7.8 Functions of two variables   77
  7.9 Color in MATLAB   79
  7.10 Parametric plots   83
  7.11 Revolution surfaces   84
  7.12 Printing graphics   84
  7.13 Exercises   87

8 Programming in MATLAB   **91**
  8.1 Basic programming structures   91
  8.2 Script files   93
  8.3 Functions   94
  8.4 Debugging programs   99
  8.5 Exercises   103

9 Numeric analysis   **111**
  9.1 Infinitesimal analysis   112
  9.2 Nonlinear equations and optimization   113
  9.3 Differential equations   115
  9.4 Exercises   117

II   SIMULINK   **121**

10 Fundamentals   **123**
   10.1 Building a SIMULINK scheme   124
   10.2 Analysis of the scheme   127
   10.3 Exercises   129

11 SIMULINK schemes   **131**
   11.1 Using the mouse   131
   11.2 Structural properties of the blocks   134
   11.3 A nonlinear example   136
   11.4 Sources and sinks   139
   11.5 Some nonlinear blocks   142
   11.6 Exercises   145

12 Simulation in SIMULINK   **153**
   12.1 Simulation problems   153
   12.2 Integration methods   157
   12.3 System with discontinuities   159
   12.4 Time-varying systems   161
   12.5 Exercises   163

13 Multi-variable systems   **167**
   13.1 Multi-variable schemes   167
   13.2 Connections   170
   13.3 An example: a planar manipulator   172
   13.4 Exercises   174

14 Group operation   **179**
   14.1 Using the group option   179
   14.2 Exercises   182

15 Discrete-time systems   **185**
   15.1 An example: the repopulation of a lake   185
   15.2 Exercises   188

16 Hybrid systems   **193**
   16.1 Hybrid schemes   193
   16.2 Exercises   196

17 Advanced topics   **198**
   17.1 System analysis   198
   17.2 Using Mask   203
   17.3 Customizing SIMULINK   207

17.4 Exercises ... 207

18 Blocks library ... **209**
18.1 Sources ... 209
18.2 Sinks ... 213
18.3 Discrete ... 216
18.4 Linear ... 219
18.5 Nonlinear ... 222
18.6 Connections ... 229
18.7 Extras ... 230

III   Control System Toolbox ... **235**

19 Models for LTI systems ... **237**
19.1 LTI system representations ... 237
19.2 Conversions among representations ... 242
19.3 Continuous to discrete conversion ... 247
19.4 Controllability and observability ... 254
19.5 Commands for other properties ... 258
19.6 Model order reduction ... 261
19.7 System connections ... 265
19.8 Exercises ... 270

20 Time domain response ... **278**
20.1 Unforced response ... 280
20.2 Step response ... 281
20.3 Impulse response ... 283
20.4 Response to any input ... 284
20.5 Exercises ... 285

21 Frequency domain response ... **289**
21.1 Bode plots ... 290
21.2 Nyquist plots ... 292
21.3 Nichols plots ... 293
21.4 Gain and phase margins ... 297
21.5 Exercises ... 300

22 Root locus ... **306**
22.1 Root locus plot ... 306
22.2 Exercises ... 310

23 State feedback   **313**
   23.1 Feedback gain design   314
   23.2 The observer   315
   23.3 Linear quadratic optimal control   317
   23.4 Kalman filter   321
   23.5 Separation principle   324
   23.6 Exercises   324

24 Discrete functions   **328**
   24.1 Discrete to continuous conversion   328
   24.2 Properties   329
   24.3 Model reduction   331
   24.4 Time domain response   331
   24.5 Frequency domain response   334
   24.6 State feedback design   339
   24.7 Exercises   340

25 Utility functions   **343**
   25.1 Modeling   343
   25.2 Time response   343
   25.3 Frequency response   344
   25.4 Lyapunov and Riccati equations   345

IV   Appendices   **347**

A Advanced graphic functions   **349**
   A.1 Graphic objects in MATLAB   349
   A.2 Root   353
   A.3 Figure   354
   A.4 Axes   355
   A.5 Line   358
   A.6 Surface   360
   A.7 Image   361
   A.8 Text   361

B Graphic user interface design   **363**
   B.1 Controls   363
   B.2 Menu   376
   B.3 Using the mouse   379

C S-function   **386**
   C.1 Writing an S-function   386
   C.2 Continuous-time systems   389

C.3   Discrete-time systems   390
C.4   Hybrid systems   391
C.5   Observations   392

References   **393**

Index   **397**

Chapter 1

Introduction

MATLAB software has become more and more popular in all engineering fields so that, today, it can be considered as the world standard for the simulation and analysis of linear and nonlinear dynamic systems, and as the most versatile numeric analysis toolbox. The name MATLAB is an acronym for MATrix LABoratory. This immediately suggests two considerations: its basic elements are matrices, so that it is naturally devoted to linear algebra applications; moreover the word 'laboratory' conjures the idea of *work in progress*, i.e. emphasizes its ability for educational and research purposes. SIMULINK is a MATLAB toolbox designed for the dynamic simulation of linear and nonlinear systems as well as continuous- and discrete-time systems. It combines a graphic user friendly interface with an advanced integrator engine which allows the user to choose between six differential equation integrator methods. The Control System Toolbox is a set of MATLAB functions which together form a library of almost all the procedures and methods used in classic and modern control theory.

This book, which is a manual for MATLAB, SIMULINK and Control Toolbox, is aimed at undergraduate and graduate students and at those academic and industrial researchers who work with dynamic systems and numerical analysis problems.

The methodological approach used in the book is different from that used in manuals and other MATLAB-based books. The distinguishing feature of the volume is its large number of worked examples and exercises. These, along with careful organization of the topics discussed, enable the reader to proceed from the basic MATLAB commands up to the more sophisticated Control System Toolbox procedures and to the optimized SIMULINK scheme, avoiding a boring and useless list of functions.

Advanced techniques are presented throughout the book in small typeface, and these can be skipped at a first reading.

The book contains twenty-four chapters and three appendices. MATLAB is described in the first eight chapters, SIMULINK in the following nine chapters and the Control System Toolbox in the final seven chapters.

1

The part devoted to MATLAB assumes that the reader has no familiarity with the package. In fact, Chapter 1 begins by explaining how to insert data from the keyboard and external files. It introduces the reader to the basic concepts of scalars, vectors and matrices and their treatment in MATLAB, it describes some utility commands (such as `help`), and it presents advanced topics such as handling formatted files and changing data formats.

Chapter 2 introduces MATLAB commands to generate interval vectors, to build up matrices from submatrices, to extract blocks from matrices; moreover some special matrices and character string variables are described in this chapter.

Operations on scalars and matrices are discussed fully in Chapters 3 and 4; arithmetic, transcendental and logical operators are presented for scalar functions, while matrix operations cover transpose, algebraic operations, sorting, seeking for largest and smallest entries and the important command `find`. More sophisticated mathematical concepts (singular values decomposition, LU factorization, etc.) can be found in the advanced topics sections.

Polynomials are discussed in Chapter 5: how to compute their roots, values at a given set of points, derivatives and partial fraction expansion. A section dealing with interpolation concludes the chapter.

Chapter 6 is devoted to graphics. This chapter deals with two-dimensional and multiple graphics, axis scaling, representation of complex data and histograms. The second part of the chapter addresses 3D graphics, including a discussion on the use and definition of color maps. The final section describes how to print MATLAB graphics. A thorough knowledge of the material presented in Chapters 3 and 4 is required in order to fully appreciate the powerful graphic capabilities of MATLAB; this chapter is, moreover, closely linked with the first two appendices.

The MATLAB programming environment is described in Chapter 7, along with the debugging functions introduced since the release 4.0.

Finally, Chapter 8 illustrates how the ODE integrators work, how to find zeros of functions, compute derivatives and definite integrals and how to solve simple optimization problems.

The SIMULINK toolbox is presented in the second part of the book. Chapter 9 describes how to build a SIMULINK scheme and how to run the simulation.

In Chapter 10 the basic elements of the SIMULINK scheme design are discussed. Particular attention is given to mouse operations, to block properties, 'Sources' and 'Sinks' blocks and the most commonly used nonlinear blocks.

In Chapter 11 the six SIMULINK integrator methods are presented and the integration problems are summarized via suitable examples. Moreover, the reader is shown how to simulate discontinuous, time-varying and time-delayed systems.

Chapter 12 is devoted to the design of multi-input multi-output schemes in SIMULINK.

The 'Group' operation is investigated in Chapter 13. This important SIMULINK feature allows the user to *group* SIMULINK schemes into a single block which may in turn be used to build up more complex schemes.

In Chapters 14 and 15 discrete-time and hybrid-time schemes are presented respectively. It is pointed out how all the synchronization problems are handled by SIMULINK.

The advanced SIMULINK functions are collected in Chapter 16. The reader is shown how to compute the linearized model, how to find trim points and how to discretize a given system. Moreover, the more sophisticated *group* function 'Mask' is presented.

Finally, in Chapter 17 all the SIMULINK blocks are presented in brief, along with their parameters and some hints about their use.

The third part of the book describes how to use the Control System Toolbox for analysis and control design of linear time invariant (LTI) systems. Many exercises encourage the reader to also use SIMULINK schemes for a complete and fast understanding of the behavior of the system considered.

Chapter 18 deals with the possible MATLAB representations for LTI systems, both single and multi-input/output. Conversion from continuous-time to discrete-time models and the inverse conversion are also considered and some theoretical considerations that are required are recalled. The procedures for detecting controllability and observability properties, and for achieving model reduction are detailed by means of examples drawn from practical electrical and electromechanical systems.

In Chapter 19 the problem of computing the time domain response of an LTI system is considered.

The procedures available for frequency domain analysis are discussed in Chapter 20. Examples dealing with the use of Bode, Nyquist and Nichols plots for analysis and controller design help the reader to understand the use of the related commands.

The whole of Chapter 21 is devoted to the tools for the root locus representation and its use in control design.

The Control System Toolbox commands for state feedback control techniques, observer and Kalman filter design are detailed in Chapter 22.

Discrete-time commands are presented in Chapter 23, whereas Chapter 24 concludes the third part of the book by presenting some of the utility commands of the toolbox.

The first appendix deals with Handle Graphics Objects, the second with the design of a graphical user interface (GUI) and the third with the 'S-function', i.e. the SIMULINK script file.

The book has been typeset in LaTeX by the authors.

Part I

MATLAB

Chapter 2

Fundamentals

The aim of this chapter is to give the reader a first glance at the MATLAB environment. To accomplish such a task, simple input and data-storing operations will be presented. Before proceeding further, the reader should know how to quit a MATLAB session: it is sufficient to type quit, or exit, or, as in any Windows-based package, to use the key combination **Alt–F4**.

As stated in the Introduction, the basic element in MATLAB is the matrix: indeed, a scalar variable is actually a 1×1 matrix, while vectors are simply either $1 \times n$ or $n \times 1$ matrices.

Matrices can be entered in four different ways:

1. from the keyboard, as a sequence of numbers;
2. loaded from external files;
3. generated by built-in functions;
4. generated by M-files (subroutines and user-defined functions).

In this chapter we will address only points 1 and 2, the other points being discussed in following chapters.

Actually, there is one more way to generate matrices: they can be the result of a MEX-file, that is, a C or FORTRAN procedure, compiled and linked in such a way as to be embeddable in the MATLAB environment. However, we will not discuss this point in the book. The interested reader can refer to the '*External Interface Guide*' booklet in the set of MATLAB manuals.

2.1 Keyboard input

Throughout this section we will use the symbol ⟨CR⟩ to denote pressing the Carriage Return (or Enter) key. This convention will be abandoned from the following section on, when the reader has hopefully acquired the minimal acquaintance with the program that makes this explicit indication redundant.

A scalar variable is defined in an obvious way:

$$A=2 \ \langle CR \rangle$$

gives the variable 'A' the value '2'. The variable is stored in the computer memory until explicitly deleted with the command `clear A`; `clear A B C` deletes the variables `A, B` and `C`; `clear` alone deletes all the variables in the current *workspace* (the computer memory). By the way, note that MATLAB variable names are case sensitive, i.e. A and a are two **different** variables.

Entering a vector from the keyboard is a very simple task. It is sufficient to surround a list of numbers separated by a blank character or a comma with square brackets, as we do when writing by hand:

$$A=[1 \ 2 \ 3 \ 4] \langle CR \rangle \quad \text{or} \quad A=[1,2,3,4] \langle CR \rangle$$

defines the vector A; note that we have not predefined the size of the vector, we have simply entered the values, without any preliminary declaration. This is one of the most attractive characteristics of MATLAB, and will hold even for matrices.

To define a matrix we need a character to 'start a new line': this character is ';' (the semicolon):

$$A=[1,2,3,4;5,6,7,8] \ \langle CR \rangle$$

defines the matrix

$$A = \begin{pmatrix} 1 & 2 & 3 & 4 \\ 5 & 6 & 7 & 8 \end{pmatrix}$$

A more intuitive way of achieving the same result is the use of the carriage return key $\langle CR \rangle$:

$$A=[1 \ 2 \ 3 \ 4 \ \langle CR \rangle$$
$$5 \ 6 \ 7 \ 8] \ \langle CR \rangle$$

Now we address the problem of defining a vector again; if we need a column vector, one possible way is

$$A=[1;2;3;4] \ \langle CR \rangle$$

A different strategy is to transpose a row vector. The transpose operator in MATLAB is the apostrophe '; then the previous result can be obtained with the following line

$$A=[1 \ 2 \ 3 \ 4]'; \ \langle CR \rangle$$

Note the use of the semicolon character ';'. Obviously in this case its role is no more to start a new line: indeed, at the end of a command the semicolon is used to avoid displaying the result of the command on the screen. The reader could consider this feature redundant, and, indeed, at this point in our discussion, to have the results of our operations given on the screen may be better, but later we will present commands generating high-dimension matrices, for instance 1000×1000, so that to avoid the fruitless scrolling of $1\,000\,000$ figures on the screen becomes crucial.

By the way, when dealing with high-dimension vectors it is useful to know how to continue any command on the next line. For this purpose we use three dots (ellipses):

$$A=[1\ 4\ 5\ 3\ 6\ 87\ 3\ 4\ 8\ \text{-6}\ 33\ 5];\ \langle CR \rangle$$

and

$$A=[1\ 4\ 5\ 3\ 6\ 87\ \dots\quad \langle CR \rangle$$
$$3\ 4\ 8\ \text{-6}\ 33\ 5];\ \langle CR \rangle$$

give the same result.

2.2 Input from external files

Let us suppose that a Pascal, FORTRAN, BASIC or C program has produced the ASCII file 'IN.DAT' with the following data:

```
1.00     2.01   3.98    9.67   -2.89
4.12   -12.65   7.99   23.43    0.01
```

In such a case the command **load IN.DAT** reads the file and stores the data in a variable with the same name as the file (without extension): in our case the result will be

$$\text{in} = \begin{pmatrix} 1.00 & 2.01 & 3.98 & 9.67 & -2.89 \\ 4.12 & -12.65 & 7.99 & 23.43 & 0.01 \end{pmatrix}$$

The dual command is **save** *FileName VarName*, which stores the variable *VarName* in the file *FileName*.MAT; **save** *FileName* stores all the variables of the current workspace in the file *FileName*.MAT and **save** alone stores all the variables in the default file MATLAB.MAT. Finally, if we want to save more variables (but not all of them) on a single file we can resort to the most general form, which uses blank characters to separate the variables to save:

save *FileName VarName1 VarName2 VarName3 ...*

It is important to point out that **save** stores the variables in MATLAB format, i.e. the variables are then readable only by MATLAB. The advantage is that such a file is more compact, and is read more quickly by MATLAB; moreover, it allows retention of the identity of the variables: we have seen above that the whole ASCII file IN.DAT is stored in the single variable **in**; thus we should store each variable in a single ASCII file, with obvious disadvantages for disk space and organization, or we should 'unpack' the different pieces of information from a single matrix variable.[1] When dealing with a MATLAB formatted file this problem is solved automatically: **load** *FileName* (note that the extension is not needed, MAT is assumed by default) loads the variables stored in *FileName*.MAT in the current workspace, while retaining the variable names.

In the case where we want to save the variables in ASCII format in order to read them from outside MATLAB, for instance with a word processor or from other programs, the option **-ascii** must be used:

$$\text{\textbf{save} } \textit{FileName.Extension VarName } \texttt{-ascii}$$

The further option **-double**, stores data in double-precision format; finally, **-tabs** separates the columns with a tabbing character.

The reader should bear in mind that, in any case, whichever screen display format, MATLAB works with double-precision data.

The remaining part of this section, as with all material in smaller font throughout the book, is more advanced and may be skipped at a first reading.

We have just remarked that MATLAB internal data representation is always in double-precision format. Some comments about screen data representation are now in order. By default, only the first five digits of a number are visualized. The command **format** allows us to change this default according to the following table

`format short`	5 digits, fixed point
`format long`	15 digits, fixed point
`format short e`	5 digits, floating point
`format long e`	15 digits, floating point
`format hex`	hexadecimal
`format bank`	2 digits after point
`format rat`	approximated to a fraction
`format +`	only $+$, $-$ and blanks

Moreover, **format compact** displays matrices in a more compact way, by suppressing extra line feeds, while **format loose** returns to the default situation (rows of a matrix are separated by a line feed). Remember that in no case is the internal data representation influenced by this command. **format** may be typed from the keyboard or selected from the 'Options' item in the MATLAB window command menu (Figure 2.1).

---

[1] We will discuss the 'unpacking' commands in Section 3.3.

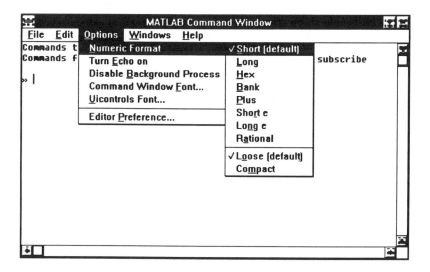

Figure 2.1 The format menu.

We conclude this section with a brief description of some low-level functions for input/output operations. Actually these are C procedures whose main feature is the ability to work with binary files; but this topic would take us too far from the context of this book, so the interested reader is advised to refer to the manuals. We will focus our attention only on some functions for handling formatted ASCII files; the way these functions work will be illustrated by an example that may occur frequently when dealing with the problem of using MATLAB to process data to be acquired from other software packages.

Suppose we are given three ASCII sequences of values resulting from any program and stored in the file RES.OUT according to the following structure:

```
Execution successful.
   The results are:
   x1    x2    x3
  0.0   3.2   4.3
  0.1   2.8   4.1
  0.2   5.9   4.0
  0.3   1.3   3.9
   ⋮     ⋮     ⋮
```

To read the numerical values, MATLAB must ignore the standard sentences 'Execution successful.', 'The results are:' and the three variable names 'x1', 'x2' and 'x3'. The `load` command could not solve this problem; thus, according to our present knowledge of MATLAB, the only solution would be to use a text editor to delete the alphabetic characters and then to use load. The use of low-level input/output functions gives a more efficient solution as follows.

First, the file must be opened, indicating its name and the operation to be performed on it ('r' to read data, 'w' to write data, 'a' to append data, i.e. write data at the end of an existing file without losing previous data, and 'r+' for both read and write operations). The following line:

$$fid=fopen('res.out','r')$$

opens the file RES.OUT for reading operations and returns the pointer fid to it. Next, we must skip (read and ignore) eight words and then read the numerical data. The command to use is fscanf, which reads formatted files; its syntax is

$$A=fscanf(fid,'format')$$

The meaning of this command is: read the data in the file pointed by fid with the format specified in the string 'format' and store the result in the variable A. The formats are those defined in standard C:

- '%s' for sequences of characters up to the first blank;

- '%d' for decimal numbers;

- '%e', '%f', '%g' for floating point numbers.

To read (and skip) the initial words we must use the format '%s'; moreover, we know that we must read exactly eight words, then we may use the further option of the same command:

$$A=fscanf(fid,'%s',8)$$

reads the first eight words and stores the result in A; the pointer is advanced by eight positions and placed at the the beginning of the next word (the figure 0.0); now we use the format '%f' to interpret the following data as numerical values, and store the result in a matrix with three rows and an unknown number of columns:

$$A=fscanf(fid,'%f %f %f',[3 Inf])$$

Finally, we transpose the matrix to obtain three columns and close the file. The latter operation is done by the command

$$status=fclose(fid)$$

To close all the files we use

$$status=fclose('all')$$

The value of the variable status is 0 in the case of success, -1 if some failure happens. The whole procedure given is here for the sake of completeness:

```
fid=fopen('res.out','r')
fscanf(fid,'%s',8);
A=fscanf(fid,'%f %f %f',[3 Inf]);
A=A';
status=fclose(fid)
```

As for the operators fread, fwrite, fseek, ftell, the reader can refer to the manuals.

2.3   Utility commands

One of the most useful commands in MATLAB is `help`, which returns an on-line MATLAB manual. If we type `help` with no parameters a list of the main MATLAB topics appears on the screen, while `help` *TopicName* returns a list of functions related to the topic and `help` *FunctionName* describes how the specific function works. For instance, let us type `help`: the result will be

```
HELP topics:

c:\matlink        - Establish MATLAB session parameters.
matlab\general  - General purpose commands.
matlab\ops        - Operators and special characters.
matlab\lang       - Language constructs and debugging.
matlab\elmat     - Elementary matrices ...
matlab\specmat  - Specialized matrices.
matlab\elfun      - Elementary math functions.

    .
    .
    .

For more help on directory/topic, type "help topic".
```

Then, typing `help elfun`, the list of the elementary mathematical functions (which we will discuss in Chapter 4) will appear. Finally, the result of `help sqrt` will explain how to use the function `sqrt` (square root).

Some standard DOS commands are available in MATLAB. For instance `cd`, `dir` changes the current directory and returns the files in the current directory, respectively; **type** *FileName* is mainly oriented to M-files (see Chapter 8) and prints the (ASCII) file *FileName*.M on the screen; the DOS-like command **more on** avoids scrolling on the screen. The current path for MATLAB files is visualized with the command **path**. **what** is similar to `dir`, but visualizes only .M and .MAT files in the current directory. A sort of dual command is **which** *FunctionName*, which displays the directory to which the function *FunctionName*.M belongs. Finally, the 'arrow up' key ($\uparrow$) allows quick data input: when pressed it returns the previously typed instructions; it may also select different commands: if we type a character (or a group of characters) and then arrow up, only the previous commands beginning with that character (or group of characters) are visualized. Here is an example. Type

```
a=5;
b=1;
c=[1 2 3]';
t=[0 1;1 0];
```

Now press $\uparrow$: the instruction defining the variable t is displayed, then that defining c, and so on backwards. But if we type a$\uparrow$ then `a=5;` appears directly.

The MATLAB analogy of the DOS command CLS is clc, which clears the screen (its meaning is CLear Current screen). This command does not delete the current workspace. By the way, we have discussed the operator clear to delete variables from the workspace, but this operation does not necessarily imply a reorganization of the memory space. In long MATLAB sessions with a large computational burden it may be effective to use the command pack periodically, which saves the current workspace on a temporary file, empties the RAM from variables and compiled functions and reloads data from the temporary file, thus executing a 'garbage collection' to optimize memory management.

2.4   Exercises

Exercise 2.1 Input the following vectors:

$$a = \begin{pmatrix} 3 & 4 & -5 & -10 & 1 \end{pmatrix}, \quad b = \begin{pmatrix} 1 \\ -1 \\ 3 \end{pmatrix}$$

Exercise 2.2 Define the variable A as the fourth-order identity matrix.

Exercise 2.3 Define the column vector v with five 0s and one 1, then save it in the ASCII file *V.DAT*, quit MATLAB, replace, by using a text editor, the final 1 with 2. Finally, restart MATLAB and reload the file. Check the result.

Chapter 3

Vector and matrix manipulation

This chapter presents some commands to generate vectors and matrices automatically, and explains how to build up matrices from submatrices and how to extract submatrices from given matrices.

3.1 Elements of a matrix

Matrix entries are available through indices; to extract from the matrix A the element on the ith row and jth column and store it in the variable a, one uses

```
a=A(i,j)
```

One of the leading features of MATLAB is that it does not require one to define the dimension of a matrix in advance; for instance the following case can be encountered:

```
x=[-1.3 sqrt(3) (1+2+3)*4/5]
```

which returns:

$$x = (-1.3000 \quad 1.7321 \quad 4.8000)$$

then entering the command

```
x(5)=x(1)
```

the result is

$$x = (-1.3000 \quad 1.7321 \quad 4.8000 \quad 0 \quad -1.3000)$$

without explicitly updating the dimension of the vector (or the matrix). Note that the lack of information about the fourth entry $x(4)$ makes MATLAB assume the value zero by default. On the other hand, the instruction `x(4)=[]` eliminates the entry x_4 from the vector, whose dimension is now reduced to 4. It is important to stress that `x=[]` empties the vector, but does not remove the variable from the workspace: to accomplish such a task we must use `clear x`, as we saw in the previous chapter.

Moreover, vectors and matrices may be juxtaposed. Let

$$A = \begin{pmatrix} 1 & 2 \\ 3 & 4 \end{pmatrix}, \quad B = \begin{pmatrix} 5 & 6 \\ 7 & 8 \end{pmatrix}$$

then the commands

$$C=[A\ B], \quad D=[A;B]$$

give

$$C = \begin{pmatrix} 1 & 2 & 5 & 6 \\ 3 & 4 & 7 & 8 \end{pmatrix}, \quad D = \begin{pmatrix} 1 & 2 \\ 3 & 4 \\ 5 & 6 \\ 7 & 8 \end{pmatrix}$$

The dual operation is to extract submatrices from a given matrix, and this topic will be discussed in Section 3.3.

The dimensions of a matrix are returned by the command `size`. For instance, if C is the matrix defined above, then

$$[m,n]=size(C)$$

gives `m=2`, `n=4`. For the dimension of a vector it suffices to apply the instruction `length`. If we apply `length` to a 4×4 matrix, the result will be 16, that is, the total number of entries in the matrix.

If we now glance at what we have done up to now, we find that the number of variables we have used has become significant: we have defined the variables A, B, C, D, x, m, n. Generally, an interactive MATLAB session produces the desired results and a fairly large number of dummy variables; the most common among the latter is the predefined variable `ans` (answer), which stores the result of the last command, unless this was not explicitly assigned to a variable. For instance, when we write

$$h=length(x);$$

the length of the vector x is computed and assigned to the variable h; but if we write

$$length(x);$$

the length of the vector x is computed and stored in the predefined variable `ans`.

Now a further problem arises: how many and which are the variables in the current workspace? The command `who` gives a list of them.

If we are also interested in the memory occupation, we can digit `whos`, which returns the list of variables along with their sizes.

3.2   Interval representation

In this section we discuss how to represent intervals in MATLAB. An interval is a set of numbers whose characteristic information consists of the first and the last number and, eventually, the step used between them. For instance, we can replace the vector $(1, 2, 3, 4, 5, 6, 7, 8, 9, 10)$ with the more compact information 'from 1 to 10 advancing by 1'.

The most commonly used operator in such cases is the character ':' (colon), which is used to generate equally spaced vectors: for instance `x=1:5` returns the row vector $x = (1, 2, 3, 4, 5)$. It is possible to specify increments different from one by using the form

$$x=x\_min:Delta:x\_max$$

where `x_min` and `x_max` are the lower and upper bounds of the interval and `Delta` is the step size, as in the case where

$$x=0:pi/4:pi$$

resulting in

$$x = (0.0000, 0.7854, 1.5708, 2.3562, 3.1416)$$

Here `pi` is the MATLAB predefined variable with value π.

Negative step size is also allowed: for instance `x=5:-1:1` gives $x = (5, 4, 3, 2, 1)$.

A different way of obtaining the same result is to use the function `linspace`, but in this case we fix the number of points in the interval rather than the step size; its syntax is

$$linspace(x\_min,x\_max,\#Points)$$

Hence, `k=linspace(-pi,pi,4)` defines

$$k = (-3.1416, -1.0472, 1.0472, 3.1416)$$

If the third parameter is omitted, MATLAB assumes one-hundred points by default: `k=linspace(-pi,pi)` creates a vector with one-hundred entries.

It is not always useful to represent a numeric interval with a vector of linearly equally spaced numbers. It is well known that, in music, the same notes on different octaves are produced by vibrations of the acoustic means with double frequency; then a frequency representation of the piano's seven octaves is best obtained by doubling the step size rather than by using a constant linear step. This consideration holds true in general for all the phenomena described in the frequency domain.[1] The command `logspace` works like `linspace`, that is, with the syntax

$$logspace(x\_min,x\_max,\#Points)$$

---

[1] See Bode plots, Section 21.1.

but the points are *logarithmically spaced* according to powers of 10, and the default number of points is fifty. The initial and final interval values are indeed the powers of 10 of the actual numbers represented: `logspace(1,2)` represents the interval $[10^1, 10^2]$ by using fifty logarithmically spaced points.

We can highlight the relationship between `linspace` and `logspace` by means of the following example: `x=logspace(1,2)` and `y=linspace(1,2,50)` define two vectors such that $x_{i+1} = 1.0481x_i$ and $y_{i+1} = y_i + 0.0204$; noting that $\log_{10} 1.0481 = 0.0204$ the link between the two operators is readily found.

3.3   Matrix manipulation operations

A powerful feature of MATLAB is its ability to extract submatrices from given matrices; this is accomplished by indexing the matrix *by a vector*. It has already been stated that given two integers i and j, `A(i,j)` gives the (i,j)th entry of the matrix A. On the other hand, when i and j are vectors, then `A(i,1)` gives the entries $A(i(1),1),\ldots,A(i(n),1)$, while the command `A(i,j)` returns the submatrix extracted from A by considering the rows indexed by the vector i and the columns by j. For instance, if we are given a 5×6 matrix A, then `B=A(1:3,1:3)` returns the 3×3 submatrix with the first three rows and the first three columns of A. Moreover, the vectors may also have negative step size; thus if we let

$$A = \begin{pmatrix} 1 & 2 & 3 & 6 \\ 3 & 5 & 8 & 12 \\ 2 & 0 & 1 & 9 \end{pmatrix}$$

then `B=A(3:-1:1,1:2:4)` gives

$$B = \begin{pmatrix} 2 & 1 \\ 3 & 8 \\ 1 & 3 \end{pmatrix}$$

To index all the rows or columns the character ':' is used; for instance `B=A(1:3,:)` extracts the first three rows and all the columns of the matrix A.

Finally, if a and k are n-dimensional vectors and the entries of k are 0s and 1s only, then `a(k)` gives only the entries of a with the same position as the 1s of k: if $a = (2, 4, 45, 22)$ and $k = (0, 1, 0, 1)$, then `a(k)` returns

ans=

4   22

This behavior is frequently involved when dealing with logical operations (see Section 5.4).

3.4 Special matrices

Here is a list of functions to generate the most frequently used matrices in MAT-LAB:

eye	identity matrix
zeros	zero matrix
ones	$a_{ij} = 1, \forall\, i, j$
diag	diagonal matrix

eye returns the identity matrix (by the way, the vowel 'I' has not been used here since it is used for the imaginary unity, thus a word with the same pronunciation has been used): eye(4) generates the 4×4 identity matrix, eye(m,n) returns an $m \times n$ matrix with 1s on the main diagonal and 0s elsewhere; finally, eye(size(A)) retains the same size as the matrix A.

zeros and ones produce matrices with all elements 0s and 1s respectively; when invoked with a single argument they return square matrices: for instance zeros(3) results in

$$\begin{pmatrix} 0 & 0 & 0 \\ 0 & 0 & 0 \\ 0 & 0 & 0 \end{pmatrix}$$

On the other hand, zeros(m,n) (and ones) produce $m \times n$ matrices: ones(1,5) gives $(1\,1\,1\,1\,1)$. Finally, zeros(size(A)) gives a null matrix with the same size as the matrix A.

diag works in two different ways. If its input argument is a vector, it returns the square matrix whose diagonal entries are the elements of the vector: A=diag([1 2 3 4]) gives

$$A = \begin{pmatrix} 1 & 0 & 0 & 0 \\ 0 & 2 & 0 & 0 \\ 0 & 0 & 3 & 0 \\ 0 & 0 & 0 & 4 \end{pmatrix}$$

If, on the other hand, the argument of diag(A) is a matrix, it works in a dual fashion, i.e. it extracts the main diagonal and returns a vector.

A second optional parameter allows the user to assign the elements of the vector to different diagonals: diag(a,k), where a is an n-dimensional vector and k an integer value, returns a square matrix with dimensions $n + |k|$ and the vector a on the kth upper-diagonal if $k > 0$ or on the kth lower-diagonal, when $k < 0$. Dually, when applied to matrices it returns the kth diagonal.

3.5   Character strings

A character string in MATLAB is a set of alphanumeric characters surrounded by apostrophes, as in `'play'`. MATLAB represents a character string as a vector of characters, then `a=['p','l','a','y']` again returns the above word. To insert an apostrophe inside a string a double apostrophe must be used: `a='Albert''s lute'` returns *a=Albert's lute*.

As with any vector variable, each character inside the string is accessed through indices: if, for instance, `a='cost'`, then typing `a(1)='h'` we obtain *a=host*, while `a=['g',a]` gives *a=ghost*; and `a(2:2:4)=[]` gives *a=got*.

Since a word (string) is actually a (row) vector of characters, a (column) vector of words will be a character matrix, hence *any word in the vector should consist of the same number of characters*, possible blanks included: storing the words 'today', 'it will', 'rain' is done by the following lines:

```
A=['  today'
   'it will'
   '   rain'];
```

A simpler way of accomplishing the same result is to use the command `str2mat`, which groups different strings into a matrix by adding the necessary blank characters *at the end* of each word:

```
A=str2mat('today','it will','rain')
```

The most useful operators in string treatment are `strcmp` and `findstr`. The first is actually a logical operator (see Chapter 4) and, when used with the syntax `strcmp(a,b)`, returns 1 if the strings *a* and *b* are the same, 0 otherwise; the second command works with the same syntax, `findstr(a,b)`, but returns the position of the string *b* into the string *a*. Consider the sentence (Shakespeare, 'Henry the Fifth', Act II) `x='A night is but small breath and little pause to answer matters of this consequence'`; the position of 'd' is sought using

```
findstr(x,'d')
```

while the position of the word 'breath' is given by `findstr(x,'breath')`. Note that the *starting* index of the word is returned.

Some operators are, moreover, available to obtain conversion between numbers and strings:

- `int2str(a)` Converts the integer *a* into a character string.
- `num2str(a)` More general than the previous command, it converts the value *a* into a character string.
- `str2num(s)` It is the dual of the previous commands, it converts the string *s* into the corresponding numerical value.

3.6   Advanced matrix manipulation techniques

In this section we will study MATLAB operators which change the 'shape' of the matrices to which they are applied. Although seldom used, knowledge of them can support powerful simplification when dealing with special situations.

- `rot90` rotates the matrix to which it is applied by 90° counterclockwise: for instance, if

$$A = \begin{pmatrix} 1 & 2 & 3 \\ 4 & 5 & 6 \\ 7 & 8 & 9 \end{pmatrix}$$

then `rot90(A)` results in

$$\text{ans} = \begin{pmatrix} 3 & 6 & 9 \\ 2 & 5 & 8 \\ 1 & 4 & 7 \end{pmatrix}$$

`rot90(A,k)` rotates by $90k$ degrees.

- `fliplr` and `flipud` exchange the columns and the rows, respectively, of the matrix they have as the argument; this is equivalent to post- and pre-multiplying the matrix by $\text{rot90}(I)$, where I is the identity matrix. With the matrix A of the above example, `fliplr(A)` and `flipud(A)` return:

$$\begin{pmatrix} 3 & 2 & 1 \\ 6 & 5 & 4 \\ 9 & 8 & 7 \end{pmatrix} \qquad \begin{pmatrix} 7 & 8 & 9 \\ 4 & 5 & 6 \\ 1 & 2 & 3 \end{pmatrix}$$

- `reshape` is the most powerful command, it allows the user to change the *shape* of a matrix. Indeed, `reshape(A,m,n)` produces an $m \times n$ matrix whose entries are taken columnwise from the matrix A. Suppose we are given an ASCII file `DATA.PAS` created by an external program, for instance a Pascal code; let the file consists of a six-element column that we want to store by rows in a 3×2 matrix. Then it will suffice to type:

```
load data.pas
A=(reshape(data,3,2))'
```

3.7   Exercises

Exercise 3.1 Define the vector $x = (-3, -4, 2, 1, 0, 2, 3, 5, 10)$ and compute:

1. `length(x);`

   2. `size(x)`;

   3. `x(12)=-x(3)`.

Exercise 3.2 Create the vector $x = (1, 2, 3, 4, 5)$ and, by manipulating it, define the vector $y = (1, 3, 4, 5, 8)$.

Exercise 3.3 Build up the row vector

$$(-1, -0.75, -0.5, -0.25, 0, 0.25, 0.5, 0.75, 1)$$

   1. by entering the value from the keyboard;

   2. by using the command `linspace(-1,1,9)`;

   3. by means of the operator ':'.

Exercise 3.4 Define the vector

$$x = \begin{pmatrix} 1 & 2 & 3 & 4 & 5 & 6 & 20 \end{pmatrix}$$

and using this vector create the vector

$$y = \begin{pmatrix} 1 & 2 & 3 & 4 & 5 & 6 & 20 & 20 & 6 & 5 & 4 & 3 & 2 & 1 \end{pmatrix}$$

   Hint. Use either the operator ':' with negative stepsize or the operator `fliplr`.

Exercise 3.5 Define the vectors

$$a = \begin{pmatrix} 0 & 3 & 6 & 9 \end{pmatrix}$$

$$b = \begin{pmatrix} 45 & 40 & 35 & 30 \end{pmatrix}$$

and build up the vector

$$c = \begin{pmatrix} 0 & 45 & 3 & 40 & 6 & 35 & 9 & 30 \end{pmatrix}$$

   Hint. Define two dummy vectors `i1` and `i2` to store the first four even and odd integers respectively.

Exercise 3.6 Define the matrix

$$A = \begin{pmatrix} 1 & 0 & 6 & -3 \\ -1 & 2 & 0 & 2 \\ 0 & 3 & -1 & -2 \\ -6 & 0 & 4 & 1 \end{pmatrix}$$

and compute:

$$A(:,2), \quad A(1:3,2:4), \quad A([2\ 4],3:4), \quad A(:)$$

Exercise 3.7 Given the matrix

$$A = \begin{pmatrix} 1 & 2 & 3 \\ 4 & 5 & 6 \\ 7 & 8 & 9 \end{pmatrix}$$

extract the submatrix with entries a_{ij}, with row indices $i = 2, 3$ and column indices $j = 2, 3$.

Exercise 3.8 Define the matrices

$$A = \begin{pmatrix} 2 & 0 & 0 \\ 0 & 4 & 0 \\ 0 & 0 & 6 \end{pmatrix}, \quad B = \begin{pmatrix} 1 & 5 & 6 & 7 \\ 5 & 2 & 0 & 0 \\ 6 & 0 & 3 & 0 \\ 7 & 0 & 0 & 4 \end{pmatrix}$$

and then the block-diagonal matrix

$$C = \begin{pmatrix} A & 0 \\ 0 & B \end{pmatrix}$$

Exercise 3.9 To plot the graph of the function depicted in Figure 3.1 two vectors of the same length are needed, the first to store the independent variable (abscissa), the second for the values of the function. Define these two vectors using equally spaced abscissae with step 0.1.

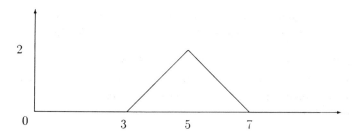

Figure 3.1 Triangular impulse.

Exercise 3.10 Given the vector $a = (3\,2\,1)$, build up the lower horizontal companion matrix associated with it, i.e. the matrix

$$\left(\begin{array}{ccc} 0| & 1 & 0 \\ 0| & 0 & 1 \\ \hline -1 & -2 & -3 \end{array} \right)$$

Exercise 3.11 This is the first of a set of exercises that address the problem of treating musical structures by computer. First of all, we must specify the musical environment we that we are going to discuss. We will always refer to the tempered system according to the equal temperament (i.e. the notes obtained with the keys of a common piano keyboard) and we will consider, within a single octave, both the chromatic scale (the one obtained by pressing both the white and the black keys) and the diatonic scale in C major tonality (i.e. the white keys only).

The objective of this first exercise is to code the notes by means of numeric values in both kinds of scale.

Solution. The diatonic scale can be coded by using integer values from 0 to 6; then define the vector

```
nameDiat=str2mat('c','d','e','f','g','a','b');
```

Now, given a vector `noteDiat` of n integers between 0 and 6, the corresponding notes are obtained with

```
nameDiat(noteDiat+1,:)
```

Note that the use of the operator `str2mat` is not strictly needed in this case, but this solution is more general than defining `nameDiat` as a simple vector: it allows different names for the notes (in Italian, C is Do, in French it is Ut).

Analogously, for the chromatic scale it is sufficient to refer to integer values between 0 and 11, and define the vector:

```
nameChrom=str2mat('c','c#','d','d#','e','f','f#');
nameChrom=str2mat(nameChrom,'g','g#','a','a#','b');
```

In this case we must use the operator `str2mat`.

Why it is convenient to code the notes starting from 0 and not from 1 (which would simplify indexing the vectors `name...`) will become clear in what follows.

Chapter 4

Scalar operations

Up to now we have discussed how to enter and manipulate data. In this chapter we will start to deal with arithmetic and logical operations on the variables by introducing scalar operations. In the next chapter operations defined in standard matrix algebra will be considered. By 'scalar operations' we mean not only those whose operands are scalar variables, but also those in which the operands are treated as scalars. For instance, the operator sqrt computes the square root, then sqrt(2) returns 1.414 21, but it is possible to use vectorial arguments: sqrt([1 2 3]) results in (1 1.414 21 1.732 05); the same result could be obtained in a less efficient way with [sqrt(1) sqrt(2) sqrt(3)]. All the operators described in this chapter work in this way, i.e. they can be applied to vectors, and the result is the one we would obtain by applying the operator elementwise (unless explicitly stated: for instance the product operator * *does not* work elementwise).

It is now useful to introduce some MATLAB predefined variables:

- ans is the result of any operation not assigned to a variable;
- eps is the actual working precision;
- computer is the type of computer;
- pi is π;
- i,j is the imaginary unit ($\sqrt{-1}$);
- Inf is infinity;
- NaN is Not-a-Number (for instance the result of 0/0);
- clock is the machine clock;
- cputime is the CPU time elapsed;
- date is the machine date;
- flops is the number of floating point operations;
- realmax is the largest number of floating point operations;
- realmin is the smallest floating point value;
- nargin is the number of function input arguments;
- nargout is the number of function output arguments.

25

The meaning of some of these variables is apparent; the others will be explained later.

4.1 Arithmetic operations

Algebraic operations are performed by using the usual operators +, -, *, /. There are, moreover, the operators ^ (power function), sqrt (square root) and \ (computes a 'reverted' division, i.e. 3\2 returns $2/3 = 0.666\,67$). The latter operator may appear useless for scalar operations, but it exhibits its importance in matrix operations, as will be shown next.

The operators described in this section obey the rules of matrix algebra. This can cause confusion when these operators are applied to vector variables. For instance, the product ay of a scalar a by a vector y is computed by a*y and returns the correct result, i.e. a vector whose ith entry is ay_i; on the other hand, the product of two vectors is defined in algebra as the inner product (or scalar product) and is so defined in MATLAB. Thus, if we have $x = (x_1, x_2, x_3)$ and $y = (y_1, y_2, y_3)^T$, x*y returns the inner product $x_1y_1 + x_2y_2 + x_3y_3$, as will be explained in full detail in the following chapter; if we want a vector whose entries are the product of the homologous entries, we must prefix the operator with the character '.' (dot): x.*y returns ans = (x_1y_1, x_2y_2, x_3y_3). The operators ./ for elementwise division and .^ to raise each element of a vector to a power are defined similarly. For the latter operator, note that the exponent may be either a scalar or a vector with the same length as the basis vector: [2 3 4].^2 returns (4 9 16), while in the second form, each element is raised to the power of the corresponding exponent: [2 3 4].^[3 2 1] gives $(2^3\ 3^2\ 4^1) = (8\ 9\ 4)$.

4.2 Elementary and transcendental functions

MATLAB has the following elementary functions:

- *Rounding*
  - round rounds to the closest integer;
  - fix rounds towards 0;
  - floor rounds down to integer;
  - ceil rounds up to integer.
- *Rational approximations*
  - rem reminder of an integer division;
  - rat rational expansion;
  - rats rational approximation.

- *Integer factorization*
  - gcd greatest common divisor;
  - lcm least common multiple.
- *Complex arithmetic*
  - real real part;
  - imag imaginary coefficient;
  - conj complex conjugate;
  - abs absolute value or complex modulus;
  - angle phase angle or complex argument.
- *Signum*
  - sign Heaviside or signum function.

The meaning of rounding and truncation operators can easily be understood.

Rational approximation functions produce an approximation of their argument by means of an integer ratio. For instance

$$\text{rats(9.22)}$$

returns ans=461/50. Note that the result is a character string.

It is also possible to prescribe the length of the string, whose default value is 13; for instance

$$\text{rats(9.22,5)}$$

would impose a *maximum* output length of five characters, and the result would be ans=46/5.

rats can also be used to compute the result of a sequence of algebraic operations in the rational field, returning an element of the same field: for instance the result of rats(1-1/2+1/3-1/4+1/5) is ans=47/60; note that the denominator is the least common multiple of all the denominators.

As discussed above, the objective of rats is substantially the presentation of the results on the screen; on the other hand, rat actually computes the partial fraction expansion to obtain a rational approximation. Indeed, given any real number x, it can always be approximated at will by the rational number defined by the truncated expansion:

$$x \approx d_1 + \cfrac{1}{d_2 + \cfrac{1}{d_3 + \cdots + \frac{1}{d_k}}}$$

The function rat has the following syntax:

$$\text{[n,d]=rat(x,tol)}$$

The result consists of two integers n and d such that their ratio satisfies the relationship

$$\left| x - \frac{n}{d} \right| \leq \text{tol} |x|$$

The default value for the tolerance is 10^{-6}.

rat can also be invoked without output arguments, and in this case it returns the truncated expansion:

$$\text{rat(sqrt(2))}$$

returns

```
1 + 1/(2 + 1/(2 + 1/(2 + 1/(2 + 1/(2 +
1/(2 + 1/(2)))))))
```

The integer factorization operator m=gcd(a,b) returns the greatest common divisor m between two integers a and b. Used in the form [m,c,d]=gcd(a,b) it also returns two integers c and d solving the Diophantine equation $m = ac + bd$. lcm works in the same way.

As for the operators in the complex field, we must stress that MATLAB has two predefined characters i and j for the complex unity. The complex number $z = 2 + j3$ is defined either by z=2+j*3, or by z=2+i*3, or by z=2+3i or, finally, by z=2+3j. Note that the MATLAB predefined variables are not protected, thus their values can be redefined: if, for instance, the variable i is given a value (this is often the case with for cycles, for i=1:n, which we will discuss in Chapter 8), for instance $i = 2$, then z=2+i*3 results in $z = 8$; to restore the original value we must type

$$\text{i=sqrt(-1)}$$

or simply delete the variable from the current workspace with clear i, thus resetting the variable to its predefined value.[1] A further remark is in order: we can avoid typing the product operator only in the case of numeric constants, not variables: MATLAB accepts the expression z=2+3j to define the complex number $z = 2 + j3$, but if we write

```
a=3;
z=2+aj
```

MATLAB considers the symbol aj to be the name of a new variable and answers with the error message

```
??? Undefined function or variable aj
```

---

[1] Note, however, that both z=2+3i and z=2+3j always return the correct result.

The solution in this case is

```
a=3;
z=2+a*j
```

The exponential and logarithmic operators are

- `pow2` base 2 exponential function;
- `exp` base *e* exponential function;
- `log` natural logarithm;
- `log2` base 2 logarithm;
- `log10` base 10 logarithm.

Finally, trigonometric operations are computed using the following commands:

- `sin` sine;
- `cos` cosine;
- `tan` tangent;
- `asin` inverse sine;
- `acos` inverse cosine;
- `atan` inverse tangent;
- `atan2` four quadrants inverse tangent;
- `sinh` hyperbolic sine;
- `cosh` hyperbolic cosine;
- `tanh` hyperbolic tangent;
- `asinh` hyperbolic inverse sine;
- `acosh` hyperbolic inverse cosine;
- `atanh` hyperbolic inverse tangent;
- `sec` secant;
- `csc` cosecant;
- `cot` cotangent;
- `asec` inverse secant;
- `acsc` inverse cosecant;
- `acot` inverse cotangent;
- `sech` hyperbolic secant;
- `csch` hyperbolic cosecant;
- `coth` hyperbolic cotangent;
- `asech` inverse hyperbolic secant;
- `acsch` inverse hyperbolic cosecant;
- `acoth` inverse hyperbolic cotangent.

A remark about the inverse tangent is now in order: `atan(x)` returns a value within the interval $[-\pi/2, \pi/2]$, while `atan2(y,x)` computes the 'four quadrants' inverse tangent, i.e. it returns a value in the interval $[-\pi, \pi]$; x and y are the cosine and the sine of the angle.

Finally, there are conversion operators among different coordinate systems:

- cart2pol Cartesian→polar (or cylinder);
- pol2cart polar (or cylinder)→Cartesian;
- cart2sph Cartesian→spherical;
- sph2cart spherical→Cartesian.

In the 2D case

$$\texttt{[theta,rho]=cart2pol(x,y)}$$

converts the pairs (x_i, y_i) in Cartesian coordinates into polar coordinate pairs (θ_i, ρ_i); the values of the θ_i are in radians. In the 3D case the conversion is in cylinder coordinates:

$$\texttt{[theta,rho,z]=cart2pol(x,y,z)}$$

Spherical coordinate conversion is performed by

$$\texttt{[phi,theta,rho]=cart2sph(x,y,z)}$$

again, ϕ and θ are expressed in radians. The inverse conversion works in the same way.

4.3   Examples

4.3.1   Elementary functions

As an example to summarize what we have discussed up to now, we consider the problem of tabulating the function $\log x$ in the interval $[1, 5]$ with step 0.1.

It is sufficient to give the sequence of commands:

```
x=(1:.1:5)';
y=log(x);
[x y]
```

The result is:

```
ans=
1.0000   0
1.1000   0.0953
1.2000   0.1823
1.3000   0.2624
1.4000   0.3365
   ⋮
```

Note that the vector x has only been transposed for better visualization of the result on the screen, to display the table columnwise.

4.3.2   Functions of functions

A more general example is: tabulate the values of $e^{3t} \sin 5\pi t$ in the interval $t \in [-2, 2]$ using forty-five points. The instructions are

```
t=linspace(-2,2,45)';
y=exp(3*t).*sin(5*pi*t);
[t y]
```

Note the use of the operators * and .* in the second line; the latter is needed because both `exp(3*t)` and `sin(5*pi*t)` return vectors.

4.3.3   Rational functions

As the last example of this section, we require tabulation of the rational function[2]

$$f(s) = \frac{3s^2 + 5s + 7}{s^3 + 5s^2 + 7s + 12}$$

with $s = j\omega$, ad $\omega \in [10^{-2}, 10^2]$; it is convenient to use logarithmically spaced abscissae, otherwise the tabulation would be too sparse at the lowest bound of the interval and too dense at the upper one. Let us use fifty points.

The following instructions solve the problem:

```
omega=logspace(-2,2);  the arguments are the exponents
s=j*omega;
x1=3*s.^2+5*s+7;  tabulates the numerator
x2=s.^3+5*s.^2+7*s+12;  tabulates the denominator
x=x1./x2;  computes the values of the function
x=abs(x);  computes the modulus of the function
[s x]  displays the result
```

4.4   Logical and relational operators

The following relational operators are defined:

<	less than
<=	less or equal
>	greater than
>=	greater or equal
==	equal
~=	not equal

---

[2]In Section 21.1 the function **bode** will be presented in order to solve the same problem in a more efficient way.

and the logical operators

&	and
|	or
xor	exclusive or
~	not

The value FALSE is indicated by 0; thus the operation `2+2~=4` gives 0; these operators can be applied to matrices: if we want to find the elements of the matrix

$$A = \begin{pmatrix} 1 & 2 & 3 & 4 \\ 5 & 6 & 7 & 8 \end{pmatrix}$$

that can be divided by 2 (without remainder), it suffices to apply

$$P=(\text{rem}(A,2)==0)$$

which returns

$$P = \begin{pmatrix} 0 & 1 & 0 & 1 \\ 0 & 1 & 0 & 1 \end{pmatrix}$$

This example gives birth to some considerations. Note that the operator **rem**, previously defined on scalar arguments, has been applied here to a matrix and its result is again a matrix. The reader could now think of an incongruity: the operator equality test **==** has a matrix on the left- and a scalar on the right-hand side; what does this mean? To give an answer to this question we must discuss the matrix treatment in MATLAB in full detail. So we suspend discussion of logical operators, which we will resume in Section 5.4, when talking about *logical functions*, and begin a new chapter to deal with matrix operations.

4.5   Exercises

Exercise 4.1 For each pair of coefficients (a, b) of the equation $ax + b = 0$ given in Table 4.1, compute the solution x subject to the constraint $x \in \mathcal{Z}$, where $\mathcal{Z}$ is the ring of integers. When the equation does not admit solution in $\mathcal{Z}$, solve the problem

$$\min_{x \in \mathcal{Z}} |ax + b|$$

Exercise 4.2 Compute $x \bmod 3$ for $x = (1, 5, 3\,12, 22, 64)$.

Exercise 4.3 Given $x = 3 + j5$, $y = -2 + j4$, $z = j3$, compute:

Table 4.1 Coefficients

a	2	5	8	13	5	0.1
b	4	3	28	−33	72	$\sqrt{2}$

1. $x + y$, $x - z$, $(x + y)z$;
2. $|x|$, $1/y$, z^2;
3. $\log x$, e^y, $|x/y|$.

Exercise 4.4 Solve the system

$$x^2 + y^2 = a$$
$$x/y = b$$

for $a = (1,\, 4,\, 3)$ and $b = (1,\, \sqrt{3},\, 0.5)$.

Hint. Solve the problem in polar coordinates.

Exercise 4.5 Verify Euler's formula

$$e^z = e^{x+jy} = e^x(\cos y + j \sin y)$$

with different complex numbers z.

Exercise 4.6 Store the abscissa and values of the function $|\sin t|$ depicted in Figure 4.1 in two vectors.

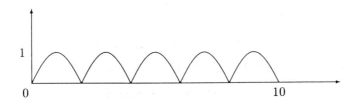

Figure 4.1 $|\sin t|$.

Exercise 4.7 Verify that the bilinear transformation

$$w = \frac{z + 2}{z + 4}$$

maps circles onto circles.

Hint. Build up circles z in polar coordinates and verify the result with the command `plot(w)`.

Exercise 4.8 Define the vector x with 31 values of the interval $[1/2, 2^5]$ equally spaced according to base 2 logarithms.

Exercise 4.9 Refer to the vector x in Exercise 3.1 and compute

$$\texttt{L=(x>=2), L=(x<3), L= (x<3), L=(x<4 \& x>-4)}$$

Exercise 4.10 Given the vector $x = (1, 34, 2, -12, 56, 7, 0, 9)$, display only values larger than 5.
Hint. Refer to Section 3.3.

Exercise 4.11 In Exercise 3.11 we discussed note coding. Now we start to manipulate sequences of notes. We will address three musical procedures in this exercise: transposition, retrogradation and inversion with respect to a given note. These techniques have been widely used in music by Guillaume de Machaut, J.S. Bach and Arnold Schönberg, to cite only some composers.

In music, transposition means to shift a sequence of notes upwards or downwards by a given interval,[3] retrogradation is simply reversing the sequence, i.e. reading the sequence from the end to the beginning, while inversion inverts the intervals with respect to the given note: for instance, the inverse of the sequence {C, C#, D} with respect to C is {C, B, A#}.

Given the sequence {E, D, C, F, F, E}, compute the sequences obtained:

- transposing two tones upwards;
- using retrogradation;
- inverting with respect to G.

Solution. The given sequence is coded with the vector

$$\texttt{notes=[4 2 0 5 5 4]}$$

Transposition by two tones (four semitones) is obtained by

$$\texttt{notesTransp=rem(notes+4,12);}$$

Note the modulo-12 sum. To obtain the retrograde:

$$\texttt{notesRet=fliplr(notes);}$$

Finally, inversion with respect to the note n is done with $2n - notes$ mod 12; hence, with respect to G (code: 7) we have:

$$\texttt{notesInv=rem(2*7-notes,12);}$$

Of course, the (names of the) notes can be displayed by using the matrix `nameChrom` defined in Exercise 3.11.

---

[3]Remember that in music an interval is defined as the distance in pitch from one note to another.

Chapter 5

Matrix operations

In Chapter 1 we have stated that the basic element in MATLAB is the matrix: even a scalar variable is a 1×1 matrix. This implies that the scalar operations discussed in the previous chapter are immediately extendable to matrices.

Consider, for instance, the matrix

$$A = \begin{pmatrix} 1 & 3 \\ 4 & 2 \end{pmatrix}$$

then C=exp(A) results in the matrix

$$C = \begin{pmatrix} e^1 & e^3 \\ e^4 & e^2 \end{pmatrix}$$

By the way, the reader must note that the result is computed elementwise, hence this operation **does not give** the matrix exponential, which is defined as

$$e^A = \sum_{n=0}^{\infty} \frac{A^n}{n!}$$

In this chapter we will devote our attention to strict matrix operations, i.e. operations defined according to standard matrix algebra rules.

5.1   Transpose

We have already mentioned this operation: it is performed, both for matrices and for vectors, using the character ' (apostrophe); the following lines

```
A=[1 2 3;4 5 6;7 8 0];
B=A';
```

result in

$$A = \begin{pmatrix} 1 & 2 & 3 \\ 4 & 5 & 6 \\ 7 & 8 & 0 \end{pmatrix}, \quad B = \begin{pmatrix} 1 & 4 & 7 \\ 2 & 5 & 8 \\ 3 & 6 & 0 \end{pmatrix}$$

Actually, the operator ' gives the *complex conjugate transpose*, thus `[1+j,1-2*j]`'
gives

$$\begin{pmatrix} 1-j \\ 1+2j \end{pmatrix}$$

It is well known that this is performed to induce in the complex field a norm starting
from the inner product, according to the expression

$$\|x\|^2 = \; <x^H, x>$$

In the case of complex vectors or matrices, to obtain a pure transpose, without
conjugation, the operator .' should be used. For instance, `[1+j,1-2*j].'` gives

$$\begin{pmatrix} 1+j \\ 1-2j \end{pmatrix}$$

5.2 Algebraic operations

The basic algebraic operators are `+,-,*`; for 'matrix division' MATLAB uses two
different symbols:

- $X = A \backslash B$ computes $X = A^{-1} \cdot B$, solving the system $A \cdot X = B$ (**left**
  division);
- $X = A/B$ computes $B \cdot A^{-1}$, solving $X \cdot A = B$ (**right** division).

The power function is obtained using the operator ˆ. The inverse matrix is
computed either with `B=inv(A)` or with `B=A^(-1)`.

The syntax rules are those of standard algebra: sum and difference are only
allowed on matrices of the same dimension; the only exception is that MATLAB
allows sum and difference between a matrix A and a scalar a. In this case the
scalar is transformed in a matrix of suitable size whose entries are *all* equal to the
scalar (note that this is quite different from what we would expect in this case, i.e.
$A + aI$, where I is the identity matrix).

As usual, in the matrix product the number of columns of the first matrix and
the number of rows of the second must be the same. Also the product between
a scalar and a matrix follows the usual algebraic definition, i.e. it gives a matrix
whose entries are the entries of the matrix times the scalar.

We comple⟍
A\B gives A^{-}⟍
lution to the⟍
entries for e⟍
right divisi⟍

⟋ors: left division
least square so-
. with k nonzero
ix. Analogously,

5.3 M⟍

In this s⟍
assume

In what follows we

• m⟍
⟋

⟋e vector x; when this
or whose elements are
⟍ce the largest element

⟋argest elements if X is a
⟋rgest value, i.e. an integer
⟍, in the matrix case I is a

⟋nd B are two matrices with
⟋ whose entries are the largest
⟍, i.e. the ijth element of Y is

When applied to matrices, it

⟋ich generate the sorting in a
⟋eturns the integer vector **ind**,
⟋ch that **x(ind)** gives the sorted
⟋or $x = (41\ 11\ 12\ 12\ 50)$; then
⟍1 50) and $ind = (2\ 3\ 4\ 1\ 5)$. The
⟋uild up with the second entry of
⟋finally the fifth; in other words, **y**
⟋ed entries (the value 12 in our ex-
⟋the matrix case [Y,Ind]=sort(X)
⟋ion matrix.

ampl⟍,⟋
sorts columnwise a⟍⟍

• **sum(x)**, **mean(x)** are the sum a⟍⟍ e mean value of the entries in the
vector x, respectively; these commands, too, as with those given above, work
columnwise, returning row vectors when applied to matrices.

---

[1]See Section 5.5.

- `rank(A)` computes the rank of the matrix A, i.e. the largest number of linearly independent rows or columns.
  It is possible to specify a tolerance: `rank(A,tol)` gives the rank up to the tolerance `tol` by computing the number of singular values exceeding `tol`.
- `det(A)` returns the determinant.
- `poly(A)` returns the coefficients, in descreasing power order, of the characteristic polynomial of the matrix A, $p(\lambda) = |\lambda I - A|$; later on, in Chapter 6, we will see that the same command, when called with a vector argument, has an utterly different meaning.
- `trace(A)` gives the trace of the matrix A, i.e. the sum of the entries on the leading diagonal.
- `norm(X,argument)` computes the norm of a matrix or a vector according to the following table, in which `A` is a matrix and `x` a vector:

Command	Operation	Comment
`norm(A,1)`	$\|A\|_1 = \max_j \sum_{i=1}^{m} \mid a_{ij} \mid$	largest column sum
`norm(A,2)`	$\|A\|_2 = \max_i \sqrt{\lambda_i(A^H A)}$	largest singular value, $\bar{\sigma}(A)$
`norm(A,inf)`	$\|A\|_\infty = \max_i \sum_{j=1}^{n} \mid a_{ij} \mid$	largest row sum
`norm(A,'fro')`	$\|A\|_F = \left(\sum_{ij} \mid a_{ij} \mid^2 \right)^{\frac{1}{2}}$	Frobenius norm
`norm(x,p)`	$\|x\|_p = \left(\sum_{i=1}^{n} |x_i|^p \right)^{\frac{1}{p}}$	p-norm
`norm(x)`	$\|x\|_2 = \sqrt{x^H x}$	Euclidean norm
`norm(x,inf)`	$\|x\|_\infty = \max_i |x_i|$	largest modulo
`norm(x,-inf)`	$\|x\|_\infty = \min_i |x_i|$	smallest modulo

here the star superscript H denotes the conjugate transpose.
`norm(X)` is equivalent to `norm(X,2)` both for matrices and vectors.
- `kron(A,B)` gives the Kronecker tensor product, i.e. returns a matrix whose ijth block is the product of the ijth entry of A and the matrix B, $(A \otimes B)_{ij} = a_{ij} B$.

These functions cover a wide range of applications: from the simple operation of searching for the extreme entries of a vector with `max` and `min` functions, to normalizing a vector with `x/norm(x)`, up to computing the dimension of a subspace spanned by a set of vectors which requires stacking of the vectors into a matrix and computing the rank: `rank([x1,x2,...,xn])`. The commands `min`, `max` and `sort` with two output parameters turn out to be particulary useful; if we want compute the largest entry of a matrix A along with two pointers to its position, h, k, we can use

```
[m1,i1]=max(A);
[m,k]=max(m1); h=i1(k);
```

A further, more complex, example follows. Suppose we need to sort the columns of a matrix *with respect to the first column;*[2] for instance we are given the matrix

$$R = \begin{pmatrix} 1 & 2 \\ 5 & 1 \\ 3 & 3 \\ 2 & 4 \end{pmatrix}$$

Using `A=sort(R)` would result in

$$A = \begin{pmatrix} 1 & 1 \\ 2 & 2 \\ 3 & 3 \\ 5 & 4 \end{pmatrix}$$

Thus the columns have been ordered independently, and the respective association is lost. To retain the association we can use

```
[S,i1]=sort(R);
A=R(i1(:,1),:);
```

which produces

$$A = \begin{pmatrix} 1 & 2 \\ 2 & 4 \\ 3 & 3 \\ 5 & 1 \end{pmatrix}$$

The matrix exponential and related functions available in MATLAB are

- `expm(A)` computes e^A;
- `logm(A)` computes $\log(A)$;
- `sqrtm(A)` only works on positive definite matrices and computes the 'square root' of the matrix, i.e. gives a matrix U such that $A = U^H \cdot U$.

---

[2]This is usually the situation in a table prepared for look-up, in which the first column must be sorted in strict increasing order.

In particular, to compute the matrix exponential function e^A, MATLAB offers three more operators, **expm1**, **expm2**, **expm3**, which compute the result via Padé approximation, Taylor approximation and eigenvalue computation respectively.

Further matrix operations are obtained with the command **funm**, whose syntax is **funm(X,'*function*')**; it simply applies the specific function to the matrix X, as in the case **funm(X,'sqrt')**, which is equivalent to **sqrtm(X)**.

5.4   Logic operations

In Section 4.4 we have seen an example of a relational operator applied to a matrix: now we are going to discuss this topic in detail.

As mentioned before, 0 indicates FALSE, while 1 or any other number different from zero indicates TRUE; then, if A and B are matrices whose entries are either 0 or 1, **A&B** returns a matrix whose entries are 1 if the corresponding entries of A and B are 1. There are more interesting logical operators: for instance **any(x)**, where **x** is a vector of 0s and 1s gives 1 if *at least one entry of* **x** is 1, while **all(x)** performs the dual operation, i.e. returns 1 if *all the entries* of **x** are 1.

If the argument is a matrix, these operators work columnwise: for instance if A is a matrix, **all(A<0.5)** returns a vector such that each entry is 1 if all the elements of the corresponding column of the matrix A are smaller than 0.5, while **all(all(A<0.5))** returns 1 if all the entries of the matrix A are smaller than 0.5.

Finally, **find** returns the indices associated with the nonzero values of its argument. In particular, if **x** is a vector, **i=find(x)** returns the nonzero entries indices, if **A** is a matrix **[i,j]=find(A)** returns the row and column indices of nonzero elements. The importance of this function becomes clear when we remember that logical operators result in matrices of 1s and 0s: looking for nonzero entries can be interpreted as looking for a condition that is satisfied; for instance the indices of the entries greater than 2 of the matrix A are computed by

$$[\text{i},\text{j}]=\text{find}(\text{A>2})$$

Note that **A(i,j)** *does not* return a matrix whose entries are those of the matrix A larger than 2: if

$$A = \begin{pmatrix} 1 & 3 & 0 \\ 4 & 5 & 1 \end{pmatrix}$$

then

$$[\text{i},\text{j}]=\text{find}(\text{A>2})$$

produces the vectors $i = (2\ 1\ 2)$, $j = (1\ 2\ 2)$. This means that the entries a_{21}, a_{12} and a_{22} are larger than 2, but **B=A(i,j)** defines the matrix

$$B = \begin{pmatrix} 4 & 5 & 5 \\ 1 & 3 & 3 \\ 4 & 5 & 5 \end{pmatrix}$$

i.e. the matrix whose khth entry is `A(i(k),j(h))`, while `diag(B)` returns a vector with the entries of A larger than 2: (4 3 5).

In this case a more simple way of obtaining the same result would have been `B=A(A>2)` (see Exercise 4.10), but the operator `find` is perhaps the most powerful of the logical operators in MATLAB and its applications go well beyond those considered here.

Consider the following situation: define the vectors

$$t=0:.005:20;$$
$$y=sin(t);$$

We want to search for the values of the independent variable t such that the corresponding values of y are 0.5. Of course, no computer can compute the solutions of the equation $\sin t - 0.5 = 0$ exactly, thus we focus our attention on the points $\bar{t}$ such that $|\sin \bar{t} - 0.5| < \delta$, where $\delta > 0$ is a suitably small scalar. The choice of the scalar δ is not trivial, as too small a value could result in a wrong result: if, for instance, we thought of the smallest possible δ, we would use

$$[m,k]=min(abs(y-0.5));$$

thus we obtain a single index k, although $\sin t$ touches the line $y = 0.5$ seven times in the interval $t \in [0, 20]$. The solution is to use a different δ, for instance $\delta = 0.05$:

$$i=find(abs(y-0.5)<0.05);$$

We stress that the operations presented in this section could be performed by repeating simple tests in `for` or `while` loops, but the result would be less readable and the execution slower: one of the most frequent errors committed by the less skilled user is to reproduce existing commands by means of programming instructions, discussed in Chapter 8, where a better knowledge of the predefined commands would solve the problem best.

The complete list of relation operators is:

`any`	true if at least a nonzero element exists
`all`	true if all the elements are nonzero
`find`	looks for indices of entries satisfying a given condition
`exist`	inspects the existence of a variable
`isnan`	inspects NaN
`finite`	inspects finite elements
`isinf`	inspects infinite elements
`isempty`	true if the matrix is empty
`isieee`	true for IEEE floating point arithmetic
`isstr`	true if the variable is a string

5.5   Complements on matrices

In this section we will discuss advanced algebraic operations, in particular factorizations, decompositions, pseudo-inverse, orthogonalization and computation of the null space of a matrix.

5.5.1   Condition number of a matrix

The condition number of a matrix is a measure of the sensitivity of the solution of a linear system of equations to perturbations of the data. It is well known that the solution of the system $Ax = b$ is $x = A^{-1}b$; hence it is clear that the operator A^{-1} can either amplify or reduce uncertainties in the constant term b. If, moreover, we assume that any entry of the vector b is uncertain, it is clear that the condition number depends solely on the matrix A. Formally, the condition number defined on the Euclidean norm is the ratio between the largest and the smallest singular value of a matrix. MATLAB computes it by the command

c=cond(A)

A quicker, although less accurate, estimate of the condition number is given by

c=rcond(A)

which uses the 1-norm and returns the reciprocal of the condition number.

In both cases, the closer to 1 is the result, the better conditioned is the matrix; the closer the result is to 0 for cond (infinity for rcond), the worst conditioned is the matrix.

5.5.2   Orthogonalization and null space

The command

Q=orth(A)

stores in Q an orthonormal basis for the range of the matrix A, i.e. Q is a unitary matrix whose columns span the same subspace as the columns of the matrix A. The number of columns of Q is the rank of A.

N=null(A)

generates a unitary matrix whose columns span the null space of the matrix A, i.e. they define an orthonormal basis of the kernel of A. The number of columns of N is the nullity of A.

5.5.3   LU factorization and Gaussian elimination

It is well known that the simplest way of solving a nonsingular system in the form $Ax = b$ is to triangularize the matrix A and compute x_i by successive substitutions. The automatic procedure with which to transform the matrix A into an upper triangular matrix is named *Gaussian elimination*. The LU factorization of the matrix A returns two matrices L and U such that L is lower triangular and unitary and U is upper triangular, and such that

$$A = LU$$

It is now clear that to solve the system one can

1. compute L and U;

2. solve $Ly = b$;

3. solve $Ux = y$.

MATLAB computes the LU factorization with the function

$$[\text{L},\text{U}]=\text{lu}(\text{A})$$

Note that if during the triangularization one of the 'pivot' entries were zero the procedure should halt. A well-known trick for solving this drawback is to exchange the rows of the matrix A. This operation leads to the factorization

$$PA = LU$$

where P is a suitable permutation matrix. In MATLAB the related command is

$$[\text{L},\text{U},\text{P}]=\text{lu}(\text{A})$$

5.5.4 LUL^T and Cholesky factorizations

A symmetric positive definite matrix A can be factored as

$$A = LDL^T$$

where L is a unitary lower triangular matrix and D is a diagonal matrix with positive entries on the diagonal. Actually, this factorization can be rewritten

$$A = LD^{\frac{1}{2}}D^{\frac{1}{2}}L^T = RR^T$$

where R is an upper triangular matrix. This is known as *Cholesky factorization*, and the matrix R is said to be a *Cholesky factor* or 'square root' of the matrix A. The corresponding instruction in MATLAB is

$$\text{R}=\text{chol}(\text{A})$$

5.5.5   Row reduced echelon form

A matrix is said to be in 'echelon' form if the number of zero entries preceding the first nonzero element of each row increases row by row until only zero rows remain. The first nonzero entry of a row is said to be a *distinguished element* of the echelon matrix. The following are echelon matrices:

$$
\begin{pmatrix}
2 & 3 & 2 & 0 & 4 & 5 & -6 \\
0 & 0 & 7 & 1 & -3 & 2 & 0 \\
0 & 0 & 0 & 0 & 0 & 6 & 2 \\
0 & 0 & 0 & 0 & 0 & 0 & 0
\end{pmatrix}
\qquad
\begin{pmatrix}
0 & 1 & 3 & 0 & 0 & 4 & 0 \\
0 & 0 & 0 & 1 & 0 & -3 & 0 \\
0 & 0 & 0 & 0 & 1 & 2 & 0 \\
0 & 0 & 0 & 0 & 0 & 0 & 1
\end{pmatrix}
\tag{5.1}
$$

In particular, an echelon matrix is said to be *row reduced* if the distinguished elements are:

- the only nonzero in their respective columns;

- each equal to one.

Hence the second matrix of (5.1) is row reduced. It is possible to show that any matrix can be brought into row reduced echelon form through a sequence of elementary transformations. The MATLAB command to accomplish this task is

`B=rref(A)`

5.5.6   *QR* factorization

Associated with any vector w we can define a corresponding *Householder transformation* as the symmetric matrix

$$
H = I - \frac{1}{\beta} w w^T
$$

where $\beta = \frac{1}{2}\|w\|_2^2$. Any Householder matrix is unitary and hence preserves the length of the vectors to which is applied. For each pair of vectors a and b of the same length, there exists a Householder matrix built up on a Householder vector w that transforms the vectors into each other:

$$
Ha = \left(I - \frac{1}{\beta} w w^T \right) a = a - w \frac{w^T a}{\beta} = b
$$

In particular, a *plane rotation* is a special orthogonal transformation used to annihilate a single element of a vector. It can be shown to be equivalent to a Householder transformation. The properties of Householder transformations state that a sequence of n matrices $\{H_i\}, i = 1, \ldots, n$ can be multiplied on the right by a matrix A $m \times n$ with rank n, obtaining the upper triangular matrix:

$$
H_n \cdots H_2 H_1 A = QA = \begin{pmatrix} R \\ 0 \end{pmatrix}
$$

where R is an $n \times n$ nonsingular, upper triangular matrix and Q is an orthogonal matrix, the product of Householder matrices. This is called *QR transformation* of the matrix A. In MATLAB:

$$[Q,R]=qr(A)$$

If A does not have full rank (i.e. not all the columns are independent), it is necessary to exchange the columns to bring the r independent columns in the first r positions, where r is the rank of the matrix. Thus the factorization becomes:

$$QAE = \begin{pmatrix} R \\ 0 \end{pmatrix}$$

obtained in MATLAB by

$$[Q,R,E]=qr(A)$$

5.5.7 Spectral decomposition

Spectral decomposition is the most commonly used among the decompositions of a given matrix A. It searches for a diagonal matrix Λ with the eigenvalues of A on the diagonal and an orthonormal matrix U whose columns are the eigenvectors, such that

$$A = U\Lambda U^{-1}$$

In MATLAB:

$$[U,Lambda]=eig(A)$$

If the (real) matrix has complex conjugate vectors the eigenvalue and eigenvector matrices have complex entries. It is possible to obtain a different form such that these matrices have only real entries; in particular, complex pairs of eigenvalues are transformed into 2×2 real submatrices. The MATLAB command for accomplishing this task is

$$[V,D]=cdf2rdf(U,Lambda)$$

(read: 'Complex Diagonal **Form to** Real Diagonal **Form**'). For instance, consider the matrix

$$A = \begin{pmatrix} 1 & 2 & 3 \\ 0 & 4 & 5 \\ 0 & -5 & 4 \end{pmatrix}$$

whose spectral decomposition is

```
[U,Lambda]  =  eig(X)
          U  =

              1.0000   0.4002-0.0191i   0.4002+0.0191i
                   0        0.6479           0.6479
                   0      0+0.6479i         0-0.6479i
       Lambda  =

              1.0000          0                0
                   0    4.0000+5.0000i          0
                   0          0          4.0000-5.0000i
```

After the conversion in real form we have:

```
[V,D]  =  cdf2rdf(U,Lambda)
   V  =

          1.0000   0.4002   -0.0191
               0   0.6479         0
               0        0    0.6479
   D  =
          1    0   0
          0    4   5
          0   -5   4
```

Note that if the matrix is *defective*, i.e. A cannot be diagonalized, the command **eig** returns two matrices solving the problem $AU = U\Lambda$, but the columns of U are no longer independent.

Finally, we must highlight the fact that **eig** also returns generalized eigenvectors and eigenvalues, i.e. given two square matrices A and B, the command

$$[\text{U,Lambda}]=\text{eig(A,B)}$$

returns two matrices such that $A \cdot U = B \cdot U \cdot \Lambda$

5.5.8   Singular value decomposition

Given an $m \times n$ real matrix A, it can always be written as

$$A = U\Sigma V^T$$

where U is a unitary matrix whose dimensions are $m \times m$, V is an $n \times n$ unitary matrix and $\Sigma = \text{diag}(\sigma_1, \ldots, \sigma_n)$ where, $\sigma_i \geq 0$ for each i.

The σ_is are named *singular values* of the matrix A and are generally ordered in decreasing order: $\sigma_1 \geq \sigma_2 \geq \cdots \geq \sigma_n \geq 0$. If A has rank r, then $\sigma_r > 0$, $\sigma_{r+1} = 0$; thus a rank r matrix has r nonzero singular values.

Moreover, the following properties hold:

- $\sigma_i^2(A) = \lambda_i(A^T A)$;

- $\sigma_1(A) = \|A\|_2$.

MATLAB uses the command

$$[\text{U,Sigma,V}]=\text{svd(A)}$$

5.5.9   Hessenberg form

A matrix A is said to be in *Hessenberg form* if all its entries below the first subdiagonal are zero. The function

$$[\text{P,H}]=\text{hess(A)}$$

returns a unitary matrix P and a *Hessenberg matrix* H such that

$$A = PHP^T$$

hence we have

$$P^T AP = \begin{pmatrix} * & * & * & * & * \\ * & * & * & * & * \\ 0 & * & * & * & * \\ 0 & 0 & * & * & * \\ 0 & 0 & 0 & * & * \end{pmatrix}$$

where the character * denotes entries whose values are of no interest for the matrix structure.

5.5.10   Schur decomposition

The idea subtending the Schur decomposition is to upper triangularize a matrix A through a sequence of unitary or orthogonal transformations. This result can be achieved by transforming the matrix A in Hessenberg form, then using the QR algorithm to clean up the subdiagonal entries. Finally, we obtain a unitary matrix U and an upper triangular *Schur matrix* T such that

$$A = UTU^T$$

The corresponding MATLAB command is

$$[\text{U,T}]=\text{schur(A)}$$

which results either in the *complex Schur form* if the matrix A has at least a complex entry, or in the *real Schur form* if all the entries of A are real: in the former case the matrix T is upper triangular, with the eigenvalues of A on the diagonal, while in the latter case it will present the real eigenvalues on the main diagonal and the complex ones in 2×2 block diagonal form. The MATLAB command **rsf2csf** converts the real Schur form into the complex one.

5.5.11   Solving the problem $Ax = b$: pseudo-inversion and least square

When the matrix A is not square or full rank its inverse does not exist. One can compute the *pseudo-inverse* A^+, instead. If A is full column rank, its pseudo-inverse is defined as

$$A^+ = (A^T A)^{-1} A^T$$

otherwise, if A is not full rank, one can use the singular value decomposition to compute the pseudo-inverse

$$A^+ = V \Omega U^T$$

where $\Omega = \text{diag}(\omega_i)$ and $\omega_i = 1/\sigma_i$, $i = 1, \ldots, r$ and $\omega_i = 0$, $i > r$. In MATLAB we have:

$$X = \text{pinv}(A)$$

By using the pseudo-inverse we are now in a position to face the problem

$$Ax = b$$

even when the matrix A is singular; it is well known that singularity leads to nonuniqueness or nonexistence of the solution.

In particular, if the system of equations is over-dimensioned, i.e. if the number of unknown is larger than the number of equations (to be more precise, if the rank of the matrix A is the same as that of the matrix $[A \ b]$), then the problem admits an infinity of solutions. Two possible solutions are obtained with the operators pinv ad \: the former minimizes the norm of the solution, the latter the number of nonzero elements, i.e. it returns as many nonzero entries as the rank of the matrix, as the following example shows.

Let

$$A = \begin{pmatrix} 17 & 24 & 1 & 8 & 15 \\ 23 & 5 & 7 & 14 & 16 \\ 4 & 6 & 13 & 20 & 22 \end{pmatrix}, \qquad b = \begin{pmatrix} 1 \\ 0 \\ 1 \end{pmatrix}$$

Then the commands

```
x=pinv(A)*b;
y=A\b;
```

return

$$x = \begin{pmatrix} -0.0366 \\ 0.0480 \\ 0.0061 \\ 0.0119 \\ 0.0246 \end{pmatrix}, \qquad y = \begin{pmatrix} -0.0376 \\ 0.0429 \\ 0 \\ 0 \\ 0.0406 \end{pmatrix}$$

The norm of the vector x is $\|x\| = 0.0666$, while $\|y\| = 0.07$, but y has only three nonzero entries vs five nonzero elements in x.

When a solution does not exist, i.e. when the rank of A is smaller than that of $[A\ b]$, the problem becomes one of looking for the best approximate solution. The operator pinv aims to minimize the norm of the error:

$$\min_x \|Ax - b\|$$

while lscov minimizes a weighted norm of the error

$$\|Ax - b\|_V^2 = (Ax - b)^T V^{-1}(Ax - b)$$

Finally, nnls seeks a nonnegative solution, solving the problem

$$\min_x \|Ax - b\|$$

subject to the constraints $x_i \geq 0, \forall i$. (By the way, the same function also returns the *dual vector* w, i.e. a vector such that $w_i < 0$ if $x_i = 0$ and $w_i = 0$ if $x_i > 0$.) As a concluding example, consider the problem

$$A = \begin{pmatrix} 0.6868 & 0.5269 \\ 0.5890 & 0.0920 \\ 0.9304 & 0.6539 \\ 0.8462 & 0.4160 \end{pmatrix}, \quad b = \begin{pmatrix} 1 \\ 0 \\ 1 \\ 0 \end{pmatrix}$$

and define a weighting matrix

$$V = \begin{pmatrix} 1 & 0 & 0 & 0 \\ 0 & 10 & 0 & 0 \\ 0 & 0 & 100 & 0 \\ 0 & 0 & 0 & 1000 \end{pmatrix}$$

Then the commands

```
x=pinv(A)*b;
y=nnls(A,b);
z=lscov(A,b,V);
```

give the solutions

$$x = \begin{pmatrix} -0.8183 \\ 2.6018 \end{pmatrix}, \quad y = \begin{pmatrix} 0 \\ 1.3316 \end{pmatrix}, \quad z = \begin{pmatrix} -0.3875 \\ 2.3973 \end{pmatrix}$$

The errors are

$$Ax - b = \begin{pmatrix} -0.1910 \\ -0.2427 \\ -0.0600 \\ 0.3899 \end{pmatrix}, \quad Ay - b = \begin{pmatrix} -0.2983 \\ 0.1225 \\ -0.1292 \\ 0.5540 \end{pmatrix}, \quad Az - b = \begin{pmatrix} -0.0030 \\ -0.0078 \\ 0.2070 \\ 0.6693 \end{pmatrix}$$

Note that, due to the weighting matrix V, the error in the solution z increases at each row.

5.6   More special matrices

In Section 3.4 we presented a set of MATLAB operators to build up some common matrices; in this section we give the complete list.

compan	companion form
tril	lower triangular
triu	upper triangular
rand	random entries matrix (uniform distribution)
randn	random entries matrix (normal distribution)
hankel	Hankel matrix
hilb	Hilbert matrix
invhilb	inverse Hilbert matrix
hadamard	Hadamard matrix
magic	'Magic square'
toeplitz	Toeplitz matrix
vander	Vandermonde matrix
wilkinson	Wilkinson matrix
pascal	Pascal matrix

The syntax of these functions varies from one to another. Let us discuss only those most frequently used. As concerns the others, the reader may refer to the MATLAB manuals or consult the on-line help.

compan requires a vector as input argument: p=[1 0 -7 6], A=compan(p) creates:

$$A = \begin{pmatrix} 0 & 7 & -6 \\ 1 & 0 & 0 \\ 0 & 1 & 0 \end{pmatrix}$$

Remember that the main feature of companion matrices is that the coefficients of their characteristic polynomial are the entries of the vector p.

tril and triu work like diag on matrices: triu(A,k) extracts the upper triangular part of a matrix starting from the kth diagonal (again, $k > 0$ is upper diagonal, $k < 0$ is lower diagonal). tril works in the same way, for the lower triangular part.

rand(m,n) produces an $m \times n$ matrix with entries uniformly distributed in the interval $(0, 1)$; invoked with a single argument it produces square matrices. randn(m,n) works analogously, but the distribution is normal with zero mean and unitary variance. It should be stressed that these functions make use of different generators and different seeds.

5.7 Exercises

Exercise 5.1 Given two matrices A and B, the classic 'row-by-column' product is only one (although the most used) of the possible matrix products. Among them we mention:

- The Schur product, defined by
  $$[A \circ B]_{ij} = a_{ij}b_{ij};$$

- The Lie product, defined by
  $$[A, B]_{ij} = \sum_{k=1}^{n} [a_{ik}b_{kj} - b_{ik}a_{kj}];$$

How does one compute these products in MATLAB?

Exercise 5.2 Solve the following linear system of equations:

$$
\begin{aligned}
2x_1 - 4x_2 + 7x_3 + 4x_4 &= 5 \\
9x_1 + 3x_2 + 2x_3 - 7x_4 &= -1 \\
5x_1 + 2x_2 - 3x_3 + x_4 &= -3 \\
6x_1 - 5x_2 + 4x_3 - 3x_4 &= 2
\end{aligned}
$$

Exercise 5.3 For each of the following vector pairs:

1. $u = (2, -3, 6)$, $v = (8, 2, -3)$;
2. $u = (3, -5, 4)$, $v = (6, 2, -1)$;
3. $u = (3, -5, 2, 1)$, $v = (4, 1, -2, 5)$;
4. $u = (5, 3, -2, -4, -1)$, $u = (2, -1, 0, -7, 2)$.

- compute the inner product $< u, v >$;
- compute the (Euclidean) distance $d(u, v)$;
- verify the Cauchy–Schwartz inequality
  $$|< u, v >| \leq \|u\| \cdot \|v\|;$$

- verify the Minkovsky inequality
  $$\|u + v\| \leq \|u\| + \|v\|.$$

Exercise 5.4 Define the matrix

$$
A = \begin{pmatrix}
1 & 0 & 6 & -3 \\
-1 & 2 & 0 & 2 \\
0 & 3 & -1 & -2 \\
-6 & 0 & 4 & 1
\end{pmatrix}
$$

and try the following commands:

1. `size(A);`
2. `max(A), max(max(A));`
3. `p=poly(A);`
4. `det(A), eig(A), [v,d]=eig(A).`

Exercise 5.5 Compute the row index and the column index of the largest and smallest entry of the matrix A defined in the previous exercise.

Exercise 5.6 The signal depicted in Figure 5.1 is the output of a three-phase rectifier bridge. It is defined by considering at each time instant the largest among the three signals $y_1 = \sin t$, $y_2 = \sin(t + 2\pi/3)$ and $y_3 = \sin(t + 4\pi/3)$. Store the abscissae and values of the signal in two vectors.

1

0   10

Figure 5.1 Three-phase system.

Exercise 5.7 Given the matrices:

$$A = \begin{pmatrix} 5 & 3 & -6 \\ 7 & 2 & 0 \\ -4 & 8 & 1 \end{pmatrix}, \qquad B = \begin{pmatrix} 2+j & -3j \\ 4-2j & 5+6j \\ 8 & 1-j \end{pmatrix}$$

compute:

1. `A*A, A'*A, A.*A;`
2. `diag(A), diag(A,1);`
3. `exp(A), expm(A), sqrt(A), sqrtm(A), sqrtm(A)^2;`
4. `exp(log(A)), expm(logm(A)), funm(A,'exp');`
5. `conj(B), real(B), imag(B), B-real(B)-sqrt(-1)*imag(B);`
6. `rand(B), max(A), norm(A), sign(A).`

Exercise 5.8 Compute a basis and the dimension of the space spanned by the vectors:

1. $\quad x_1 = \begin{pmatrix} 1 \\ 1 \\ 1 \end{pmatrix}, \quad x_2 = \begin{pmatrix} 3 \\ 4 \\ 1 \end{pmatrix}, \quad x_3 = \begin{pmatrix} 5 \\ 3 \\ 9 \end{pmatrix}, \quad x_4 = \begin{pmatrix} 2 \\ -3 \\ 12 \end{pmatrix};$

2. $\quad x_1 = \begin{pmatrix} 1 \\ 2 \\ -3 \end{pmatrix}, \quad x_2 = \begin{pmatrix} 1 \\ -3 \\ 2 \end{pmatrix}, \quad x_3 = \begin{pmatrix} 2 \\ -1 \\ 5 \end{pmatrix}.$

Exercise 5.9 Build up a 4×4 matrix with random entries with normal distribution, mean 10 and variance 0.5.

Exercise 5.10 With reference to the matrix A in Exercise 5.4, compute

1. `test=(A<=1) & (A>=-2), all(test), all(all(test))`;
2. `i=find(A==max(max(A)))`.

Exercise 5.11 Find the set of all the solutions of the system

$$\begin{aligned} x_1 - 3x_2 + 4x_3 - 2x_4 &= 5 \\ 2x_2 + 5x_3 + x_4 &= 2 \\ x_2 - 3x_3 &= 4 \end{aligned}$$

- by reducing the system in *echelon* form;
- by computing a particular solution and the general solution of the homogeneous associated system.

Exercise 5.12 Which condition must the scalars a, b and c satisfy for the system

$$\begin{aligned} x + 2y - 3z &= a \\ 3x - y + 2z &= b \\ x - 5y + 8z &= c \end{aligned}$$

to admit a solution?

Exercise 5.13 From a computational point of view, spectral decomposition algorithms are very fragile in numerical analysis. Compute the eigenvalues and eigenvectors of the matrices

$$A = \begin{pmatrix} 1 & -3 & 3 \\ 3 & -5 & 3 \\ 6 & -6 & 4 \end{pmatrix}, \quad B = \begin{pmatrix} -3 & 1 & -1 \\ -7 & 5 & -1 \\ -6 & 6 & -2 \end{pmatrix}$$

Note that, actually, the matrix B *is not diagonalizable*, as becomes apparent when computing the rank of the eigenvector matrix with different tolerance values.

Exercise 5.14 Compute the singular value decomposition of the matrices defined in the previous exercise.

Exercise 5.15 In poetry, the *sestina* is a particular compositive structure used in Italy and France and based on permuting six words in six six-line stanzas (hence its name). Roughly speaking, a sestina consists of six stanzas, each of them made of six verses (actually there are seven stanzas, but the last, one of three lines only, will not be considered in this exercise); the distinguishing characteristic of the sestina is that in the total thirty-six lines the last words are bound to belong to a restricted set of six words, namely, those defined in the first stanza, according to the permutation order defined in the second stanza (the classic permutation used is the so-called 'cruciate retrogradation', but, according to Raymond Quenau, any permutation makes sense both from a mathematical and from a literary point of view). Hence, having written the first stanza, defining the words, and the second, imposing the permutation, the remaining twenty-four lines have the final words prefixed. Consider, for instance, a sestina by the Italian poet Francesco Petrarca (1304–74), specifically the twenty-second in the 'Canzoniere' collection, whose first stanza is

> A qualunque animale alberga in **terra**,
> se non se alquanti c'hanno in odio il **sole**,
> tempo da travagliare è quanto è 'l **giorno**;
> ma poi che 'l ciel accende le sue **stelle**,
> qual torna a casa e qual s'anida in **selva**
> per aver posa almeno in fin a l'**alba**.

The final words have been highlighted. A translation by Mark Musa is

> For every animal that lives on the **earth**,
> except for but those few that hate the **sun**,
> the time to toil is while it is still **day**;
> but then when heaven lights up all its **stars**,
> some go back home while some nest in the **wood**
> to find some rest at least until the **dawn**.

The next stanza is

> Et io, da che comincia la bella **alba**
> a scuoter l'ombra intorno della **terra**
> svegliando gli animali in ogni **selva**,
> non ho mai triegua di sospir col **sole**;
> poi, quand'io veggio fiammeggiar le **stelle**,
> vo lagrimando e disiando il **giorno**.

whose meaning is

> And I, from the first signs of lovely **dawn**,
> shaking the shadows from around the **earth**

awakening the beasts in every **wood**,
can never cease to sigh while there is **sun**;
then when I see the flaring of the **stars**
I start to weep and long for the gone **day**.

Note that the final words for each line are the same as those of the first stanza,
but their order is different: *terra* (earth), the final word of the first line of the first
stanza, appears in the second line in the second stanza, the second final word *sole*
(sun) becomes the fourth, etc. The third stanza will be created accordingly: *alba*
(dawn), the end-word of the first line of the second stanza becomes the final word
of the second line of the third stanza, and so on. The act of permuting words
can be described mathematically using the powers of an elementary matrix (a row
exchange, if each final word of the first stanza is stored in a matrix).

Find the permutation matrix and the final words of the next twenty-four lines.

Hint. The product of a matrix by a vector of strings (a character matrix) is
a matrix whose entries are integers that can again be converted into a vector
of strings by using the command `setstr`.

Solution. Here is the whole sestina.

A qualunque animale alberga in terra,
se non se alquanti c'hanno in odio il sole,
tempo da travagliare è quanto è 'l giorno;
ma poi che 'l ciel accende le sue stelle,
qual torna a casa e qual s'anida in selva
per aver posa almeno in fin a l' alba.

Et io, da che comincia la bella alba
a scuoter l'ombra intorno della terra
svegliando gli animali in ogni selva,
non ho mai triegua di sospir col sole;
poi, quand'io veggio fiammeggiar le stelle,
vo lagrimando e disiando il giorno.

Quando la sera scaccia il chiaro giorno,
e le tenebre nostre altrui fanno alba,
miro pensoso le crudeli stelle,
che m'hanno fatto di sensibil terra,
e maledico il dì ch'i' vidi 'l sole,
che mi fa in vista un uom nudrido in selva.

Non credo che pascesse mai per selva
sì aspra fera, o di notte o di giorno,
come costei ch'i' piango a l'ombra e al sole,
e non mi stanca primo sonno od alba;
ché, ben ch'i' sia mortal corpo di terra,
lo mio fermo desir vien dalle stelle.

prima ch'io torni a voi, lucenti stelle,

o tomi giù nell'amorosa selva,
lassando il corpo che fia trita terra,
vedess'io in lei pietà, che 'n un sol giorno
può ristorar molt'anni, e nanzi l'alba
puommi arichir dal tramontar del sole!

Con lei foss'io da che si parte il sole,
e non ci vedess'altri che le stelle,
sol una notte, e mai non fosse l'alba,
e non se transformasse in verde selva
per uscirmi di braccia, come il giorno
ch'Apollo la seguia qua giù per terra!

Ma io sarò sotterra in secca selva,
e 'l giorno andrà pien di minute stelle,
prima ch'a sì dolce alba arrivi il sole.

Translation.

For every animal that lives on the earth,
except for but those few that hate the sun,
the time to toil is while it is still day;
but then when heaven lights up all its stars,
some go back home while some nest in the wood
to find some rest at least until the dawn.

And I, from the first signs of lovely dawn,
shaking the shadows from around the earth
awakening the beasts in every wood,
can never cease to sigh while there is sun;
then when I see the flaring of the stars
I start to weep and long for the gone day.

When night drives out the clarity of the day
and our darkness brings out another's dawn,
I gaze all full of care at the cruel stars
that once created me of sentient earth,
and I curse the first day I saw the sun
which makes me seem a man raised in the wood.

I think there never grazed in any wood
so cruel a beast, whether by night or day,
as she for whom I weep in shade and sun,
from which I am not stopped by sleep or dawn;
for though I am a body of this earth,
my firm desire is born from the stars.

Before returning to you, shining stars,
or sinking back into the amorous wood
leaving my body turned to powdered earth,
could I see pity in her, for one day

can restore many years, and before the dawn
enrich me from the setting of the sun!

Could I be with her at the fading sun
and seen by no one, only by the stars,
for just one night, to never see dawn,
and she not be transformed into green wood
escaping from my arms as on the day
Apollo had pursued her here on the earth!

But I'll be under earth in a dry wood
and day will be all full of tiny stars
before so sweet a dawn will see the sun.

Chapter 6

Polynomials

Polynomials are represented in MATLAB by row vectors containing their coefficients in decreasing power order; for instance, the polynomial $p(s) = s^3 + 4s^2 + 2s + 5$ is represented by the vector p=[1 4 2 5]. Of course, zero coefficients must also be considered: $s^3 + 1$ is represented by p=[1 0 0 1].

6.1 Basic operations

We have already mentioned the *characteristic polynomial* of a square matrix A, which is computed by the instruction p=poly(A); the function poly is also used to build up the monic polynomial with a *given set of roots*: for instance, the monic polynomial with roots at $-3 \pm j2, -5$, i.e. the polynomial $p(s) = (s + 3 + j2)(s + 3 - j2)(s + 5)$, is obtained by

```
r=[-3-2*j,-3+2*j,-5];
p=poly(r);
```

The *polynomial product* is computed by the function conv:

```
a=[1 2 3]; b=[4 5 6];
c=conv(a,b);
```

gives c=[4 13 28 27 18] (indeed $(s^2 + 2s + 3)(4s^2 + 5s + 6) = 4s^4 + 13s^3 + 28s^2 + 27s + 18$).

The inverse operation is polynomial division, obtained with the command deconv. Its syntax is

```
[q,r]=deconv(a,b);
```

which returns two polynomials, $q(s)$ (quotient) and $r(s)$ (remainder) such that

$$a(s) = q(s)b(s) + r(s)$$

Obviously, if $a(s)$ is divisible by $b(s)$ the remainder $r(s)$ is zero.

The *roots of polynomials* are computed with `roots` (which actually builds up the companion matrix associated to the polynomial and computes its eigenvalues). With reference to the previous example, `roots(p)` returns

```
ans=
    -3-2*j
    -3+2*j
    -5
```

`polyval(p,x)` computes the value $p(x)$; x may be either a scalar, or a vector or a matrix. In the first case it returns the value of the polynomial at the given point, in the second case we obtain the image of the set of points in the vector, and in the third case we obtain a matrix whose entries are $a_{ij} = p(x_{ij})$. Hence, in the example on page 31 the third and the fourth row, given below,

```
x1=3*s.^2+5*s+7;
x2=s.^3+5*s.^2+7*s+12;
```

can be rewritten as

```
x1=polyval([3,5,7],s);
x2=polyval([1 5 7 12],s);
```

`polyvalm(p,A)` works like `polyval`, but in the space of (square) matrices. Remember that, given the nth degree polynomial

$$p(s) = \sum_{i=0}^{n} p_i s^i$$

and the square matrix A, $p(A)$ is defined by

$$p(A) = \sum_{i=0}^{n} p_i A^i$$

If, for instance, $p(s) = s^2 + s + 1$ and $A = \begin{pmatrix} 2 & 3 \\ 5 & 8 \end{pmatrix}$, then `polyval(p,A)` gives

$$\text{ans} = \begin{pmatrix} 7 & 13 \\ 31 & 73 \end{pmatrix}, \quad \text{that is,} \quad \begin{pmatrix} p(2) & p(3) \\ p(5) & p(8) \end{pmatrix}$$

thus computing the values elementwise, while `polyvalm(p,A)` returns

$$\text{ans} = \begin{pmatrix} 22 & 33 \\ 55 & 88 \end{pmatrix}, \quad \text{that is,} \quad A^2 + A + I$$

this is the correct result in matrix algebra.

In particular, `polyvalm(poly(A),A)` returns the zero matrix, according to the Cayley–Hamilton theorem.

A quite frequent operation is the derivative. MATLAB uses the function `polyder` to compute the derivative of a polynomial; in particular

- `q=polyder(p)` returns the derivative of the polynomial: $q(s) = d\,p(s)/d\,s$.
- `q=polyder(a,b)` returns the derivative of the product $a(s)b(s)$.
- `[q,d]=polyder(a,b)` returns the derivative of the polynomial ratio as a rational function:

$$\frac{q(s)}{d(s)} = \frac{d}{ds}\frac{a(s)}{b(s)}.$$

Another common operation when dealing with polynomials is *partial fraction decomposition*: given a rational fraction function $b(s)/a(s)$ with partial fraction decomposition:

$$\frac{b(s)}{a(s)} = \frac{r_1}{s - p_1} + \frac{r_2}{s - p_2} + \cdots + \frac{r_n}{s - p_n} + k(s)$$

the command `[r,p,k]=residue(b,a)` returns the decomposition, where the meaning of the symbols is obvious. In the case where the multiplicity m_j of the *i*th root of $a(s)$ is larger than 1, $m_j > 1$, the decomposition also includes such terms as

$$\frac{r_j}{s - p_j} + \frac{r_{j+1}}{(s - p_j)^2} + \cdots + \frac{r_{j+m_j-1}}{(s - p_j)^{m_j}}$$

`residue` also works dually, when invoked with three input and two output arguments: `[b,a]=residue(r,p,k)` returns the polynomial $a(s)$ and $b(s)$. Note that the joint use of the two working modes allows the user to obtain the real decomposition easily, even in the case of complex poles. Consider, for instance, the rational function

$$\frac{b(s)}{a(s)} = \frac{s - 1}{s^4 + 8s^3 + 23s^2 + 26s + 10}$$

The command

$$[r,p,k]=residue(b,a)$$

returns

```
r =
   -0.2600 + 0.3200i
   -0.2600 - 0.3200i
    0.5200
   -0.4000


p =
   -3.0000 + 1.0000i
   -3.0000 - 1.0000i
   -1.0000
   -1.0000


k =
   []
```

to obtain the real form it is necessary to compute the second-degree term associated to the pair $p = -3 \pm j$ by using the command

$$\texttt{[b1,a1]=residue(r(1:2),p(1:2),[])}$$

whose result is

```
b1 =

      -0.5200    -2.2000

a1 =

       1.0000    6.0000    10.0000
```

Hence the expansion is

$$\frac{0.52}{s+1} - \frac{0.4}{(s+1)^2} - \frac{0.52s + 2.2}{s^2 + 6s + 10}$$

The following list summarizes the polynomial operators

`poly`	polynomial with given roots
`roots`	roots of a polynomial
`polyval`	value of a polynomial at a given point
`polyvalm`	matrix evaluation of a polynomial
`conv`	polynomial product
`deconv`	polynomial division
`residue`	partial fraction decomposition
`polyder`	polynomial derivative
`polyfit`	polynomial interpolation

The last thing to do is to discuss `polyfit`. The next section will be devoted to this command, and to the more general problem of interpolation in MATLAB.

6.2 Interpolation

The command `polyfit` solves linear interpolation problems. Given the independent variable in a vector x and the values to interpolate in a vector y, the command `p=polyfit(x,y,n)` returns an nth degree polynomial $p(x)$ that interpolates the pairs (x_i, y_i) by least square interpolation.

At this point, it will be useful to mention the interpolation methods available in MATLAB:

- `polyfit` Polynomial interpolation. Discussed above.
- `spline` Cubic spline interpolation. Given the vectors x and y above, and an abscissae vector x_i, the command

$$\texttt{yi=spline(x,y,xi);}$$

returns a vector y_i with the values associated with the abscissae x_i obtained through cubic spline interpolation.
- `interpft` Computes a larger number of samples starting from a vector of samples y, assuming a fixed step. Its syntax is

$$\texttt{yi=interpft(y,n);}$$

and the result is an n-dimensional vector y_i. The command transforms the original sequence through FFT, then computes an inverse transform with a larger number of points.
- `interp1` Is the general table look-up algorithm. It uses linear interpolation, and its call is

$$\texttt{yi=interp1(x,y,xi);}$$

where the symbols retain the same meaning as above, but y may be a matrix, then also y_i will be a matrix whose columns interpolate the corresponding columns of y at the same abscissa points. It is also possible to choose the interpolation method by using

$$\texttt{yi=interp1(x,y,xi,'}method\texttt{');}$$

The following methods are possible:

1. `'linear'` linear interpolation;

2. `'spline'` cubic spline interpolation;

3. `'cubic'` cubic interpolation.

All these methods require x to increase monotonically; moreover, `'cubic'` also requires equally spaced abscissae.
- `interp2` Computes a 2D interpolation. The command

$$\texttt{zi=interp2(x,y,z,xi,yi,'}method\texttt{');}$$

works like `interp1`, but uses 2D tables. The possible methods are now `'linear'` and `'cubic'` only. Again, `'linear'` is the default method.

6.3   Exercises

Exercise 6.1 Given the polynomials $a(s) = s^3 + 2s^2 + 5s + 1$ and $b(s) = s^2 + 1$, compute $a(s) + b(s)$, $a(s) - 5b(s)$, $a(s)b(s)$ and $a(s)/b(s)$.

Exercise 6.2 Determine the (monic) polynomials whose roots are

1. $-1, -3, -5$;
2. $-2, -5 \pm j2$;
3. $1, -1, j$.

Exercise 6.3 Compute the roots of the following polynomials:

1. $s^3 + 5s^2 + 7s + 2$;
2. $(s + 1)^3(s^2 + 12s + 9)$;
3. $2s^5 + j3s^2 + 1$.

Exercise 6.4 The minimal polynomial of a matrix is the lowest degree monic polynomial among the annihilators of a matrix. Compute the minimal polynomial of

$$A = \begin{pmatrix} 2 & 1 & 0 & 0 \\ 0 & 2 & 0 & 0 \\ 0 & 0 & 1 & 1 \\ 0 & 0 & -2 & 4 \end{pmatrix}$$

Exercise 6.5 The space of polynomials with given maximum degree is a vector space. The basis that MATLAB uses for this vector space is $\{1, s, s^2, \ldots\}$, but, as for any vector space, we can consider different bases. Compute the representation of the polynomial $p(s) = 2s^2 - 5s + 6$ in the basis $\{1, s - 1, (s - 1)^2\}$.

Exercise 6.6 Looking for regions in the complex plane to which all the zeros of a polynomial belong, starting from the knowledge of the coefficients of the polynomial itself, is a problem faced by generations of mathematicians. The main results have led to circular regions around the origin to which either all of the zeros or no zeros belong. Here are some examples: given the polynomial

$$p(s) = a_n s^n + a_{n-1} s^{n-1} + \cdots + a_0$$

we have:

1. the zeros belong to the disk

$$|s| < 1 + \max \left| \frac{a_k}{a_n} \right|, \quad k = 0, \ldots, n - 1;$$

2. for each pair $p > 1$, $q > 1$ such that $1/p + 1/q = 1$, the zeros belong to the disk

$$|s| < \left[1 + \left(\sum_{k=0}^{n-1} \left| \frac{a_k}{a_n} \right|^p \right)^{q/p} \right]^{1/q};$$

3. the zeros belong to the disk

$$|s| \leq \sqrt{ 1 + \left| \frac{a_0}{a_n} \right|^2 + \left| \frac{a_1 - a_0}{a_n} \right|^2 + \cdots + \left| \frac{a_n - a_{n-1}}{a_n} \right|^2 };$$

4. the zeros belong to the disk

$$|s| \leq \sum_{k=1}^{n} \left| \frac{a_{n-k}}{a_n} \right|^{1/k};$$

5. the zeros lie outside the disk

$$|s| = \min \frac{|a_0|}{|a_0| + |a_k|}, \quad k = 1, \dots, n.$$

Verify the accuracy of these estimates by using different test polynomials.

Exercise 6.7 Compute the turning points and the possible points of inflection of the functions

1. $f(x) = x^4 + 5x^3 + 1$;
2. $g(x) = \dfrac{x+1}{x^3 + 3x + 1}$.

Exercise 6.8 Compute the partial fraction decomposition of the function

$$\frac{s^2 + 3s}{s^4 + 4s^3 + 7s^2 + 6s + 2}$$

Exercise 6.9 The air density ρ (in kg m^{-3}) varies with the altitude h (in km) according to Table 6.1

Table 6.1 Air density

h	7	10	15	21	27	34	39	43	47	51	55	59	61
ρ	556	369	191	75	26.2	9.9	4.4	2.3	1.4	0.8	0.5	0.33	0.25

Compute interpolating polynomials of degree $n = 1, 2, \dots, 6$ and determine the norm of the error.

Compute, moreover, the values of air density at altitude 10, 20, 30, 40, 50 and 60 km by using the algorithms discussed in Section 6.2 (note that the values of the independent variable are not equally spaced).

Exercise 6.10 Compute the coefficients of the *Legendre polynomials* of degree $n = 1, 2, 3$ defined by the formula

$$L_n(x) = \frac{1}{2^n n!} \frac{d^n}{dx^n} \left(x^2 - 1 \right)^n$$

Chapter 7

Graphics

Graphics is one of the most advanced characteristics in MATLAB. This chapter is organized as follows: we start with the simplest 2D graphic commands to allow the reader to trace the graph of functions of one real variable quickly; then axis scaling and labels are defined; an advanced part follows, with a brief account of MATLAB graphics primitives for handling more complex situations; the 3D part is structured in the same way.

For a deeper insight on the graphic capabilities of MATLAB the reader is urged to consult Appendices A and B.

7.1 2D graphics

MATLAB allows one to trace graphics on different windows, named 'figures'. By default, MATLAB traces graphics on window no. 1; if we want to open more windows we type the command `figure(n)`, where n defines the number of the window to open. From now on MATLAB will plot graphics on the nth window, unless the user again changes window with a new `figure` command (or selects a different window by using the mouse pointer or the Windows key combination ALT–TAB). To close the nth window we use the command `close(n)`. `close all` closes all the windows, while `close` alone closes the current window only.

Let us work for the moment on the first window. The main command to trace 2D graphics is `plot`.

Let us start with a simple example: we want to represent the following sequence graphically:

$$\{0, 0.48, 0.84, 1, 0.91, 0.6, 0.14\}$$

Then it is sufficient to type:

```
Y=[0 .48 .84 1 .91 .6 .14]
plot(Y)
```

66

which results in the plot given in Figure 7.1.

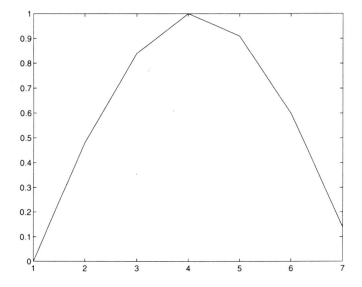

Figure 7.1 Simple graphic.

Some remarks arise from this first example. First of all, we note that axis scaling is automatic; moreover we note that the title of the window is 'Figure No. 1' to keep track of the window we are currently working on. Finally, the points are joined by solid lines.

The instruction `plot` accepts an optional string parameter (i.e. surrounded by apostrophes) to define the type and color of the plots.

We have four types of line, five types of points and eight basic colors, as summarized in Table 7.1.

If we want to plot the points of the vector Y defined above with green circles and without joining them, we will use the command

plot(Y,'og')

We now add some comments (Figure 7.2):

```
title('Point sequence')
xlabel('Abscissa')
ylabel('Ordinate')
grid
```

`title`, `xlabel` and `ylabel` define the title and axis labels; each time we retype these commands, the previous labels are replaced by the new ones. The command

Table 7.1 `plot` options

Line type		Point type		Color	
solid	–	point	.	yellow	y
dashed	--	plus	+	magenta	m
dotted	:	star	*	cyan	c
dash–dotted	-.	circle	o	red	r
		cross	x	green	g
				blue	b
				white	w
				black	k

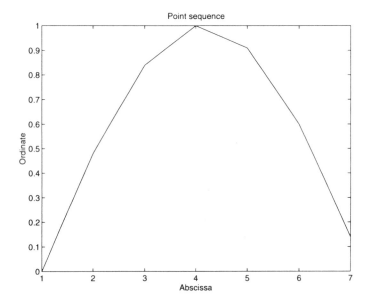

Figure 7.2 Adding text.

`grid` draws a grid on the plot; each time it is executed it reverses its status, then on typing `grid` again the gridding disappears.

To insert text anywhere in the plot there exists the command

<code>text(x,y,'text')</code>

which places the character string *text* at the point with coordinates (x, y). These three parameters may also be vectors (of same length); in this case the ith entry of the string vector (actually it is a matrix of characters, see Section 3.5) is placed at the point with coordinates (x_i, y_i). There also exists a mouse-driven version of the same command:

$$\texttt{gtext('\textit{text}')}$$

places the text at the point selected by pressing the left-hand mouse button in the graphic window.

We now present something less trivial: we want to trace the graph of the function $y = f(t)$. The command $\texttt{plot(a,b)}$ draws the graph of the pairs (a_i, b_i); then the first thing to do is to tabulate the function under consideration, as seen in Section 4.3, and then to draw its graph.

As a simple example we consider the function $y = \sin t$ (Figure 7.3):

```
t=0:.05:4*pi;
y=sin(t);
plot(t,y)
```

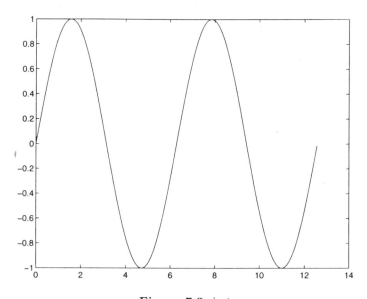

Figure 7.3 $\sin t$.

Obviously, the type and color of the graph can be assigned according to Table 7.1: the same plot traced with a red dashed–dotted line is obtained with $\texttt{plot(t,y,'r-.')}$.

By the way, tabulation is not always a trivial operation, since it may happen that the function exhibits different 'time scales'; this is, for instance, the case for the function $f(t) = e^{-10t} + \sin(0.05t)$, for which the use of a very small tabulation step stresses the contribution of the exponential term, while a large step highlights the sine function. These situations are to be solved for any particular case, possibly by using two plots. Hence, we may think of producing two graphs on two different windows, using two different timescales:

```
t1=0:.01:1;  first timescale, 0 thru 1, step 0.01
y1=exp(-10*t1)+sin(0.05*t1);
t2=0:200;  second timescale, 0 thru 200, step 1
y2=exp(-10*t2)+sin(0.05*t2);
figure(1)
plot(t1,y1)
figure(2)
plot(t2,y2)
```

The result is shown in Figure 7.4.

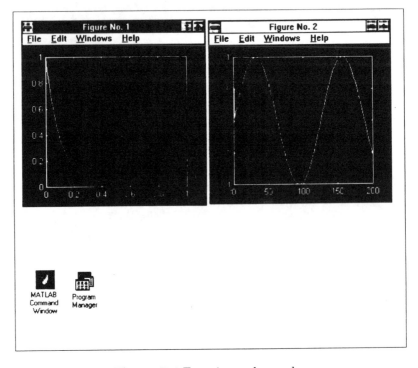

Figure 7.4 Two timescale graphs.

This last example leads us to the problem of simultaneously tracing more graphs, which will be addressed in the next section.

7.2   Multiple plots

Obviously, the best method of comparing more plots is to trace them on the same window. We have three methods for tracing multiple plots:

1. If we need to trace the graph of more plots on the same abscissa values, we can use `plot` with matrix arguments. Using `plot(X,Y)`:

   - if Y is a matrix and X a vector, then the columns (or the rows) of Y are plotted vs the values in X;
   - if X is a matrix and Y a vector, the columns (or the rows) of X are plotted vs the vector Y.

   For instance, to trace the graph of the functions $y = \sin t$ and $y = \cos t$ for $t \in [0, 2\pi]$ we type the following lines:

   ```
   t=(0:.1:2*pi)';  transpose to obtain a column vector
   Y=[sin(t),cos(t)];
   plot(t,Y)
   ```

   Again, note that axis scaling is automatic and is determined on the basis of the largest and the smallest values in the matrix Y and the vector t, for ordinate and abscissa respectively.

2. If the abscissae domains are different, we use `plot` with more arguments:

   ```
   plot(X1,Y1,X2,Y2,...,Xn,Yn)
   ```

   traces the graphs of the vector pairs (X_i, Y_i), $i = 1, ..., n$. To trace the plots of $\sin t$ for $t \in [0, 3]$ and $\cos t$ for $t \in [1, 4]$ we use (Figure 7.5)

   ```
   t1=(0:.1:3)';
   y1=sin(t1);
   t2=(1:.1:4)';
   y2=cos(t2);
   plot(t1,y1,t2,y2)
   ```

3. Finally, one may 'keep' the previous graph with the couple of commands `axis(axis)`, `hold`.

   ```
   plot(X1,Y1)
   axis(axis)
   hold
   plot(X2,Y2)
   plot(X3,Y3)
       ⋮
   plot(Xn,Yn)
   ```

   This solution is the preferred one when a graph must be emphasized against the others; then it must impose axis scaling.

   For instance, if we want to compare the behavior of the function $y = \sin t$ with that of $y = t$ and $y = e^t$, in the domain $t \in [0, 5]$, it is better to avoid method 1, which would impose the maximum y-scale $e^5 = 148.41$, thus concealing the behavior of $\sin t$; the solution is then

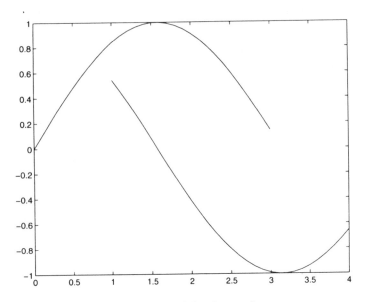

Figure 7.5 Multiple graphs.

```
t=0:.1:5;
plot(t,sin(t))
axis(axis)
hold
plot(t,[t;exp(t)])
```

When tracing graphs according to the first two methods, the default colors are, along with their order, those defined in Table 7.1, i.e. the first curve is drawn in yellow, the second in magenta, the third in cyan, and so on, rotating among the first colors of the table.

hold is a command to freeze the current definition of axis. It is a switching command, in the sense that each time we type **hold** its status is reversed, when a graph is frozen, the next **hold** unlocks the scale factors; a further **hold** freezes the scale again. **hold on** and **hold off** allows one to define the holding status explicitly.

A legend may be useful to distinguish among different plots; this is available via the command **legend**. If the current window contains N curves, the command **legend('**linetype1**','**string1**',...,'**linetypeN**','**stringN**')** creates a legend associating each line type with the corresponding string.

7.3   Axis scaling

As mentioned above, axis scaling is automatic, but it is often necessary to redefine it. To accomplish such a task, we use the command `axis`, which we have already mentioned in conjunction with the command `hold`. `axis` has three main working functions:

- `axis([xmin xmax ymin ymax])` defines the extreme values for the x- and y-axes; *after* graph plotting it is possible to change axis scaling by using this command.
- `axis(axis)` seen above, freezes axis scaling.
- `axis('auto')` restores automatic axis scaling.

`v=axis` stores the current axis limits in the vector v. If one of the limits has the value `inf`, then the corresponding extreme data value is used; it is equivalent to (local) auto-scaling; then `axis([-inf inf -inf inf])` is the same as `axis('auto')`.

Other capabilities of `axis` are:

- `axis('ij')` uses matrix coordinates, with the vertical axis directed downward (according to the index-increasing direction in a matrix).

- `axis('xy')` uses Cartesian coordinates, with the vertical axis directed upwards.

- `axis('square')` imposes a square graphic region.

- `axis('equal')` sets equal axis scale factors on both axes.

- `axis('off')` turns axis tracing off.

- `axis('on')` turns axis tracing on.

It is possible to use the command `zoom` to magnify the details of the current graph using the mouse and the left-hand mouse button to define the region to enlarge. Pressing the right-hand button returns to the previous view. Finally, `zoom off` turns off the zoom mode.

7.4   Complex data

To trace curves in the complex plane we must use *a single* vector of complex numbers: if X is a complex vector, `plot(X)` is the same as `plot(real(X),imag(X))`.

To trace more complex curves one must use method 3 given in Section 7.2 or, alternatively, method 2 by plotting imaginary coefficients vs real parts for each curve. Since, often (think of the roots of polynomials or root loci), the curves in the complex plane are not connected, but rather grouped into 'branches', it is convenient to specify tracing by points rather than using solid lines.

7.5 More 2D graphs

This is the last section on basic commands for 2D plotting. It introduces all the remaining functions to trace 2D plots using representations that are different from the linear-axis Cartesian coordinates discussed up to now.

semilogx plots the x-axis in $\log_{10}$ scale, and the y-axis in linear scale; its parameters are the same as those of plot.

semilogy works like the previous command, but the role of the x- and y-axes is reversed.

loglog works like the two previous commands, with both axis logarithmically scaled.

polar(rho,theta) traces graphics in polar coordinates; the anomaly θ must be in radians, but the axes are traced in degrees; moreover, it is impossible to use multiple arguments (unless using method 3).

bar, stairs trace bar and staircase graph, respectively; invoked with output arguments, such as [Xb,Yb]=bar(X,Y), they do not trace plots, but return vectors such that the command plot(Xb,Yb) traces bar or staircase plots. This is useful, for instance, when continuous- and discrete-time plots are to be represented simultaneously on a single figure. It is possible to use multiple arguments only according to method 1, if the vectors to trace are given as column vectors.

hist computes and traces histograms. In particular, hist(x) computes and traces the histogram related to the data in vector x, with the x-axis scaled on the vector values and the related occurrence frequencies on the y-axis. The histogram is computed and plotted by dividing the x-axis into segments; a second optional parameter may be either an integer, defining the number of bins to trace, or a vector, defining the intervals to highlight. For instance, suppose we want to test the accuracy of the uniform random generator rand. By definition, a vector of samples of a stochastic variable with uniform distribution must result in a flat histogram; then we can use the following lines to trace a twenty-bin histogram:

```
n=100;
y=rand(n,1);
hist(y,20)
```

By increasing the number n the histogram becomes more flat (Figure 7.6). The polar coordinates version of this command is rose.

fplot traces the graph of given functions. Its two most interesting forms are:

fplot('*function*',[xmin xmax])

and

[x,y]=fplot('*function*',[xmin xmax])

By using the first form the function *function* is plotted in the abscissa interval $[x_{min}, x_{max}]$. The second form, on the contrary, does not result in plotting

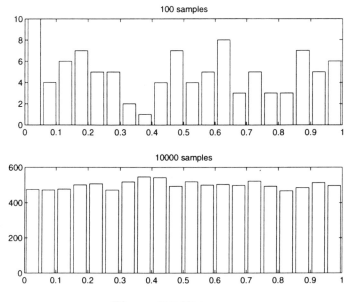

Figure 7.6 Histogram.

actions, but returns two vectors of abscissae and values (or a matrix of values, if the function under consideration is a vectorial one) that can then be plotted by using `plot(x,y)`.

7.6 Plotting more graphs on the same window

We have explained how to open more graphic windows, but sometimes it is necessary to visualize more plots that, although referred to nonhomogeneous quantities, depend on the same variable. Think, for instance, of Bode plots, which represent completely different entities (modulus and phase angle of complex rational functions), but their behavior depends on the same variable (the set of imaginary numbers). Instead of using two windows and trying a wayward alignment between them (consider, moreover, that by selecting one window we are forced to deselect the other), we can make use of the instruction `subplot`. Its syntax is

$$\texttt{subplot(m,n,p)}$$

which splits the current window into `m` rows and `n` columns and selects the `p`th subwindow; the windows are ordered from left to right and from top to bottom; then the window at row i and column j is denoted by the number $p = (i-1)n + j$. To split the current window into four subwindows (two rows and two columns) and select the lower left one (row 2, column 1) we use

```
subplot(2,2,3);
```

The following graphic commands will affect this subwindow. To select the one up on the right we use

```
subplot(2,2,2);
```

Finally, to restore a single window we type:

```
subplot(1,1,1);
```

As an example, the graphic given in Figure 7.6 has been traced by the code:

```
subplot(2,1,1)
n=100;
y=rand(n,1);
hist(y,20);
title('100 samples')
subplot(2,1,2)
n=10000;
y=rand(n,1);
hist(y,20);
title('10000 samples')
```

7.7   3D graphics

3D graphics are surely more complex than 2D graphics. It is sufficient to think that a 3D object may appear in completely different ways by simply changing the point of view. We can divide 3D objects into three wide categories:

- surfaces generated by mathematical functions of two variables $z = f(x, y)$;
- curves or surfaces of which a parametric description is known $x = x(t)$, $y = y(t)$, $z = z(t)$;
- surfaces of revolution generated by the rotation of a curve about a fixed axis.

Before discussing these three possibilities, a digression about the definition of the view point, i.e. the point from which we look at the 3D graph, may be of interest. MATLAB allows one to define the view point angle, but not the distance of the 'eye' from the plotted object. This is done through two angles, namely the *azimuth* θ and the *elevation* ϕ, which we can think of as longitude and latitude respectively. Referring to the picture in Figure 7.7, we see that θ is measured starting from the y *negative semiaxis*. The default values are $\theta = -37.5°$ and $\phi = 30°$. It is possible to change these values by using the command view; it works essentially in two ways:

- view(az,el) sets the view point as az (θ) and el (ϕ) in degrees.

- view([x,y,z]) allows one to define a view point $P(x, y, z)$ in Cartesian coordinates; since, as mentioned before, it is not possible to define the observation distance, the length of P is ignored.

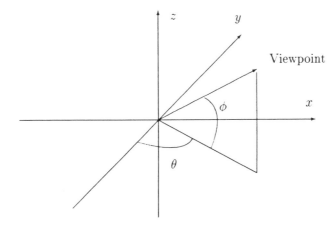

Figure 7.7 Reference frames.

7.8 Functions of two variables

Given a real function of two real variables $z = f(x, y)$, we use the command **mesh** or the command **surf** to plot its graph. The difference between the two commands is that the former traces the graph using line segments joining the z values, while the latter also paints the regions, named 'faces', bounded by these segments; their syntax is the same, so we will describe the **mesh** command only. An alternative to a 3D plot is a 2D representation of the surface through contour curves, using the command **contour**.

The first operation is to build up, by using the command **meshgrid** two matrices X and Y defining the domain of the function under consideration; the X matrix stores copies of the x interval arranged by rows, the Y matrix contains copies of the y values stored by columns. In other words, given two vectors with the x and y intervals defining the domain,

```
[X,Y]=meshgrid(x,y)
```

creates the matrix X with as many rows as the length of the vector y, and each row is a copy of the vector x; analogously, the Y matrix has as many columns, copies of the vector y, as the number of elements of x. Thus the generic point with coordinates (x_i, y_j) can be computed by $(X(i, j), Y(i, j))$.

Then, if we want to compute the values of the function $z = \sin x \cos y$ for $x \in [0, 4]$ and $y \in [-2, 1]$ we can use the following lines:

```
x=0:.1:4;
y=-2:.1:1;
[X,Y]=meshgrid(x,y);
Z=sin(X).*cos(Y);
```

We have now built up the domain and the values of the function; we now only need to trace the graph.

The 2D contour plot is obtained by

$$contour(x,y,Z)$$

Note that, as the plot is 2D, the first two input arguments are the vectors defined before; thus in this case the instruction meshgrid is only used to compute the values of the function. In the case of 3D plots, the matrices X and Y are also used for plotting: the command

$$mesh(X,Y,Z)$$

traces the 3D plot shown in Figure 7.8.

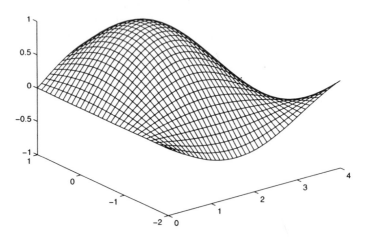

Figure 7.8 3D plot.

It is, moreover, possible to trace both the surface and its contour curves (in the x–y plane) with the command meshc (or surfc). Finally, meshz evidences the reference plane tracing lines joining the zero-altitude points with the x–y plane (by the way, an analogous command surfz does not exist).

7.9   Color in MATLAB

The instruction `mesh` and its alternative forms discussed in the previous section accept a fourth matrix C as an optional parameter. Its rule is to select the colors used in plotting the elements defining the surface.

At this point a short discussion about the use of color to represent a vector will be useful. The reader should remember that measuring a set of objects means defining a correspondence between the members of the set and an ordered set. If the ordered set is the set of (positive) integers, we are 'counting' the set: if we say that there are three books on a desk, we are associating the number 'one' to the first book, 'two' to the second book and 'three' to the third book. If the set is the field of the real numbers we obtain the measure in the usual sense, and using the letters of the alphabet we could speak of line segments whose lengths are 'a', or 'b', and so on. Obviously, length 'b' does not imply a double length with respect to 'a': if the ordered set is the diatonic scale, the note D does not correspond to a double vibration frequency with respect to the note C. The key point is only the definition of *order* on a set. In particular, we may order a set of colors in an arbitrary way and define a measure by colors. Thus if we choose the order 'green–white–red' we can speak of 'green entries' in a vector, i.e. those whose dimensions are the smallest, while 'red entries' are the largest and 'white' ones are the others.

Although the ordering is arbitrary, it is impracticable to insist on this point: a complete random ordering would force such a complex interpretation of the data as to result in a useless measure.

One possible way of overcoming this drawback is to make use of sortings that we are used to as tactile ones: sensations of cool and warm make us associate the red color to a high-temperature level, white to a higher one, while blue and black are related to lower temperatures. MATLAB is equipped with a set of *color maps*, that is, a set of matrices defining an order among different colors. Among them some deserve mentioning: the `hot` map defines the 'warm' colors black–red–yellow–white, along with intermediate tones; the `cool` map defines the cold tones (different blue tones); the `gray` map refers to different gray tones. To activate these maps, it is sufficient to type the command `colormap(`*map*`)`: typing `colormap(hot)` the 'warm' colors are the current ones; any graphic 3D command introduced after this command will paint the smallest elements in black, and the largest in white.

After the above considerations, we are able to give an interpretation of a color as a measure. In standard plots the measure is defined by an axis with numerical labels attached to it, while by using colors we may think of a line segment painted with different colors as 'proportional' to y-axis values. When using standard axes, MATLAB automatically scales the axes, although we can rescale the plot (see Section 7.3). Analogously, when using colors MATLAB scales the colors, associating the first color of the current map with the smallest value to be represented, and the last color with the largest. Moreover, it is possible to 'rescale' the colors: the command `caxis` achieves this task. Suppose we want to represent a set of data whose values belong to the interval $[0, 10]$; typing

```
caxis([0 5])
```

data with values greater than five are not represented; on the other hand, typing

```
caxis([0 1000])
```

the graphic uses only the 'low' colors of the current map.

Now we can go back to the command **mesh**. If the optional parameter C is not given explicitly, MATLAB assumes $C = Z$ by default; hence the colors used are taken from the current map and are proportional to the z-values.

The relevance of the parameter C increases dramatically in the representation of functions of three variables. A classic Cartesian approach would require a 4D space; the reader may think of the problem of representing the temperature distribution on the surface of a 3D body, where each point to trace is characterized by a triple of values defining its spatial location and a fourth number representing the temperature. A possible representation uses the triplet to plot the point in 3D space and the fourth data to define the color of the point. This is done by defining a C matrix containing data associated with the temperature. By the way, in this case it is better to use **surf** rather than **mesh** in order to obtain a more readable plot.

The following example, with a high-resolution screen with 256 colors, gives a rather appealing result.

We consider the plane defined by the equation $z = x + y$, and let the x–y domain be the set $x \in [0, 5]$, $y \in [0, 5]$; a fourth variable w to be represented attains its maximum at the center of the x–y rectangle $[0, 5] \times [0, 5]$, and decays exponentially; for instance w can be described by $w = \exp(-(x - 2.5)^2 - (y - 2.5)^2)$. We now want to represent this situation graphically.

A possible solution to this exercise is obtained by the following sequence of MATLAB commands:

```
x=0:.1:5;[X,Y]=meshgrid(x);
Z=X+Y;
W=exp(-(X-2.5).^2-(Y-2.5).^2);
surf(X,Y,Z,W);colormap(hot);caxis([-.11 1]);shading interp
```

The result is plotted in Figure 7.9.

Note that **meshgrid** invoked with a single argument assumes $x = y$. The command **shading interp** defines the transition between adjacent colors (in the present case by interpolation, obtaining a fading effect), and will be described briefly later in this section. The automatic scaling of the color axis would associate the white color with the largest value ($w = 1$) of w and black with the smallest ($w = \exp{-12.5} \approx 3.7 \times 10^{-6}$); in our example the command **caxis** redefines the color axis scaling giving the color black to the value -0.11, so that the points with the smallest w-value (always positive) will not be painted in black, but in a dark color.

We conclude this section with a description of the color tones handled by MATLAB. This topic is strictly related to the definition of color maps, which are, as mentioned above, ordered sets of colors. The three basic colors in MATLAB are red, green and blue. A single color is defined by a triplet of numbers between 0 and 1 that characterizes the relative intensity of the red–green–blue components; red is, for instance, defined by the triplet $(1, 0, 0)$, while $(0.5, 0, 0)$ defines a darker tone of red. According to the theory of colors, black and white are not, strictly speaking, true colors: the former is a complete absence of color, the latter is due to the simultaneous presence of all the colors of the visible spectrum; consequently their MATLAB representation will be $(0, 0, 0)$ and $(1, 1, 1)$ respectively. A color map is defined as a three-column matrix, each row of which defines

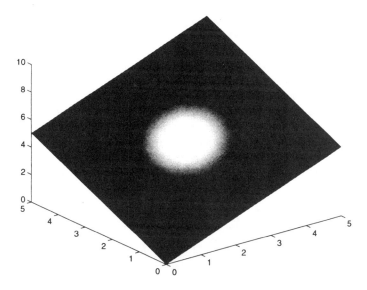

Figure 7.9 Using colors

a color. The predefined MATLAB maps have sixty-four rows, thus they define sixty-four colors. Then the following question arises: how, with a 256 (or more) color screen, are the remaining colors handled or, in other words, how are the sixty-four predefined colors shaded off? The reader should remember that both **mesh** and **surf** plot 3D graphics, but only the latter paints the 'faces', i.e. the small surfaces bounded by the mesh gridding. The command **shading** is devoted to painting the faces; it may produce three different results:

- **shading flat** paints each face with a single color, without shading.

- **shading faceted** (this is the default value for this command) works as the previous option, and adds black mesh lines.

- **shading interp** shades off the colors, painting each face with a mixture of colors resulting from a linear interpolation of the edge colors.

A color map can be visualized by using the instruction **pcolor**, as follows

$$\texttt{pcolor}(map)$$

This plots a checkerboard with colors defined by the color map *map*. It is possible to use the syntax **pcolor(X,Y,C)**, which traces colors defined in the map C on a grid defined by X and Y. Indeed, **pcolor** is simply a form of **surf** with a viewpoint 'from above', i.e. an elevation of 90°.

The use of **pcolor** is not restricted to the simple visualization of color maps, but can be used in more complex contexts. Consider, for instance, the problem of tracing the

Mandelbrot set, one of the most complex mathematical objects, and the one that has given the most powerful impulse to the study of fractal geometry.

It is defined as follows. Consider the complex number sequence $\{z_k\}$ recursively defined by the relationship $z_{k+1} = z_k^2 + c$, $z_0 = 0$, where c is a prescribed complex number. As c varies we obtain a family of sequences, and it is possible to prove that any member of this family (i.e. any sequence) diverges (when k goes to infinity) if there exists an index h such that the modulus of z_h reaches the value 2. By varying c in the complex plane the corresponding sequences may diverge (with a different speed) or not diverge. The Mandelbrot set is the set of points c such that the corresponding sequence does not diverge. To plot this set the first problem is the selection of a region in the complex plane in which c must take values: we could think of a circle of radius 2 centered at the origin (as stated before, outside this circle the sequences must diverge), but it is possible to show that the only interesting zone is the rectangle $-2 < \mathrm{Re}\{c\} < 0.5$, $-1.25 < \mathrm{Im}\{c\} < 1.25$.

The following algorithm traces the Mandelbrot set.

1. Choose a value for c and let $z_0 = 0$.

2. Repeat: compute $z_{k+1} = z_k^2 + c$ until a maximum iteration number (say 50) is reached or an index h exists such that $|z_h| \geq 2$.

3. If the modulus of the last element computed in the previous step is less than 2 then plot the point c in the complex plane. Go to step 1.

Moreover, the divergence speed can be taken into account by using different colors: it is sufficient to use colors that are proportional to the index h defined at step 2. As explained above, this leads to the definition of a matrix W, which associates the value h with each point in the complex plane. The following lines translate the previous considerations in MATLAB form. The algorithm makes use of the iteration operator `for`, which will be discussed in Section 8.1

```
[X,Y]=meshgrid(-2:.015:.5,-1.25:.015:1.25);
C=X+j*Y;
W=100*ones(size(C));
Z=zeros(size(C));
for n=1:50,
Z=Z.*Z+C;
h=find(abs(Z)<2);
if ~isempty(h),
W(h)=n*ones(size(h));
else
break
end;
end;
clear C Z
pcolor(X,Y,W)
```

The reader may alter the color map and use the instruction **shading** to obtain different effects. The result is shown in Figure 7.10

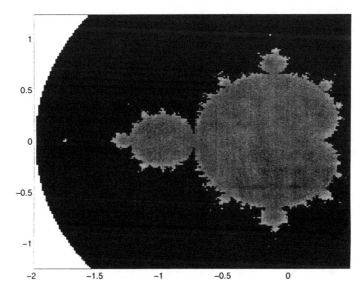

Figure 7.10 Mandelbrot set.

7.10   Parametric plots

It is well known that a parametric description allows an easier representation of curves and points in a 3D space than does a description through functions of two independent variables. For instance, a straight line may be described implicitly as the intersection of two planes or explicitly as the locus of points depending linearly on a single parameter. Moreover, we know that it is always possible to obtain (at least locally) a parametric representation of a curve, while if we want to describe a curve by points, without analytic expressions, a command to plot single triplets of points in a 3D space is mandatory. This command is

```
plot3(x,y,z)
```

Its syntax is the same as that of `plot`, including options about line type and colors, and multiple plots. For instance a 3D logarithmic helix is traced by (see Figure 7.11)

```
t=0:.1:5*pi;
r=exp(t/10);
x=r.*cos(t);
y=r.*sin(t);
z=t;
plot3(x,y,z);
```

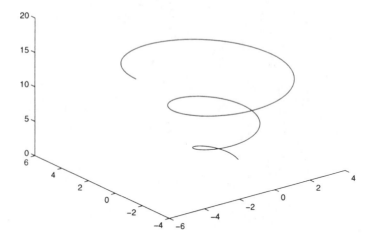

Figure 7.11 Logarithmic helix.

7.11   Revolution surfaces

The function `cylinder(r)` plots the revolution surface generated by the rotation of the curve stored in the vector r. For instance, to plot the cone resulting from the rotation of the line $y = x$, $x \in [0, 2]$ we use

$$\texttt{cylinder(0:.1:2)}$$

It is useful to store the set of points describing the surface. To this end we type

$$\texttt{[X,Y,Z]=cylinder(r);}$$

which returns three matrices X, Y and Z that can be input to **surf** or **mesh** to trace the surface. A last remark is in order: **cylinder** traces surfaces assuming angular steps of 18° for each point in r; in other words, it uses twenty points for each circle. To change this number it is possible to use a second optional parameter, n: `cylinder(r,n)` draws n points for each circumference.

7.12   Printing graphics

MATLAB uses two basic ways of printing graphics: by using the Windows environment or by managing the printing task itself.

In the first case it is sufficient to use the **Print** option from the **File** menu of the graphic window to print the current figure on the printing device currently defined in Windows. Alternatively, it is possible to copy the figure into the Windows **Clipboard**, either as a metafile or as a bitmap, by using the **Edit** menu and setting the copy option. The figure thus copied is available in other Windows applications by using the standard **Cut** and **Paste** commands.

If we prefer MATLAB to manage the printing process, we type the command `print` at the MATLAB prompt. When used without parameters, this command sends the graph in the current window to the default printing device. Note the difference from the previous case: the default device is not the one currently defined in Windows, but rather the default MATLAB device, defined with the command `printopt`. It is, moreover, possible to select different output devices and print options. The complete form of the command `print` is:

`print -d`*device* *-options FileName*

The command creates the file defined by the parameter *FileName* according to the printing device selected with the first parameter. If the file name is omitted, the output is directed to the printer.

MATLAB handles different PostScript printers:

`-dps`	black and white PostScript printers
`-dpsc`	color PostScript printers
`-dps2`	black and white Level 2 PostScript printers
`-dpsc2`	color Level 2 PostScript printers
`-deps`	black and white Encapsulated PostScript printers
`-depsc`	color Encapsulated PostScript printers
`-deps2`	b/w Level 2 Encapsulated PostScript printers
`-depsc2`	color Level 2 Encapsulated PostScript printers

Level 2 files are smaller than standard ones, but not all PostScript printers accept them. The following printers are handled indirectly, through the Ghostscript postprocessor:

`-dlaserjet`	HP LaserJet printers
`-dljet2p`	HP LaserJet IIP printers
`-dljet3`	HP LaserJet III printers
`-dcdeskjet`	HP DeskJet 500C printers, 1bit/pixel
`-dcdjcolor`	HP DeskJet 500C printers, 24 bit/pixel
`-ddeskjet`	HP DeskJet and DeskJet Plus printers
`-dpaintjet`	Color HP PaintJet printers
`-dln03`	DEC LN03 printers
`-depson`	9 or 24 pins printers
`-deps9high`	9 pins, triple resolution printers
`-depsonc`	Epson LQ-2550, Fujitsu 3400/2400/1200 printers
`-dgif8`	Color GIF-formatted (8 bit) files

The command

$$\texttt{print -dmfile } FileName$$

saves on the disk the files needed to regenerate the plot, which can be replotted in MATLAB by typing the file name as a command.

With this command both a .MAT file storing the data of the current plot and an .M file to assign the data to the graphic objects are created (see Appendix A).

Finally, it is possible to make use of the printing devices defined by Windows:

`-dwin`	current Windows printer
`-dwinc`	current color Windows printer
`-dmeta`	metafile clipboad
`-dbitmap`	bitmap clipboard
`-dsetup`	changes the current Windows printing options

Finally, as for the option (the second parameter of the `print` command), the most important option is `-append`, which adds the current graphic to an existing graphic file. If not specified, the current graphic replaces the previous one.

Strictly related to `print` is the command `orient`, which defines the orientation of the graphic during the printing process. In particular

- `orient portrait` is the default value and prints the figure centered on the page with aspect ratio 4/3;
- `orient landscape` prints the figure longitudinally, so as to fill the whole page;
- `tall` prints like 'portrait', but the figure is stretched to fit the whole page.

7.13 Exercises

Exercise 7.1 Plot the graph of all the curves defined in the exercises of the previous chapters.

Exercise 7.2 Verify that the function $\cos x$ in a neighborhood of the origin is approximated by the partial sums of its McLaurin series expansion

$$\cos x = \sum_{n=0}^{\infty}(-1)^{n}\frac{x^{2n}}{(2n)!}$$

by plotting the graph of the function $\cos x$ and those of its first five partial sums.

Exercise 7.3 Trace the disks in the complex plane defined in Exercise 6.6 and plot the roots of the test polynomials used.

Exercise 7.4 *Lucas's theorem* states that each convex polygon containing the roots of a polynomial also contains the roots of the derivative of the polynomial. Moreover, for real coefficients polynomials a better (non convex) estimate is possible: the union of disks. The circles whose diameters are line segments joining pairs of complex conjugate roots are named *Jensen circles*. According to *Jensen's theorem*, each complex zero of the derivative lies inside (or on) at least a Jensen circle of the polynomial.

Verify the accuracy of these estimates using different test polynomials.

Exercise 7.5 Verify that the series of functions

$$\pi - 2\sum_{k=1}^{\infty}\frac{\sin(kx)}{k}$$

converges to the 'sawtooth' function depicted in Figure 7.12.

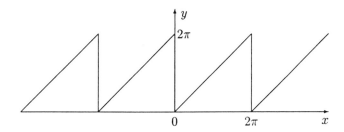

Figure 7.12 Sawtooth function.

Exercise 7.6 Compare the plots, in the interval $x \in [-1, 1]$, of *Chebyshev's polynomials* of degree $n = 0, 1, 2, 3, 4$ defined by the formula

$$T_n(x) = \cos(n \arccos(x))$$

Then, by using the command `polyfit`, find the coefficients of these polynomials.

Exercise 7.7 Verify *Gershgorin's theorem*: the eigenvalues of a real or complex $n \times n$ matrix lie in the union of the circles

$$|s - a_{ii}| \le r_i \quad i = 1, \ldots, n$$

where

$$r_i = \sum_{j=1, j \ne i}^{n} |a_{ij}|$$

and also in the union of the disks

$$|s - a_{jj}| \le \rho_j \quad i = 1, \ldots, n$$

where

$$\rho_j = \sum_{i=1, i \ne j}^{n} |a_{ij}|$$

Exercise 7.8 It is known that a unitary 'sphere' in $\mathbf{R}^n$ is the locus of points

$$S = \{x \in \mathbf{R}^n : \|x\| < 1\}$$

Test the unitary spheres in $\mathbf{R}^2$ defined by the norms

1. $\|x\|_1 = |x_1| + |x_2|$,
2. $\|x\|_2 = \sqrt{x_1^2 + x_2^2}$,
3. $\|x\|_\infty = \max(x_1, x_2)$,

by generating $10\,000$ random complex pairs (x_1, x_2) with uniform distribution in the interval $[-1, 1] \times [-1, 1]$ and tracing on the x_1-x_2 plane only those points belonging to each sphere (using a smaller number of points, we may trace the point inside the sphere).

Hint. It is better to avoid the command **norm**.

Exercise 7.9 The trajectory described by a point on a circle that rolls without slipping along a straight line is a curve named a *cycloid*. If we choose as a parameter the rotation angle ϕ, the parametric equations of the cycloid are

$$x = R(\phi - \sin \phi)$$
$$y = R(1 - \cos \phi)$$

where R is the radius of the circumference.

Trace the cycloid and deduce the equations of the 'reduced' cycloid, i.e. when the point describing the curve is inside the circle, and the 'expanded' cycloid, with the point outside the circle; then plot these curves.

Exercise 7.10 The *hypocycloid* and the *epicycloid* are described by a point on a circle rolling inside or outside another fixed circumference. In particular, the hypocycloid with radii ratio 4 : 1 is named the *asteroid* and its parametric equations are

$$x = R\cos^3\phi$$
$$y = R\sin^3\phi$$

while the epicycloid with radii ratio 1 : 1 is called the *cardioid* and is described in polar coordinates by

$$\rho = 2R(1 - \cos\theta)$$

Trace asteroids and cardioids for different values of the radius R.

Exercise 7.11 Plot the graph of the function

$$f(x, y) = -\frac{5000}{740\sqrt{x^2 + y^2} - 100x^2 - 100y^2 - 6369} - 1$$

Exercise 7.12 With reference to the previous exercise, use the command surf and try different color maps; finally, generate a random color map (a 64×3 matrix) and try it.

Exercise 7.13 Trace the plot of the function defined in polar coordinates by $\rho = |\cos\theta|$ using the command polar and then using plot, after a conversion in Cartesian coordinates.

Exercise 7.14 In 1751 the English musician William Hays proposed a method to compose music automatically. It was sufficient to dip a tooth-brush into an ink-

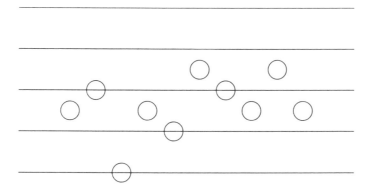

Figure 7.13 MATLAB score.

pot, then to pass a finger over the bristles to squeeze the ink onto the staves. The 'composition' was completed by adding bar-lines according to a rythm generated at random by a pack of playing cards.

Throughout the eighteenth century composers such as Bach, Haydn, Händel and, particularly, Mozart were amused by automatic compositions by means of dice (Mozart wrote a treatise on the 'automatic composition' of minuets).

Using the same principle and the command **rand**, neglecting the duration of the notes (the rhythm), trace score and notes in a MATLAB window, as depicted in Figure 7.13

Exercise 7.15 Verify that the sequence of functions

$$f_n(x) = \frac{x^2}{1 + nx^2}$$

converges uniformly to the function $f(x) = x^2$ and that the sequence

$$g_n(x) = \frac{nx}{1 + n^2 x^2}$$

converges to the zero function, but not uniformly.

Hint. Remember that a sequence of functions $\{f_n(x)\}$ converges uniformly to a limit function $f(x)$ in a set X if

$$\lim_{n \to \infty} |f_n(x) - f(x)| = 0, \ \forall x \in X$$

Exercise 7.16 Solve the following constrained optimization problem graphically:

$$\min x^2 + y^2 + z^2$$

subject to

$$x^2 + 2y + 4z - 14 = 0$$

Hint. Trace the constraint surface and paint it according to the objective function.

Chapter 8

Programming in MATLAB

Up to now we have used MATLAB in interactive mode, typing each command from the keyboard. This way of operating is fast and useful, and uses the computer like a sort of extended calculator to obtain results quickly, but it does face two problems: what if we need to repeat the same computation more times and are not willing to retype the same instructions? Moreover, what if we need to repeat the same commands more times with different values of the parameters involved? Is it possible in this case to simply change the parameters without repeating the instructions?

The answer to these questions is yes; in particular, the first point will be addressed by describing MATLAB control flow structures, while describing how to design MATLAB programs will answer the second question, leading to the development of MATLAB commands that will enrich the basic function library. In the following section the commands to control the logic flow of the instructions will be presented, in the next two sections we will deal with the design of MATLAB programs and functions, while in the last section the debugging tools available in MATLAB are discussed.

8.1 Basic programming structures

It is known from the theory of structured programming that any program can be implemented by a programming language having the three basic operators: 1) sequence, 2) selection, 3) iteration. Let us consider how MATLAB implements these operators.

- The *sequence* in MATLAB is simply given by the lexicographic order of the instructions;
- the *selection* is obtained with the classic operator

91

```
if condition,
instruction set 1
elseif condition,
instruction set 2
else
instruction set 3
end
```

- the *iteration* is obtained with the operators

```
for i=1:n,
instructions
end
```

and

```
while condition
instructions
end
```

As for the command if, it can be used in any form, i.e. like a simple if, like an if, else or like an if, eleseif, else and is always closed by end.

The counter used in the for loop can actually assume a more general form than the one presented above. If we remember that the operator : (colon) defines a vector, we realize that the syntax for i=A is possible, where A is a matrix and the counter index i takes values columnwise from the matrix. For instance a (rather inefficient) way to compute $y = \sin t$ for $t \in [0, 2]$ is

```
T=0:.1:2;
for t=T,
y=[y,sin(t)];
end
```

The execution of a loop can be halted asynchronously with the command break. For nested loops, break stops the execution of the inner loop only.

Since loop execution is rather slow, it is better avoided where possible; in this context the command find described in Section 5.4 often offers a more elegant and efficient alternative to loops. For instance, if we want to set the entries larger than 10 of the matrix A to the value NaN (to avoid their representation on a plot), the worst solution is

```
[m,n]=size(A);
for i=1:m
  for j=1:n
    if A(i,j)>10,
    A(i,j)=NaN;
    end
  end
end
```

a better solution is

```
i=find(A>10);
A(i)=NaN*ones(size(i));
```

and the best solution is

```
A(A>10)=NaN*A(A>10);
```

These commands may be used in interactive mode, but their natural habitat is batch mode, within MATLAB programs.

8.2 Script files

A **Script file** is a MATLAB procedure written with any text editor and consisting of a set of MATLAB commands to be executed when invoked from MATLAB.[1] Script files work directly on the variables in the current workspace and do not permit the use of local variables. They are also named M-files as they are to be saved with extension .M. A script file can be viewed like a sequence of MATLAB instructions written outside MATLAB and then executed in MATLAB as a single block of commands. For example, let us write the following lines with any editor:

```
% An M-file to compute the Fibonacci numbers
f=[1 1]; i=1;
while f(i)+f(i+1)<1000
    f(i+2)=f(i)+f(i+1);
    i=i+1;
end
plot(f)
```

then save these lines in the file FIBNO.M; then, typing from MATLAB the command `fibno`, the commands are executed and the first sixteen Fibonacci numbers are computed and plotted. By the way, anything after the character % up to the end of the line is considered to be a comment. After the execution of the script file, the variables `f` and `i` are present in the workspace.

Of course, at this point the problem of both syntax and semantic errors arise, which we could detect only when executing the file. More desirable would be to verify at least the syntax consistency in creating the script file, but this of course would require the file to be created *inside a MATLAB session*. To do this we can use the MATLAB command `diary` *FileName*. This command saves everything appearing on the screen, including errors and intermediate results, on

---

[1]To edit a script file or a function, one can use the options *Open M-file...* or *New ... M-file* from the MATLAB main window menu **File**. The text editor to be used may be specified by the option *Editor preference...* from the **Options** menu (usually the standard Windows **Notepad** is sufficient).

the file *FileName*, until a `diary` command is typed again, halting the recording. The file thus created must be modified with a text editor in order to be used as a script file next. As an example we show a possible list of the file FIBNO.M created by using the command `diary fibno.m` and typing the instructions interactively:

```
f=[1 1];i=1

i =

    1

while f(i)+f(i+1)<1000
f(i+2)=f(i)+f(i+1);
i=i+1;
end
plit(f)

??? Undefined function or variable plit.

plot(f)
diary
```

Three points are worth mentioning: 1) in the first line, if we do not use the character ';' at the end of the instruction, the value of the variable *i* is visualized both on the screen and on the file; 2) the error message due to incorrect typing (`plit` instead of `plot`) is reported in the file; 3) the concluding command `diary` (needed to close the file) is stored on the file. These three points should be corrected with a text editor to obtain the script file presented before.

8.3   Functions

A `function` is created in MATLAB like a script file, but it should start with the following heading:

    function [*output variables*]=*FunctionName*(*input variables*)

All the temporary variables used inside the function are **local ones**, then, after the execution of the function, they disappear from the workspace.

As an example, here is how MATLAB implements the function `trace`:

```
function t = trace(a)
%TRACE  Sum of diagonal elements.
%   TRACE(A) is the sum of the diagonal elements of A,
%   which is also the sum of the eigenvalues of A.
```

```
%          Copyright (c) 1984-94 by The MathWorks, Inc.
```

```
t = sum(diag(a));
```

In this example we see that the second through fourth lines are typed like comments (prefixed by %): when comments are introduced immediately after the function heading they define the on-line help of the function itself, i.e. typing `help trace` the result will be

```
TRACE  Sum of diagonal elements.
   TRACE(A) is the sum of the diagonal elements of A,
   which is also the sum of the eigenvalues of A.
```

We have encountered various functions with many input and output variables. MATLAB stores the number of input parameters in the variable `nargin`, and that of output parameters in `nargout`. An example is the function `rank` which, as stated before, computes the rank of the input matrix. From a theoretical point of view, the rank can be computed as the number of nonzero singular values, but, from a practical point of view, the number of singular values larger than a very small quantity will suffice. The function `rank`, depending on the number of parameters, may assume a default value for this quantity or accept it as an input parameter:

```
function r = rank(x,tol)
%RANK   Number of linearly independent rows or columns.
%   K = RANK(X) is the number of singular values of X
%   that are larger than MAX(SIZE(X)) * NORM(X) * EPS.
%   K = RANK(X,tol) is the number of singular values of X
%   that are larger than tol.

%          Copyright (c) 1984-94 by The MathWorks, Inc.

s = svd(x);
if (nargin == 1)
    tol = max(size(x)) * max(s) * eps;
end
r = sum(s > tol);
```

Analogously, functions that exhibit different behaviors depending on the number of input and output parameters, such as `residue`, make use of the values of `nargin` and `nargout`.

One of the most powerful commands in MATLAB programming is `eval`, which is able to consider a string as a MATLAB command. For instance,

```
t=pi/2;
c='y=sin(t)';
eval(c);
```

sets y to the value 1. A less trivial example is the following. We have seen that the basic MATLAB element is a 2D array, i.e. a matrix. But a 3D matrix could be implemented as a set A1, A2, ..., An of 2D matrices. Unfortunately, in this way we would lose the possibility of referring to an entry through integer indices. eval is able to overcome this difficulty. Suppose that the $a_{i,j,k}$ entry is represented by Ak(i,j); then the assignment $x = a_{i,j,k}$ is implemented with

$$\text{eval}(['x=A',\text{num2str}(k),'(i,j);'])$$

Note the square brackets surrounding the argument of the command, required because eval accepts only a single string as input parameter.

Now we give a more complex and significant example of how to use the command eval. Consider the problem of seeking the greatest common divisor (GCD) and the least common multiple (lcm) of two polynomials. Mathematically, the problem is stated as follows: given two polynomials $a(s)$ and $b(s)$, find a polynomial $g(s)$ and two coprime polynomial pairs $(p(s), q(s))$ and $(r(s), z(s))$ such that the following equalities are satisfied:

$$
\begin{aligned}
ap + bq &= g \\
ar + bz &= 0
\end{aligned}
\tag{8.1}
$$

Equation (8.1) is known as the *Diophantine equation.* The problem can be solved by using an extended version of the Euclidean algorithm to look for GCD in the integer ring. The scheme of the algorithm is summarized in the following steps:

- Build up the vector of polynomials

$$F(s) = [a(s)\ b(s)]$$

  and, by means of elementary operations, set the term $F_2(s)$ to zero.
- Define a polynomial matrix $V(s)$, initialized to the identity matrix, and perform the same operations on it; at the end of this phase the matrix has the structure

$$V(s) = \begin{pmatrix} p(s) & r(s) \\ q(s) & z(s) \end{pmatrix}$$

By detailing the previous steps we have the following algorithm:

1. Let $V = I_2$.
2. Select the lowest degree polynomial in F; if one of the polynomials is zero go to Step 6.
3. Let λ be the ratio between the largest degree coefficient of the polynomial selected in the previous step and the largest degree coefficient of the other polynomial, and let k be the degree difference between the two polynomials.

4. Subtract the product λs^k times the lowest degree polynomial from the largest degree polynomial and perform the same operation on the columns of the matrix V.
5. Return to step 2.
6. If the zero polynomial is F_1 then exchange the columns of both F and V.

The use of polynomial matrices is not trivial in MATLAB, since a polynomial is a vector itself; hence a polynomial matrix is a matrix of vectors. In what follows we propose a possible solution to the problem. Actually, the decision process is based on the columns of the vector F alone; then, rather then 'vectorizing' the matrices we have preferred to handle the matrices with `eval`. The problems to face are: subtracting polynomials with different degrees at step 4 and computing the largest degree of polynomials represented by vectors whose first entries could be zero after elementary operations. Before analyzing the proposed solution, the reader is invited to solve the problem by him or herself, possibly skipping the rest of the section to go directly to the next section, where (s)he will study a tool to verify the correctness of the produced code.

Finally, here is the solution to the problem. The data structure is the following: the F vector is implemented with two variables F1 and F2, the matrix V by using the four variables V11, V12, V21 and V22. To remove the leading zero entries in a vector we write the following function:

```
function azr=rmz(a)
% a1=rmz(a) removes the leading
% zero elements of a vector until
% a possible scalar variable remains
azr=a;
while (azr(1)==0)&(length(azr)>1),
   azr(1)=[];
end
```

To execute the elementary operations, which are the core of step 4, the function to use is

```
function f=polyop(f1,f2,la,k);
% F=polyop(F1,F2,L,K), where F1 and F2 are
% polynomial coefficients (row vectors),
% L is a scalar and K an integer, computes
%          F(s)=F1(s)-L*s^k*F2(s)
fp=la*[f2,zeros(1,k)];
l1=length(f1);
lp=length(fp);
if (l1>lp),
     fp=[zeros(1,l1-lp) fp];
```

```
    elseif (lp>ll),
        f1=[zeros(1,lp-ll) f1];
    end
    f=f1-fp;
```

Finally, the main program is

```
function [g,l,p,q,r,z]=polygcd(a,b);
% The function [g,l,p,q,r,z]=polygcd(a,b)
% returns the GCD g(s) and the lcm l(s) of
% the polynomials a(s) and b(s);
% moreover two coprime polynomial pairs
% can be returned, (p(s),q(s)) and (r(s),z(s))
% such that
%      a(s)p(s)+b(s)q(s)=g(s)
%      a(s)r(s)+b(s)z(s)=0
F1=a;F2=b;
V11=1;V21=0;V12=0;V22=1;
while (any(F1~=0)&any(F2~=0))
k=abs(length(F1)-length(F2));
[n,i]=max([length(F1),length(F2)]);
one=num2str(i);two=num2str(3-i);
eval(['la=F',one,'(1)/F',two,'(1);']);
eval(['F',one,'=polyop(F',one,',',F',two,',',',...
num2str(la),',',num2str(k),');']);
eval(['F',one,'=rmz(F',one,');']);
eval(['V1',one,'=polyop(V1',one,',',V1',two,',',',...
num2str(la),',',num2str(k),');']);
eval(['V1',one,'=rmz(V1',one,');']);
eval(['V2',one,'=polyop(V2',one,',',V2',two,',',',...
num2str(la),',',num2str(k),');']);
eval(['V2',one,'=rmz(V2',one,');']);
end
if F1==0,
    t1=F1; F1=F2; F2=t1;
    t1=V11; V11=V12; V12=t1;
    t1=V21; V21=V22; V22=t1;
end
g=F1;
p=V11;r=V12;q=V21;z=V22;
l=conv(a,r);
```

As test polynomials the reader can use $a(s) = s^3 - 2s^2 - s + 2$ and $b(s) = s^2 - 2s$. The results must be $g(s) = -s+2$, $p(s) = 1$, $q(s) = -s$, $r(s) = s$ and $z(s) = -s^2+1$.

Now let us present some more MATLAB commands. We have stated that in a MATLAB function all the internal variables are local, thus the variables are available outside the function only when passed as output parameters; analogously, inside a function only input parameters can be used as external variables. This can be a limitation in various respects: first, some functions have a fixed number of parameters, as in the case of the differential equation integrators (ode23 and ode45), which integrate only functions of the form $\dot{x} = f(t, x)$, and do not allow a parameterization of the function in the form $f(t, x, \pi)$, where π is a vector of parameters; moreover, passing the variables as parameters slows down the execution of the function. To overcome this problem, the command global is available in MATLAB, which works like the FORTRAN declaration COMMON: it makes a variable globally visible. The instruction global A must be typed in MATLAB *before* calling the function, and must also be declared inside the function itself. Every modification to the variable performed inside the function will be reflected on the current MATLAB workspace, but not on other functions, unless the latter declares the same variable to be global. Finally, note that clear A *does not remove* the global variable A (it simply masks it); it is necessary to use clear global A.

The command echo *FunctionName* on, allows one to see the lines of a function during run time, and it can be used for debugging purposes.

The command pause(n) suspends the execution for n seconds.

The command input allows one to enter data in interactive mode during run time: for instance the instruction:

```
n=input('How many points?  ')
```

causes the quoted sentence to appear on the screen and stores the number typed from keyboard in the variable n.

8.4 Debugging programs

It is well known that in writing a program every possible source of error becomes a sure source of errors, so that someone says that a program that works well at the first trial contains many interlaced errors that compensate each other! Let us discuss how to detect and correct the errors. We will deal with errors in interactive mode first, then errors in script files and finally errors in M-functions, in particular with logic errors, the so-called *run-time errors*, since the syntax errors are more simple to check and correct.

Working in interactive mode, the easiest way to detect errors is to visualize the intermediate results for each operation (it is sufficient to dispense with the semicolon at the end of the instruction). This should of course be done with care, to avoid an endless flow of numbers on the screen, especially when working with high-dimensional matrices and vectors. Often in these cases the keen use of find, all and any results is advantageous. Remember that MATLAB indicates the command in which the malfunction has taken place, then the first thing to do is to invoke the on-line help of the function itself (this makes it clear that, in writing a function, it is very important to provide it with an accurate on-line help, describing at least the usage and input/output parameters).

We have stated before that a script file is nothing but a batch session, thus the approach to debugging for the script files is the same used in interactive mode; the only difference is that we must wait for the last line of the file to be executed before we can correct the errors.[2] It is possible to suspend the execution of the script files, give the control to the keyboard, execute some operations in interactive mode (for instance to control partial results) and resume the file execution. This is done with the command `keyboard` from inside the file. When MATLAB encounters this command in a script file, it switches to interactive mode, changing the standard prompt to K>> to signal that the execution is paused. To resume the execution we must type `return` (N.B.: not press the RETURN key, but type `return`).

The most frequent result of a MATLAB programming process is the creation of a function file, which we can think of as a new command added to the MATLAB basic library. This is why closer attention is devoted to developing function files correctly; to accomplish this task we have a dedicated tool, unable to work in interactive mode and for script files: the debugger. One of the main advantages of an M-function is to use local temporary and dummy variables, so that, from the user point of view, execution of a function only affects input/output parameters (and `global` variables, of course). During the error checking phase this results in a limitation: the designer of the function would prefer complete visibility of all the variables, 'masking' temporary variables only in a later phase.

Obviously, a possible solution is intensive use of the command `keyboard` described above, but this is not the most efficient way to test a MATLAB function. Using the debugger we can interrupt the execution of a function in prescribed points, inspect called functions and change variable values in a 'structured' way.

Let us examine the use of the debugger in detail. Suppose we have the file `plteig.m` that computes the eigenvalues of a matrix A and plots them on the complex plane. We know that if at least one of the eigenvalues is complex then the command `plot(l)` suffices, where the vector l contains the eigenvalues. But if all the eigenvalues are real a possible solution is to add a small imaginary part when plotting. This is done by the file `addimg.m`. Here are the two files

```
function l=plteig(A)
% Give a square matrix A, the command
%            l=plteig(A)
% plots on the complex plane the eigenvalues
% and stores them in the variable l.
l=eig(A);
if ~any(imag(l)),
```

---

[2]The execution can be stopped by using the key combination CTRL-C

```
% if does not exist at least a complex eigenvalue
% then adds a fictitious imaginary part
l1=addimg(l);
else
l1=l;
end
% the variable l1 is used for the plot only,
% the eigenvalues are stored in l.
plot(l1,'o')

function y=addimg(x);
% Given a vector or a matrix x, the command
%         y=addimg(x);
% returns y=x+j*eps (eps is the predefined variable
% indicating the machine working precision).
y=x+j*eps;
```

Now let us examine how to start a debugging session. The only thing to do is to define breakpoints and to run the function plteig. To do this we type

$$\text{dbstop in plteig}$$

Then we execute plteig(A) (where the matrix A has been defined in advance). MATLAB answers with

```
6 l=eig(A);
K>>
```

Note that the execution is suspended just before the first executable statement, the eigenvalue computation, as evidenced by the prompt that has changed, assuming the form described above when discussing the command **keyboard**. On-line help and comments are ignored. Note that the debugger indicates the line numbers of the instruction to be executed. It is indeed possible to break the execution at prescribed line numbers different from the first executable statement: to break the function at line 10, we type

$$\text{dbstop in plteig at 10}$$

to see the line numbers one uses

$$\text{dbtype}$$

which results in our case in

```
1    function l=plteig(A)
2    % Give a square matrix A, the command
3    %              l=plteig(A)
4    % plots on the complex plane the eigenvalues
5    % and stores them in the variable l.
6    l=eig(A);
7    if ~any(imag(l)),
8    % if does not exist at least a complex eigenvalue
9    % then adds a fictitious imaginary part
10   l1=addimg(l);  .
11   else
12   l1=l;
13   end
14   % the variable l1 is used for the plot only,
15   % the eigenvalues are stored in l.
16   plot(l1,'o')
```

We can now continue the execution with

dbcont

which will stop at line 10, where the function addimg is called. Note that if we now inspect the workspace with the command who we find the variables A, l and l1, which are now visible and can be changed by the usual MATLAB assignment instructions. In particular, the variable l1 is still empty, as its value has not been computed yet.

If we now entered dbcont again, the program will proceed until the end, since no more breakpoints have been defined. If we prefer to advance by a single instruction each time, it would be unacceptable to define a breakpoint for each line, and thus we can resort to the command dbstep. At this point we are going to execute line 10, which is the function call addimg. Typing dbstep the instruction would be executed like a single command, and the execution would stop at line 11 of DISEIG.M. If, on the other hand, we also want to inspect the variable inside the function addimg, we enter

dbstep in

(By the way, dbstep may advance by n steps by using the form dbstep n). Now MATLAB warns us that it is going to execute line 6 of addimg. If we now inspect the workspace with who we find only the variables x and y, i.e. the *local* variables of the current function. To inspect the previous workspace we type

dbup

A further **who** evidences that the workspace has changed. In this way we can, for instance, verify the correct parameter exchange between different functions: we display 1, then we go back to the workspace of **addimg** with

```
dbdown
```

and visualize **x**; if no error has occurred in passing parameters, the two variables have the same value. Now we resume the execution with **dbstep**. It is now clear that in more complex cases we can lose ourselves in a bifurcation tree. In this case, the following turns out to be very useful:

```
dbstack
```

which tell us 'where we are now and how we got here'. Moreover, the number of breakpoints in each function may increase dramatically; the command

```
dbstatus FunctionName
```

lists the breakpoints defined in the function. They can all be canceled with

```
dbclear all in FunctionName
```

or we can remove a single breakpoint at line *n* by adding **at n**; finally, we can remove all the breakpoints in all the functions with

```
dbclear all
```

Note that the latter operation can also be performed by removing all the functions from memory with

```
clear functions
```

Finally, to leave the debugging mode we use

```
dbquit
```

8.5   Exercises

Exercise 8.1 By using the **while** operator, compute the value of the function $\sin x$ for $x = 0.01, 0.02, \ldots, 1$ radians using the series

$$\sin x = x - \frac{x^3}{3!} + \frac{x^5}{5!} - \frac{x^7}{7!} + \cdots$$

Consider only the terms necessary to guarantee accuracy up to the fourth digit and compare the result with the one given by the MATLAB function **sin**.

Exercise 8.2 Write a function to compute the value of the general bilinear transformation

$$w = \frac{az + b}{cz + d}, \quad ad - bc \neq 0$$

and, with $a = 1$, $b = 2$, $c = 1$, $d = -3$ compute and plot on the complex plane the images of

1. the circles through $z_1 = -1$ and $z_2 = 3$;
2. the straight line $\Re(z) = 0$;
3. the circles through the point $z = 3$;
4. the circles through $z_1 = 1$ and $z_2 = -3$.

Exercise 8.3 The studies on *chaos theory* are based on recursion, that is, on mathematical equations such as

$$x_{k+1} = F(x_k), \quad k = 0, 1, 2, \ldots$$

which generate a sequence of points $\{x_n\}$, called the *orbit of x_0 with respect to F*. The most important points of the orbits are the *fixed points*, i.e. the points x_0 such that $F(x_0) = x_0$; moreover x_0 is said to be *periodic* if for some $n > 0$ we have $x_n = x_0$; finally, x_0 is said to be *eventually fixed* or *eventually periodic* if it is neither fixed nor periodic, but a point of its orbit is fixed or periodic respectively, i.e. if an integer $n > 0$ exists such that $F(x_n) = x_n$ or integer $n, m > 0$ exist such that $F(x_n) = x_m$.

Study the orbits of

1. $x_0 = 1 + \sqrt{5}$ with respect to $F(x) = x^2 - x - 4$;
2. $x_0 = 0$ with respect to $F(x) = x^2 - 1$;
3. $x_0 = 0$ with respect to $F(x) = -\frac{3}{2}x^2 + \frac{5}{2}x + 1$;
4. $x_0 = 0$ with respect to $F(x) = x^2 - 2$;

Exercise 8.4 One of the best-known families of functions in chaos theory is the quadratic map

$$Q_c(x) = x^2 + c$$

for which the asymptotic orbits of $x_0 = 0$ vary dramatically by varying the real constant c. An *orbit diagram of x_0* is a plot with the values of c on the x-axis and the asymptotic values of the orbits of x_0 on the y-axis. The number of iterations needed to consider the orbit in steady-state with sufficient accuracy varies with the value c (and x_0); in our exercise one hundred iterations will suffice.

Trace the orbit diagram of $x_0 = 0$ of the quadratic map varying c in the interval $[-2, 0.25]$ in the following way:

1. ten points in the interval $-0.75 < c < 0.25$;
2. ten points in the interval $-1.25 < c < -0.75$;

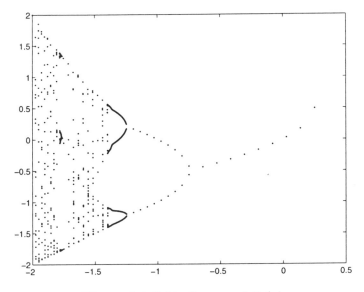

Figure 8.1 Orbit diagram of $Q_c(x)$.

3. fifty points in the interval $-1.4 < c < -1.25$;
4. ten points in the interval $-1.75 < c < -1.4$;
5. twenty-five points in the interval $-1.78 < c < -1.76$;
6. ten points in the interval $-2 < c < -1.78$.

The result should be like that shown in Figure 8.1.

Exercise 8.5 Write a subroutine to represent a sequence of integers like notes on the staff (in treble clef) and generalize the result of Exercise 7.14 writing the function music.m whose input parameter is the number of notes and which returns their representation on the staff.

   Solution. The first thing to do is to draw the pentagram, then the notes; accidentals are to be considered, i.e. sharp notes should occupy the same position as natural notes, but are to be prefixed by the sharp symbol #. Moreover, notes like the middle C need additional lines (leger lines). Let us start tracing the staff with the function

```
function pentagr(n)
axis([0 n+1 -5 17])
axis('off')
hold on
for i=2:2:10
plot([0 n+1],[i i],'-');
end
```

The input parameter is the length of the sequence to trace.

To handle the problem of sharp notes, it is convenient to avoid the 0–11 coding, and associate a decimal part to the note to be sharpened, so that the integer part gives the position and the decimal part the presence of a sharp:

```
function plotnot(note)
basenot=[0:0.5:2 3:0.5:6];
posnot=basenot(note+1);
plot(fix(posnot),'o','MarkerSize',10)
sharp=(rem(posnot,1) =0);
sharpos=find(sharp);
text(sharpos-0.5,fix(posnot(sharpos)),'#');
C=find((note==0)|(note==1));
plot([C-.3;C+.3],...
[fix(posnot(C));fix(posnot(C))],'-');
```

Note that the last two lines are used to trace leger lines on the central C and C#. The option 'MarkerSize' in the command plot defines the size of the circle to be traced and will be discussed in detail in Appendix A.

By using these functions to write the program music.m becomes a trivial operation.

Exercise 8.6 Continuing to deal with music, consider the more complex problem of automatically generating two voice canons. The canon is a musical structure based on the principle of two-part imitations, a practice that is basic in the counterpoint technique and finds applications in most general contexts from the most complex fugues to the less articulated rondels. Perhaps, the best way to describe the canon is to mention the rondel (or round), the simplest among imitative forms. The rondel is familiar to all those who have sung 'Fraier Jaques', or 'Fra' Martino' or 'Three Blind Mice' in a choir: a voice starts the melody and, when the end of the melodic phrase has been reached, a second voice starts singing the same melody, in unison or one octave far, while the first voice repeats its part. Then, also a third and a fourth voice start singing at fixed time intervals. The voices may catch each other an undefined number of times, and finally conclude the song in turn, starting from the first voice.

The canon has a similar structure, but with two further degrees of freedom: the interval between the voices is not confined to unison or octave but may be any degree of the scale; moreover it is not necessary for the second singer to wait for the end of the phrase sung by the first singer. These two new variables give birth to harmonic problems unknown in the rondel: when two people sing on the same note, the interest is focused on melody, rhythm and timbre (in increasing complexity order); when two voices sing different notes simultaneously, the first problem is (or, more exactly, was) to avoid dissonance. The definition of dissonance is too complex a problem to be discussed here, but, for the sake of simplicity, we will

Table 8.1 Consonance table

	C	C#	D	D#	E	F	F#	G	G#	A	A#	B
C	1	0	0	1	1	1	0	1	0	1	0	0
C#		1	0	0	1	1	1	0	1	0	1	0
D			1	0	0	1	1	1	0	1	0	1
D#				1	0	0	1	1	1	0	1	0
E					1	0	0	1	1	1	0	1
F						1	0	0	1	1	1	0
F#							1	0	0	1	1	1
G								1	0	0	1	1
G#									.1	0	0	1
A										1	0	0
A#											1	0
B												1

confine ourselves to an axiomatic definition of a minimal set of consonant intervals to avoid too poor a harmonic structure. The 'admissible' intervals are summarized in Table 8.1, whose columns refer to the lower voice, the bass voice, and whose rows refer to the upper voice, the soprano; the value 1 means consonance, 0 stands for dissonance.

Having defined the consonance, let us turn to the problem of generating two-voice canons with a fixed number of notes *num*, with a time delay of *rit* notes and an interval between voices of *int* halftones.

The simplest way to achieve such a task is to try all the possible combinations of *num* notes within an octave, accepting those generating a canon with consonant intervals and rejecting each canon with at least a dissonance. To do this we suggest coding each melodic line as a *num* digits number in base 12.

Solution. In principle, the following algorithm solves the problem:

1. input *num*, *rit* and *int*;

2. let the vector *note*, of length *num*, be zero;

3. let the test variable *continue* be true;

4. while a consonant sequence is not encountered

5.   increase the vector *note*

6.   if the new sequence is consonant

7.   then print the sequence

8.   if the user want more sequences, the variable *continue* becomes true

Step 5. is implemented by the following function:

```
function y=incr12(x,i);
if i>0,
    x(i)=x(i)+1;
    if x(i)>11,
        x(i)=0;
        x=incr12(x,i-1);
    end
end
y=x;
```

the calling sequence of this function is

$$x=incr12(x,n)$$

It increases the number represented in the vector x (of length n) by one.

To analyze the interval consonance, define the symmetric matrix *Acc*, based on the values given in Table 8.1:

$$Acc = \begin{pmatrix}
1 & 0 & 0 & 1 & 1 & 1 & 0 & 1 & 0 & 1 & 0 & 0 \\
0 & 1 & 0 & 0 & 1 & 1 & 1 & 0 & 1 & 0 & 1 & 0 \\
0 & 0 & 1 & 0 & 0 & 1 & 1 & 1 & 0 & 1 & 0 & 1 \\
1 & 0 & 0 & 1 & 0 & 0 & 1 & 1 & 1 & 0 & 1 & 0 \\
1 & 1 & 0 & 0 & 1 & 0 & 0 & 1 & 1 & 1 & 0 & 1 \\
1 & 1 & 1 & 0 & 0 & 1 & 0 & 0 & 1 & 1 & 1 & 0 \\
0 & 1 & 1 & 1 & 0 & 0 & 1 & 0 & 0 & 1 & 1 & 1 \\
1 & 0 & 1 & 1 & 1 & 0 & 0 & 1 & 0 & 0 & 1 & 1 \\
0 & 1 & 0 & 1 & 1 & 1 & 0 & 0 & 1 & 0 & 0 & 1 \\
1 & 0 & 1 & 0 & 1 & 1 & 1 & 0 & 0 & 1 & 0 & 0 \\
0 & 1 & 0 & 1 & 0 & 1 & 1 & 1 & 0 & 0 & 1 & 0 \\
0 & 0 & 1 & 0 & 1 & 0 & 1 & 1 & 1 & 0 & 0 & 1
\end{pmatrix}$$

(Note that symmetry of the matrix is allowed since we have not included augmented intervals in the table.) The solution is given by the following function:

```
function mel=canon(int,rit,num,Acc,nameChrom)
mel=zeros(1,num);
continue=1;
while continue
    mel=incr12(mel,num);
    if all(diag(Acc(mel(1+rit:num)+1,...
        1+rem(mel(1:num-rit)+int,12)))),
      nameChrom(rem(mel,12)+1,:),
      nameChrom(rem(mel+int,12)+1,:),
      continue=input('Continue? (1/0)');
    end
end
```

The input parameters are the imitation interval, delay, length of the canon and the matrices `Acc` and `nameChrom`, the latter presented in Exercise 3.11, is required to return a readable output.

Alternatively, we can trace the notes on a pentagram. The program `plotnot.m` is described in the previous exercise, and can be used to trace the lower voice, while the soprano part must be modified to allow the upper voice to cover the next octave; to this end the transposition is no longer computed modulo 12. Here is the modified program:

```
function mel=canon(int,rit,num,Acc,nameChrom)
mel=zeros(1,num);
continue=1;
while continue
      mel=incr12(mel,num);
      if all(diag(Acc(mel(1+rit:num)+1,...
         1+rem(mel(1:num-rit)+int,12))))),
        clf
        nameChrom(rem(mel,12)+1,:),
        nameChrom(rem(mel+int,12)+1,:),
        pentagr(num+rit);
        plotnot(mel);
        plotnot2(mel,int,rit);
        hold off;
        continue=input('Continue? (1/0)');
      end
   end
```

where the function `plotnot2` is

```
function plotnot2(note,int,rit)
notetrasp=note+int;
basenot=[0:0.5:2 3:0.5:6];
basenot=[basenot basenot+7];
posnot=basenot(notetrasp+1);
plot((1:length(notetrasp))+rit,fix(posnot),'o','markersize',10)
sharp=(rem(posnot,1)~=0);
sharpos=find(sharp);
text(sharpos-0.5+rit,fix(posnot(sharpos)),'#');
C=find(notetrasp>20);
plot([C-.3;C+.3]+rit,[fix(posnot(C));fix(posnot(C))],'-');
```

Figure 8.2 shows one of the possible results of the following command:

```
canon(4,2,6,Acc,nimiCrom)
```

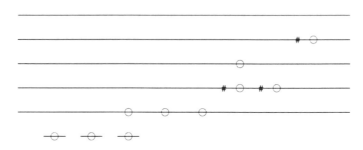

Figure 8.2 MATLAB-generated canon

The same result is shown in Figure 8.3 on two staves to improve readability; this is obtained by modifying the lines to plot notes on the pentagram as follows:

```
subplot(2,1,2)
pentagr(num+rit);
plotnot(mel);
subplot(2,1,1)
pentagr(num+rit);
plotnot2(mel,int,rit);
```

Figure 8.3 Canon score.

Obviously, these musical exercises have no musical value; they can only be considered as MATLAB exercises: in music (and not only in music) the rules imposed must be violated as soon as the necessity to violate them arises: most of the canons generated by this program are trivial, hence boring, none of them having a viable reason for its existence. More interesting results can be obtained by also introducing rules about the generation of single melodic lines, irrespective of harmonic considerations (for instance, diatonic intervals work fine from a melodic point of view, but are harmonically forbidden), but in any case it is better not to exceed 10–12 notes for each canon in order to avoid completely meaningless results.

Chapter 9

Numeric analysis

There is a class of operators in MATLAB that are mathematical problems numerically. They are operators in the strict mathematical sense, since they work on functions rather than on matrices, and can be divided into the following three classes:

- numeric derivatives and integrals;
- optimization and treatment of nonlinear equations;
- solution of differential equations.

Let us refer to the predefined function

$$\text{humps}(x) = \frac{1}{(x-0.3)^2 + 0.01} + \frac{1}{(x-0.9)^2 + 0.04} - 6$$

whose *function file* is

```
function y = humps(x)
y=1./((x-.3).^2+.01)+1./((x-.9).^2+.04)-6;
```

and whose graph, given in Figure 9.1, is obtained with:

```
x=-1:.01:2;
plot(x,humps(x))
```

Generally speaking, the function file defining the function to be given as input to the operators discussed in this chapter must be able to return vectors; this is why dot operators have been widely used in writing HUMPS.M, since they work elementwise (see Section 4.1); thus if x is a vector, humps(x) returns a vector.

111

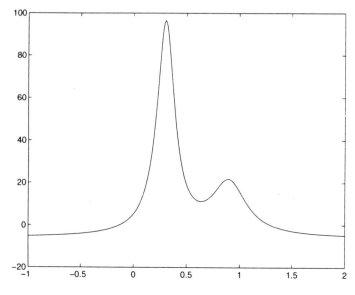

Figure 9.1 *humps(x).*

9.1   Infinitesimal analysis

To compute the derivative MATLAB uses the prime difference operator `diff`, which computes the difference between adjacent elements of a vector: $\mathrm{diff}(x) = [x(2) - x(1), x(3) - x(2), ..., x(n) - x(n-1)]$; then a first-order approximation of the derivative of $y = f(x)$ is obtained with

$$\texttt{Dy=diff(y)./diff(x);}$$

Note that the vector Dy has length $n - 1$.

When the argument of `diff` is a matrix, the differences are computed column-wise. The form `diff(x,n)` computes the nth difference (remember that the nth difference operator is recursively defined by $\Delta_n(x) = \Delta_1(\Delta_{n-1}(x))$, where Δ_1 is the prime difference operator).

For two-variable functions it is possible to use the operator `gradient` in the form

$$\texttt{[dZx,dZy]=gradient(Z,dX,dY);}$$

Z is a matrix with the values of the function to be differentiated, while dX and dY may either be scalars, defining uniform spacing between Z samples along the x and y directions, or vectors containing the points at which the derivatives are to be computed. When omitted, they are assumed to be scalars and unitary.

The results dZ_x and dZ_y are matrices whose entries are the partial derivatives $\partial Z/\partial x$ and $\partial Z/\partial y$.

When invoked in the form `dY=gradient(Y,dX)` it returns the approximate derivative for functions of one real variable.

Numeric integration is obtained with the command `quad`. Used with the syntax

$$s=\text{quad('}\textit{FunctionName'}\text{,a,b)};$$

it computes the integral

$$s = \int_a^b f(x)dx$$

with a relative error smaller than 10^{-3}. For instance, the area of the function humps in $[0, 1]$ is computed by `s=quad('humps',0,1)`.

It is possible to change the relative error by using a fourth optional parameter `tol`; finally, the evolution of the integration process can be displayed on the screen, setting the further parameter `trace` to a nonzero value; considering again the previous example, if we want a relative error smaller than 0.01% and a plot of the antiderivative function, we type

$$\text{s=quad('humps',0,1,1e-4,1);}$$

The command `quad8` works like `quad`, but uses a different integration algorithm.

9.2 Nonlinear equations and optimization

The MATLAB commands to compute the minimum and the zeros of a function of one or more variables are described in this section. Further commands devoted to the solution of optimization problems require the *Optimization Toolbox* and will not be treated in this book.

The function `fzero('`*FunctionName*`',x0)` computes the (real) zero closest to x_0 of the function defined by the function file *FunctionName*.M.

Here is an example:

$$\text{xz1=fzero('humps',0)} \quad \textit{zero close to } x = 0$$
$$\text{xz2=fzero('humps',1)}$$

results in: $xz1 = -0.1316$, $xz2 = 1.2995$.

As for `quad`, in this case it is possible to specify a value for the tolerance as a third parameter and to follow the evolution of the algorithm with a fourth parameter.

Zeros of functions of more variables may be found with the command

$$\text{x=fsolve('}\textit{FunctionName'}\text{,x0)}$$

where the meaning of the parameters is the same, but x_0 is now a vector. A third parameter is a vector to set some optional parameters that we will describe briefly later, while a fourth parameter is a function again, the gradient of the function whose zeros are to be sought. If this parameter is not specified, a numeric evaluation of the gradient occurs.

The most general form of this command is

$$\texttt{fsolve('}\mathit{FunctionName}\texttt{',x0,options,'}\mathit{GradientName}\texttt{')}$$

options is a vector with at most eighteen entries. It is also used by the functions fmin and fmins, to be discussed later, and by the *Optimization Toolbox* commands. The entries of this vector are given here, along with their meaning:

1. Display parameter, with default value 0; when set to 1 displays intermediate results.

2. Tolerance with respect to x (domain of the function), with default value 10^{-4}.

3. Tolerance with respect to the value of the function, with default value 10^{-4}.

4. Tolerance on constraints. Used by the *Optimization Toolbox*.

5. Optimization strategy. Used by the *Optimization Toolbox*.

6. Optimization method. Used by the *Optimization Toolbox*.

7. Line-Search algorithm, with default value 0.

8. Function value. Used by the *Optimization Toolbox*.

9. When set to 1, checks the gradient function given as the fourth parameter.

10. Number of function and constraints calls.

11. Number of gradient function calls.

12. Number of constraints calls.

13. Number of equality constraints.

14. Maximum iteration number. The default value is 100 times the number of variables.

15. Used for special objective functions in the *Optimization Toolbox*.

16. Minimum variation for the numeric gradient.

17. Maximum variation for the numeric gradient.

18. Step size.

In particular, fsolve uses options(2) and options(3) to define the tolerance on the solution and accuracy of the zero, and options(5)=1 to perform the computation by using the Levenberg–Marquardt method; the default value options(5)=0 uses a modified Gauss–Newton algorithm.

The search for a local minimum of a function of one real variable is undertaken by

$$\texttt{x=fmin('}\mathit{FunctionName}\texttt{',x1,x2);}$$

which finds the minimum in the interval $[x_1, x_2]$ of the function defined in the file *FunctionName*.M.

Again, it is possible to use the options vector as a parameter, and, moreover, up to ten input parameters to be passed to the function to be minimized as input arguments; hence, the most general form of the calling sequence is

$$\texttt{x=fmin('}\textit{FunctionName}\texttt{',x1,x2,options,p1,p2,...);}$$

where the parameters p_i are passed to the function *FunctionName* as arguments; if, for instance, *FunctionName* is `fun`, then `fmin` at each evaluation step calls `fun` with the instruction `fun(x,p1,p2,...)`.

Example:

$$\texttt{fmin('cos',3,4)}$$

returns an approximated value for π.

$$\texttt{fmin('cos',3,4,[1,1e-12])}$$

displays the steps performed and returns an approximation for π of up to twelve digits.

For functions of more variables, the command `fmins` is used, with the syntax

$$\texttt{x=fmins('}\textit{FunctionName}\texttt{',x0);}$$

where x_0 is the point about which the local minimum is to be sought.

Again, it is possible to use an options vector and up to ten parameters to be passed to the function to optimize with the call

$$\texttt{x=fmins('}\textit{FunctionName}\texttt{',x0,options,[],p1,p2,...);}$$

Note the presence of the fourth empty parameter, required for consistency problems when used with other optimization procedures of the *Optimization Toolbox*.

The options used are the same as those discussed when dealing with `fmin`.

9.3   Differential equations

MATLAB has two methods for solving ordinary differential equations:

1. `ode23` 2nd-3rd order Runge-Kutta method;
2. `ode45` 4th-5th order Runge-Kutta method.

The syntax is the same for both functions and requires that the functions to be integrated are expressed in *normal form* in a function file, i.e. as a first-order system

$$\frac{dx}{dt} = f(t, x) \tag{9.1}$$

where x is a vector whose length is the order of the equation and t is the independent variable (time, for equations modeling the behavior of dynamic systems).

The simplest form for the calling sequence is

$$[\texttt{x,t}]=\texttt{ode23}(\texttt{'}xDot\texttt{'},\texttt{t0,tf,x0});$$

where $xDot$.M is the function file computing the function f in (9.1), with two input arguments t and x, t0 and tf are the initial and final integration times and x0 is the vector of initial conditions. After computation x stores the solution of the equation and t the corresponding values of the independent variable (time); this is done because the integration algorithm uses a variable step; thus the integration points are not known *a priori* and must be returned by the algorithm as output arguments.

Let us consider the van der Pol equation

$$\ddot{x} + (x^2 - 1)\dot{x} + x = 0$$

which we can rewrite as:

$$\begin{aligned}
\dot{x}_1 &= x_2 \\
\dot{x}_2 &= x_2(1 - x_1^2) - x_1
\end{aligned}$$

We write the function file vdpol.m:

```
function xdot=vdpol(t,x)
xdot(1)=x(2);
xdot(2)=x(2).*(1-x(1).^2)-x(1);
```

and let us start the simulation of the differential equation in the interval $0 \le t \le 20$ with the initial condition $x_0 = (0.25\ 1)^T$:

```
t0=0; tf=20;
x0=[0.25 1]'; % initial conditions
[t,x]=ode23('vdpol',t0,tf,x0);
plot(t,x)
```

The result is shown in Figure 9.2.

The routines ode23 and ode45 accept two more optional input arguments: tolerance on the function to be integrated, and a 'trace' indicator that, if nonzero, displays the evolution of the integration at each step.

Since the function to integrate can be given two input parameters only, t and x, further parameters to be given (more inputs, coefficients, etc.) must be defined as global variables (see Section 8.3).

As an example, we consider the more general form of the van der Pol equation:

$$\ddot{x} + a(x^2 - b)\dot{x} + x = 0$$

where a and b are two real parameters. The corresponding function file is

```
function xdot=vdpol1(t,x)
global a b
xdot(1)=x(2);
xdot(2)=a*x(2).*(b-x(1).^2)-x(1);
```

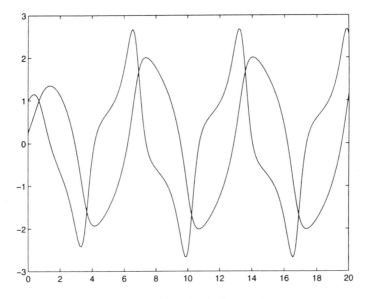

Figure 9.2 Van der Pol equation.

whose calling sequence is

```
global a b
a=1; b=1;
t0=0; tf=20;
x0=[0.25 1]';
[t,x]=ode23('vdpol1',t0,tf,x0);
plot(t,x)
```

For more complex models and quicker results, the use of Simulink is recommended. Moreover, it is known that Runge-Kutta methods are inaccurate for *stiff* problems.[1]

9.4   Exercises

Exercise 9.1 Compute the roots of the equation

$$x^2 - 3\sin x + 0.1 = 0$$

Exercise 9.2 In the simplified economic model of a perfectly competitive market, great importance is given to supply and demand curves; that on the x-axis represents the quantity of the required or produced goods in a fixed time interval and

---

[1]See Chapter 12 in Part II for a detailed discussion on this topic.

that on the y-axis represents the unitary price for the goods. The intersection of the two curves determines the *equilibrium price* of the goods. Assuming the following data for supply and demand of harvest in a market compute the equilibrium price and trace the supply and demand curves using at least thirty points.

Price ($/bushel)	Demand (thousands of bushels/day)	Supply (thousands of bushels/day)
1.25	350	190
1.40	330	220
1.50	320	250
1.60	310	270
1.75	295	300
1.90	280	320
2.00	245	350

Hint. Build up a function expressing the difference between demand and supply and find its zero, taking care to extrapolate problems.

Exercise 9.3 Compute the minimum of the function

$$f(x) = \sin x^2 \tanh x \cosh 3x$$

in the interval $x \in [3, 3.35]$.

Exercise 9.4 Compute the minimum of the function

$$f(x, y) = e^{x-y} + x^2 + y^2$$

and trace the plot about the computed value.

Exercise 9.5 A factory wants to start the production of a new kind of good. The production cost is 1000$ for each package and the marketing office foresees a monthly request of $100\,000/p^2$ packages, where p is the total sale price for each package. Compute the sale price p to maximize the profit.

Hint. Obviously, in this very simplified scheme the monthly profit is
$$(p - 1000)100\,000/p^2.$$

Exercise 9.6 Compute the minimum of the function

$$x_1^4 + 6x_1x_2 + x_1x_2^3 - 6x_1x_3^3 - x_2x_3 + 4x_3^3$$

Exercise 9.7 The Volterra–Lotka equations[2] describe a simplified model of the behavior of two conflicting species (it is often named the 'predator-prey model'):

$$\dot{x} = (a - by)x$$
$$\dot{y} = (cx - d)y$$

Study the solution of these equations starting from nominal parameters $a = 2.7$, $b = 0.7$, $c = 1$ and $d = 3$.

Exercise 9.8 The simplified model of the trajectory of a satellite on a circular orbit about a planet is

$$\ddot{r} - r\dot{\theta}^2 = -\frac{k}{r^2}$$
$$\ddot{\theta} + \frac{2}{r}\dot{r}\dot{\theta} = 0$$

where r is the distance between the satellite and the center of the planet, θ the azimuth angle of a polar coordinate 2D system on the orbit plane and k is the planet's gravity constant.

Study the motion of the satellite about the Earth for different altitudes between 350 and 500 km.

Exercise 9.9 Looking for the fixed points of a map $F(x)$ requires solution of the following equation

$$x = F(x).$$

Compute the fixed points of the maps

1. $F(x) = x^2 - x/2;$
2. $F(x) = x(1 - x);$
3. $F(x) = (2 - x)/10;$
4. $F(x) = x^4 - 4x^2 + 2;$
5. $F(x) = \pi/2 \sin x;$
6. $F(x) = -\sin x;$
7. $F(x) = \arctan x;$
8. $F(x) = \tan x;$
9. $F(x) = \log |x - 1|.$

---

[2]See also page 147.

Part II

SIMULINK

Chapter 10

Fundamentals

SIMULINK is the acronym for the words *Simulation* and *Link*. This term calls to mind the idea of simulating, i.e. the idea of 'artificially' reproducing what happens or what might happen in the real world.

SIMULINK has a very friendly user interface which also allows the use of the tool without a thorough knowledge of MATLAB (however, if one wants to use SIMULINK skillfully then he or she must be familiar with MATLAB: SIMULINK is only a toolbox of MATLAB, not a stand-alone package).

It has been designed to work in graphic environments such as Windows; hence the most natural way to include information or models in SIMULINK is to draw them. SIMULINK has a very wide range of functional blocks, and all the tasks can be reduced to suitable links of these blocks.

Before studying the SIMULINK environment it is important to understand the meaning of the achievable results; more precisely we want to stress that the result of each simulation is *almost always* exact. In particular, it is essential to dwell on the words 'almost' and 'always'.

We start with the latter. The result of each simulation is correct, since it is the consequence of the adopted model for the physical phenomenon under consideration. A simulator by itself cannot (and never will be able to) check if the model is consistent with the real world. SIMULINK works on a series of numbers and returns other numbers. There is no way of preferring one number over another; in principle all the numbers are allowable.

The real world is not so generous: a 'must' for simulator users is to check if the numbers have a physical meaning, i.e. if the model used for the simulation is consistent with the real world.

We analyze the word 'almost'. A computer, as a calculating machine, is not able to evaluate any calculus exactly; indeed it performs some approximations and round-offs that, especially in the presence of complex and repeated operations, may modify the results considerably.

This is dramatically apparent if one works with nonlinear dynamic systems.

In this case there is no method for obtaining an exact solution; there are only some approximated numerical methods which allow one to solve some particular classes of problem. SIMULINK has six different algorithms for the integration of the differential equations and hence it is able to handle a wide class of processes. However, it is very important to choose a 'good' integration method in order to avoid computational overload or incorrect results.

10.1   Building a SIMULINK scheme

Perhaps the easiest way to study a simulation tool such as SIMULINK is to work with it. We start our study of SIMULINK by means of an easy example.

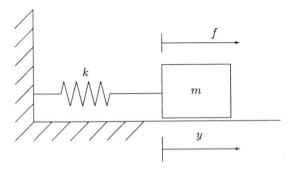

Figure 10.1 Mass-spring system.

Let us consider the mass-spring system depicted in Figure 10.1 whose input–output model is

$$m\ddot{y} + ky = f \tag{10.1}$$

where m is the mass, k the stiffness constant of the spring, y the displacement with respect to the equilibrium point and f is the applied force. If, for the sake of simplicity, all constants are assumed unitary, a possible block diagram is shown in Figure 10.2. It is interesting to compare this diagram with the corresponding SIMULINK scheme represented in Figure 10.3.

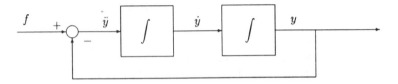

Figure 10.2 Block diagram of the mass-spring system.

The reader can see that only small differences appear, which gives an idea of how close the SIMULINK representation is to the 'standard' way of representing the block diagram. Let us go on to implement the scheme in Figure 10.3.

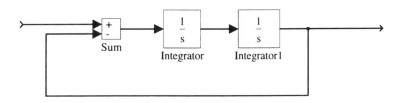

Figure 10.3 SIMULINK scheme of the mass-spring system.

From the MATLAB prompt entering the command

$$>> \texttt{simulink}$$

a window called *Simulink* appears at the top of the screen. If you select *New* from **Files** menu, a new window, called *Untitled*, is opened: this is the sheet on which we are going to draw the block diagram of our system.

Looking at Figure 10.2, one can see that we need a sum. Double-clicking with the left-hand mouse button on the *Linear* icon of the *Simulink* window opens a window which includes some blocks. The first of them is the *Sum*. Move the mouse on the *Sum*, click the left-hand mouse button and hold it pressed (notice that the cursor form is changed), drag the *Sum* in the center of the *Untitled* window and then release the mouse button. We have 'drawn' a summer on the sheet. Repeat the same operation for the first *Integrator*, putting it on the right of the *Sum*. Then repeat the same operation for the last integrator.

In Figure 10.3 the three elements are aligned. It would also be desirable if the blocks in the SIMULINK scheme were aligned. If they are not, you may click (with the left-hand mouse button) and drag in the correct position the out-of-alignment-blocks.

Now we must draw the lines which connect the blocks. Place the mouse pointer near the angle bracket (>) pointing out the *Sum* (angle brackets pointing out are the block outputs) and click the left-hand mouse button: the pointer becomes a thin cross. Without releasing the button move the mouse to the right; as you move it a line is drawn. When you reach correspondence with the angle bracket (>) pointing into the first *Integrator* (angle brackets pointing into are the block inputs); release the mouse button. If the connection has been performed correctly, the angle brackets disappear, and a line with an arrowhead shows the direction of the data flow.

You will notice that if the left-hand mouse button is not released exactly on the angle bracket pointing into the block, but near to it, the result is the same.

Otherwise, if it is released too far from the *Integrator* input the line terminates with an angle bracket, i.e. it has an 'unattached' end.

Using the same operation connect the integrators, and from the last integrator draw a horizontal line and release the button. As mentioned before, an angle bracket indicates that the line is unattached. Place the pointer in the middle of this unattached line and click with the **right-hand** mouse button: the pointer becomes a thin cross. Move the mouse to the bottom and release the right-hand button. With this operation we have drawn a new unattached line, and we have introduced a branch point.

Starting from the end of this new line, and holding the left-hand mouse button down, draw a horizontal line and stop a little way after the *Sum* block. Draw a new line to the top and at last close the feedback on the *Sum* lower angle bracket. In order to complete the scheme, place the pointer close to the *Sum* upper angle bracket, click the left-hand mouse button and draw a line to the left.

A last particular: the feedback of the scheme shown in Figure 10.2 is a negative one, whereas that of our SIMULINK scheme is positive; hence we must change the *Sum* sign in correspondence with the feedback line. Place the pointer on the *Sum* block and double-click: a dialog box such as that shown in Figure 10.4 is displayed. This dialog box, as we will see later, shows the block's *structural* properties. In our case, in the 'parameters' field there are two + signs. This means that the block has two inputs and that the output is the sum of the inputs.

Enter the string + − with no space between the type in the parameter field *List of signs:*.

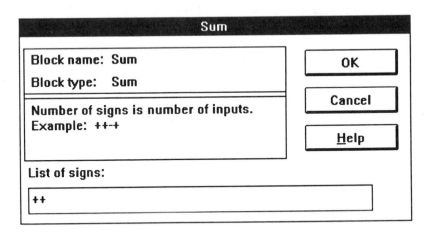

Figure 10.4 *Sum* dialog box.

This means that the block inputs still number two, but the output is the sum of the first input (the higher input) with the opposite for the second one, i.e. the block is a subtractor. Click on the *OK* button to close the window.

10.2   Analysis of the scheme

Drawing the block diagram on the screen is the first step towards analyzing a system by means of SIMULINK. A very direct way of studying the behavior of a system is to force it suitably and to extract information from its time response.

We must then insert an input signal in our scheme. In *Sources* there is a wide range of signal generators; select the *Sine Wave* block and drag it to the system input. Double-clicking on the *Sine Wave* opens its dialog box. By means of this box one can modify some characteristics of the sine wave. In particular, put the frequency equal to 10 rad s^{-1}.

After the input has been connected, you must find some 'reading' points, i.e. some points at which you are interested in the system time history. The 'output' blocks are collected in *Sinks*; select the *Scope* block and drag it to the second integrator output (see Figure 10.5).

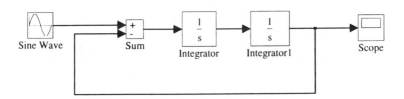

Figure 10.5   SIMULINK scheme of the mass-spring system with inputs and outputs.

A double click on the *Scope* opens a window which is a graphic representation of an oscilloscope (see Figure 10.6). The way the *Scope* works is very similar to that of an oscilloscope: the horizontal axis is the time axis, whereas the measured quantity is represented on the vertical axis.

The *Scope* vertical axis is centered on zero and varies between $+amp$ and $-amp$, where 'amp' is the magnitude of the visualization range which appears in the *Vertical Range* field.

Leaving the *Scope* opened, select *Start* from the **Simulation** menu (the same result may be obtained by means of the hot-key CTRL-T). Some moving lines appear on our oscilloscope. Suitably adjusting the scale axis factors, you are able to display the system output.

The adjustment may be done either by dragging the corresponding bar (notice that the range value is modified dynamically) or by writing the range value directly and then clicking on the corresponding bar or pressing the TAB key.

If you are interested in the time history of a different variable, e.g. the output of the first integrator, you may connect another *Scope* to it.

So, as a classic oscilloscope may successively show different variables of a plant without affecting its time evolution, the *Scope* block may analyze the system variables corresponding to different scheme points without stopping the simulation.

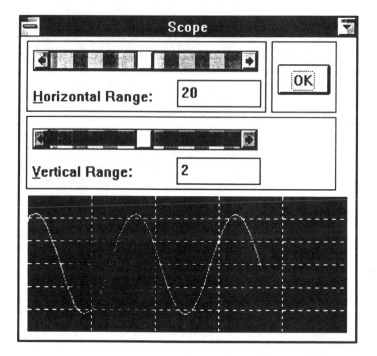

Figure 10.6 Scope.

In particular an unattached *Scope*, called a floating scope, shows the time history of the quantities flowing on the selected lines and is useful for a quick survey.

This is the only situation where one may delete a 'link' without stopping the simulation. Generally, during a simulation it is not possible to delete or to add any blocks, although it is possible to add links and to modify the 'structural' parameters of the blocks.

In order to stop the simulation, select *Stop* from the **Simulation** menu. It is worth noting that the first field of the menu has changed, along with the meaning of the hot-key CTRL-T which is equivalent to *Stop* during a simulation.

If the simulation is to be evaluated over a given time range, the extremes of this range must be inserted in the fields *Start Time* and *Stop Time* of the *Parameters'* submenu, as will be explained on page 155.

The *Scope* should be used for a qualitative analysis of the system only. For a quantitative analysis it is preferable to transfer data in MATLAB and treat them by means of MATLAB tools.

In particular, this may be obtained using as the output block the *ToWorkspace* of the group *Sinks* (see page 139). Once the simulation is stopped (and only after that moment), we find the vector yout in which the simulation results are stored in the MATLAB workspace.

10.3   Exercises

Exercise 10.1 Modify the mass-spring model described by (10.1) and then the scheme of Figure 10.3, taking into account a unitary damping due to friction.

> **Hint.** Viscous friction force is proportional to velocity. In the presence of a unitary constant of proportionality, this force may be represented by connecting the output of the first integrator (which is the velocity) with a third input of the *Sum* (with − sign).

Exercise 10.2 Analyze the response of the system described by means of the scheme in Figure 10.3 forced by a unit step.

> **Hint.** Use the *Step Input* block included in the *Sources* blocks.

Exercise 10.3 Analyze the behavior of the system described in Exercise 10.1 when it is forced by a square wave with unitary magnitude and frequency 5 rad s^{-1}.

> **Hint.** Use the *Signal Gen.* block of the *Sources* to generate the square wave.

Exercise 10.4 Analyze the behavior of the mass-spring system when the input is a sawtooth wave with unitary magnitude and frequency, both with and without the friction term.

> **Hint.** Use the previously considered schemes. The sawtooth wave signal may be obtained through the *Signal Gen.* of *Sources*.

Exercise 10.5 Analyze the response to a unit step of the system

$$\dot{x} = -x + u$$

Exercise 10.6 Analyze the response to a unit step of the system

$$\dot{x}_1 = -x_1 + x_2$$
$$\dot{x}_2 = -x_1 - x_2 + u$$

Exercise 10.7 SIMULINK does not have a ramp generator; build one.

> **Hint.** The ramp is the integral of a constant.

Exercise 10.8 If a vibrating system is forced by a vibrating perturbation with a frequency that differs from the natural frequency of the system, then the system output exhibits the beat phenomenon.

Consider the system

$$\ddot{y} + \omega y = \sin(t)$$

and analyze the system output when the parameter $\omega = 1.1$ and 0.8. What happens if $\omega = 1$?

> **Hint.** In order to build the scheme, use the *Gain* block of *Linear*.

Exercise 10.9 Analyze the time history of the system described by the equation:

$$\dot{x} = -x + \sin(10x) + \sin(t)$$

Hint. The nonlinear term may be built by means of the *Fcn* blocks (of the *Nonlinear* group). Double click on this block and type `sin(10*u)` in the parameter field.

Exercise 10.10 Analyze the time history of the system $\dot{x} = -x + x^2$ with the initial condition $x(0) = 0.2$.

Hint. In order to insert the initial condition, double click on the *Integrator* block and type the condition into the parameter field. The quadratic term may be built by means of the *Product* block of the *Nonlinear* group; connect both inputs of the block to the output of the *Integrator* block (which is the x variable).

Chapter 11

SIMULINK schemes

In the previous chapter we discussed an easy SIMULINK scheme; let us go on to see how to solve some common imperfections:

- **Out of alignment blocks:** select the block to move and, holding the left-hand button down, drag it.
- **No perpendicular lines:** place the pointer at the corner to be modified and click the left-hand button: the pointer becomes a circle. Moving the mouse, without releasing the button, the corner is stretched.
- **Delete a line:** place the pointer on the line to be deleted, click the left-hand button (the line is selected, black squares appear at its extremes), press the DEL key.
- **Move a line:** select the line and without releasing the left-hand button (the pointer becomes an arrowhead) move the line (notice that the original line still appears on the screen, with a different color, until you release the button).
- **Turn snap off:** every object on the screen moves by discrete steps; if you want to move it continuously click on it with both mouse buttons (the pointer becomes an arrow).

In the following we show how to draw complex SIMULINK schemes. In order to realize this goal, we must consider the use of the mouse in depth, the block properties and the way the most common SIMULINK blocks work.

11.1 Using the mouse

One of the first things that one discovers when working with SIMULINK is that all the mouse operations are 'context sensitive', i.e. the click of the mouse button when a block is selected produces, generally, a different result than when a line is selected. Even though this may introduce some difficulties for the beginner, after

131

a little practice it allows one to draw, to correct and to modify schemes using only the mouse.

We have seen that the SIMULINK schemes are made by functional blocks and connecting lines; sometimes it may be useful to include some labels in order to increase the clarity of the schemes. Text may be placed all over the screen. Place the pointer where you want to insert the text and click, then type your text (the ENTER key is a line feed). In order to end the text insertion, click on a different object. The text is an object on which one may perform such operations as deleting, moving, modifying, etc.

Mouse actions which are not context sensitive are selection, deletion and movement of an object. In order to select any object (line, block, etc.) click on it with the left-hand mouse button. The selected object has at its end points some little black squares ('text' objects change the background color). Select an object and to delete it press the DEL key.

In order to move an object, select it and, holding the left-hand button down, drag it.

Except for the selection, which is a 'static' operation, all other operations are 'dynamic', i.e. until the mouse button is released the operation is not active, even if the scheme is dynamically modified on the screen.

More specifically:

- **Working on blocks:**

  - *Left-hand button:* allows one to select and to move the block (to cancel a block select it and press the DEL key).

  - *Right-hand button:* allows one to duplicate the block, a copy of the block is placed at the point where the right-hand button is released (the same result may be obtained by means of the hot-keys CTRL-Ins, SHIFT-Ins).

  - *Double click:* allows one to open the block dialog box.

  - *Both buttons together:* allows one to multi-select (like SHIFT left-hand mouse button).

- **Working on lines** (it is important to stress we can define a SIMULINK line as any piecewise-linear segment which connects two blocks):

  - *Left-hand button:*

    * *clicking on a line:* selects or deletes the line (to delete it press the DEL key).

    * *clicking at a line end:* allows one to add a further element to the line.

    * *clicking on a corner of the piecewise-linear segment:* allows one to stretch the corner. If the corner is removed (i.e the discontinuity of the line is eliminated), then it will not be possible to modify the

corner further. During the stretching operation the pointer becomes a circle centered on the corner.

- *Right-hand button:* allows one to create a branch from a line (working on the free extreme of a line allows one to add elements to the line).

- *Both buttons together:* removes the snap.

- **Working on text elements:**
  - *Left-hand button:* allows one to select, to move, to cancel and to modify the text (the new text replaces the old).
  - *Right-hand button:* has the same functions as the *left-hand button*.
  - *Double click:* allows one to append text further.

- **Working on the screen**
  - *Left-hand button:* allows one to select where to put a text element, to activate multi-selection via the bounding box and to deselect any object. Moreover:
    * *placing the pointer close enough to a free output port of a block* allows one to draw an unattached line starting from the block;
    * *placing the pointer close enough to a free input port of a block* allows one to draw a line ending at the block and with an unattached starting point.
  - *Right-hand button:* has the same functions as the *left-hand button*.

Normally SIMULINK snaps the line to angles that are multiples of 45°. However, if you click on both mouse buttons the snap is turned off, i.e. the line moves continuously.

In order to move a line it must have two fixed extremes and at least two moving corners. In other words, a line may be moved if it consists of at least three segments, and it is possible to move only the central element. In particular, this implies that it is not possible to move the last piece of an unattached line after it has been drawn.

Drawing a scheme may be simplified by means of the *Reroute Line* option. This option allows one to connect the output (input) port of a block with the input (output) port of another block directly; when the connection is done SIMULINK 'makes the circuit planar', i.e. SIMULINK tries to avoid any intersection between lines and blocks, and tries to square all the corners.

In order to avoid the *Reroute Line* one may release the mouse button far from any block ports or make the connection holding the SHIFT key down.

To implement the same operation on a drawn line, select it and use the *Reroute Line* option of the **Options** menu (the same result may be obtained via the hot-key CTRL-L).

All the operations considered until now allow one to work on a single object; indeed if one selects a different object the first one is deselected. Sometimes it may be useful to select a set of objects, and to work on the set as if it were a single one, i.e. move all the objects without modifying the relative positions, or delete all of them. To this aim one may select the object, clicking with both mouse buttons (the reader will note that the corners of all selected objects are highlighted).

To remove the selection it is enough to click on any point of the screen out of the selected objects.

This is not the only way to multi-select; a different solution is to select all the objects enclose into a bounding box.

If you click on a point of the screen where there are neither blocks nor lines, and do not release the mouse button, the pointer becomes a black cross with a thin white cross inside it. Moving the mouse, a dashed box appears. When you release the button, all blocks and all lines at least partially enclosed in the box are selected.

This technique, called **selection via bounding box**, is fundamental, as we will see in Section 14.1, to the building of new blocks.

Before running any simulation, SIMULINK checks, by a warning message, whether there are unattached lines. This message may be useful for avoiding errors due to incorrect block connections. The same message may be obtained via the *Update Diagram* option of the **Style** menu (or by means of the hot-key CTRL-D).

The presence of blocks with free input or output ports is not checked.

11.2   Structural properties of the blocks

In this section we analyze in detail the fundamental elements of any SIMULINK scheme: the blocks. Every block has two set of properties: stylistic and structural.

The **stylistic properties** are modified via the **Style** menu:

Style	
Drop Shadows	
Orientation	▷
Title	▷
Font ...	
Foreground Color	▷
Background Color	▷
Screen Color	▷

In particular:

- *Drop Shadows* toggles the shadows of selected blocks on and off.
- *Orientation* sets the orientation of the selected block.
  The orientation may also be changed via the hot-key CTRL-R for 90° rotations and CTRL-F for 180°.
- *Title* sets the name of the block:

Displayed	shows block name
Hidden	hides block name
Top/Left	shows the name on the top (on the left)
	based on the block orientation
Bottom/Right	shows the name on the bottom (on the right)

The block name is changed as common text.

Usually, if you give suitable names to blocks, you increase the readability of the scheme dramatically.

In the case of very complex schemes, heavy use of block names may reduce the readability of the overall scheme. In these cases one should use the *Top/Left*, *Bottom/Right* options and *Hidden* when the block function is apparent from its icon. A further way to 'title' a block (limited to some of them, e.g. the *Gain*) is to use the parameter field suitably. However, this capability depends on the graphical resolution of the screen; sometimes the block needs resizing.

- *Font* allows one to modify the font type, the dimension and other typographic properties of the text used for the block name.
- *Color* allows one to manage colors by means of three options: *Foreground* for the text color; *Background* for the block background color; *Screen* for the screen color.

  Clever use of colors increases the readability of the scheme: it allows one to highlight the fundamental elements, or stress some important connections. However, excessive use of color should be avoided because it may be confusing for the user.

The blocks may be resized: select the block, then click with the left-hand button on one of the black squares at the block corners (the pointer becomes a diagonal line with arrows at both ends) and stretch it.

The **structural properties** allow one to describe the block functions. All the blocks have structural properties collected in the dialog box, which is opened by double-clicking on the block.

In these dialog boxes there are essentially (see Figure 11.1):

- block name;
- brief description of the behavior of the block;
- parameter fields, if they exist;
- explanation of the behavior of the block (*HELP button*).

A brief description of the structural properties of all the SIMULINK blocks will be presented in Chapter 18.

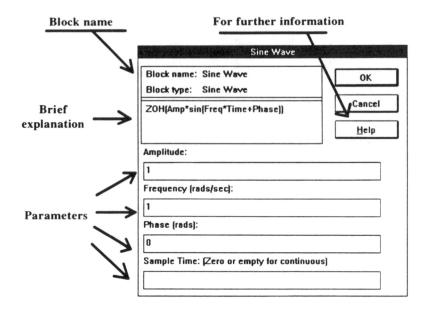

Figure 11.1 *Structural* properties of a SIMULINK block.

11.3   A nonlinear example

SIMULINK is intrinsically devoted to the modeling and simulation of nonlinear systems. This may be shown via an example.

We want to study the rolling of the ship's cross-section depicted in Figure 11.2, subject to a rolling moment.

The roll angle, i.e. the ship's slope in the transversal plane, is described by the following nonlinear second-order differential equation:

$$J\ddot{y} = u - B\dot{y} + P\left(a - r\left(1 + \frac{1}{2}\tan^2 y\right)\right)\sin y \qquad (11.1)$$

where u is the rolling moment.

In the case of a coaster container-ship, the parameters may be assumed to be

J	0.62×10^8	kg m s^2 rad^{-1}	ship moment of inertia
P	10^7	kg	ship weight
B	0.49×10^7	kg m s rad^{-1}	hydrodynamic damping
r	4.95	m	transversal metacentric radius
a	4	m	center of mass level

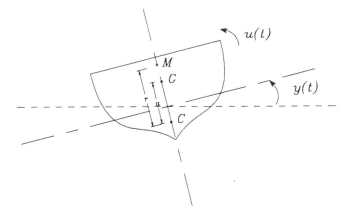

Figure 11.2 Ship cross-section.

The SIMULINK model is shown in Figure 11.3.

In order to build up this scheme, in addition to the blocks analyzed in previous sections, the *Fcn* block of the *Nonlinear* group has been used which implements a generic nonlinear function (see Section 11.5).

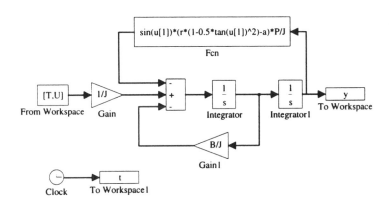

Figure 11.3 SIMULINK ship model.

Moreover, the *From Workspace* block is used as the system input, and the *To-Workspace* block as the output. A detailed description of the input and output blocks is given in the next section. In particular, the *From Workspace* block allows one to use a MATLAB matrix as the system input, whereas the *To Workspace* block stores the results in a MATLAB vector.

Note that the way to build a SIMULINK scheme is quite general, as can be deduced by comparing this nonlinear example with the mass-spring scheme described previously. This feature makes SIMULINK a powerful and extremely versatile simulation tool.

A last consideration: a block scheme, such as that of Figure 11.3, generally provides a more intuitive understanding of the physical phenomenon than does a mathematical equation such as (11.1). Indeed, looking carefully at the scheme it is possible to see that the ship's roll motion is that of a body subject to friction due to water (linear feedback), plus a further force (nonlinear feedback) depending on the roll angle that is mainly due to the interaction between the gravity and the buoyant force.

The most important element used to build up the scheme of Figure 11.3, as well as the mass-spring system, is the *integrator* block.

As a rough analysis of the motion equations, one could think of using derivative blocks to build up the scheme. Considering, for instance, the mass-spring system, whose equation is rewritten here for the sake of clarity:

$$\ddot{y} + y = f$$

one might build a simulation scheme such as that of Figure 11.4. A fundamental weakness of this scheme is that every signal is corrupted by noise. When such a signal is differentiated, the derivative of the, usually fast varying, noise will 'drown out' the derivative of the signal.

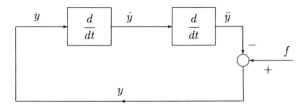

Figure 11.4 Simulation scheme based on derivative blocks.

In order to avoid these problems, simulation schemes must be built with *integrator* blocks.

The idea of using integrator blocks as basic building blocks for simulation schemes was suggested by Lord Kelvin in 1876 for the systems described by a general differential equation of the form

$$\frac{d^n y}{d t^n} = F\left(\frac{d^{n-1} y}{d t^{n-1}}, \dots, \frac{dy}{dt}, y, u, t\right)$$

The simulation scheme is obtained as follows: assuming that $d^n y/dt^n$ is somehow directly available, by integrating this signal n times using the same number of integrator blocks one obtains the quantities $d^{n-1} y/dt^{n-1}$, ..., dy/dt, y. These quantities may be used as inputs (with the external input u and eventually the time t) to evaluate $F(\cdot)$; since this quantity is just $d^n y/dt^n$, we can finally close the loop (see Figure 11.5).

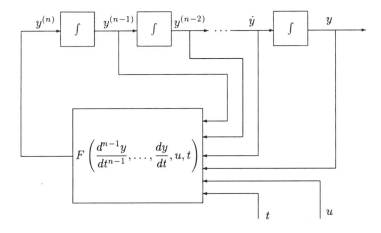

Figure 11.5 Lord Kelvin's simulation scheme.

In the presence of first-order normal form differential equations, that is, of the form $\dot{x} = f(x, u, t)$, the previous procedure involves the use of as many integrator blocks as the system's order. The output of these integrator blocks consists of the state variables.

11.4   Sources and sinks

As mentioned before, the input and output blocks in the scheme of Figure 11.3 allow one to transfer data from MATLAB to SIMULINK and vice versa. This feature contributes to making SIMULINK a very flexible environment, i.e. it may use the sophisticated MATLAB commands both to design the input signals and to analyze the system responses.

We start with the output blocks: these are included in the *Sinks* group. The first block of this group is the *Scope*. The *Scope* works as an oscilloscope, and is able to represent more than one signal (it is also possible to open more than one *Scope* at the same time).

The *Scope* graphic allows one to perform qualitative analysis only. Moreover, like a real oscilloscope, as soon as the trace reaches the right limit of the *Scope* screen, the *Scope* screen is refreshed and the trace restarts from the left limit. This implies that should the time history be reexamined, the simulation must be repeated.

The *Graph* block may be considered as an improvement of the *Scope*; indeed it uses the MATLAB graphic window to improve the graphic capabilities. Moreover, it is possible to choose the type and the color of the line and to obtain a hard-copy of the graphic.

The use of the *Graph* implies a considerable increase in the time needed for the simulation, and the adjustment of the axes scale is not as easy as for the *Scope*.

More than one *Graph* may be used in a single scheme and each of them is associated with a different `figure`. Moreover, more than one signal may be represented on the same *Graph*, and with each of them may be associated a line with different type and/or color. To this end the *Graph* supports the same options as are used for the `plot` command (see page 67): for each signal the color and the line type; the information corresponding to each signal us separated by means of a / (slash). In the presence of graphic format information for a number of signals larger than those actually connected to the *Graph*, the extra signals are ignored.

As for the *Graph*, the MATLAB graphic window is used for the *Auto-Scale Graph* and the *XY Graph*, and this increases the time needed for the simulation.

In particular, the *Auto-Scale Graph* automatically modifies the scale axes in order to represent the signals optimally, and thus it is useful in the presence of signals whose values may not be known *a priori*.

The *XY Graph* works like the *Graph*, but it may represent a generic quantity on the horizontal axis (in particular, the variable connected to the first input, i.e. the higher one, is represented on the horizontal axis). It is useful, for example, for representing the phase portrait of a dynamic system.

The *XY Graph* block works only on scalar quantities. In the presence of multi-variable signals the *Auto-Scale Graph* follows the same rules as does *Graph*.

A different solution, which allows one to overcome most of the drawbacks of the blocks previously described, is to use the *To Workspace* block. This block transfers data from SIMULINK to MATLAB: at the end of each simulation (and only at the end) there is a vector in the MATLAB workspace which contains the values assumed by the analyzed variables. Thus one may use the MATLAB commands to post-process these data suitably.

There are two parameter fields in the dialog box of the *To Workspace* block. The first one, *Variable name*, defines the name of the MATLAB variable where the data will be stored.

More exactly, being the *To Workspace* multi-variable, one may transfer to MATLAB at the same time n variables collected into a matrix: each column contains the values of a different variable, and each row represents the values assumed by all the variables at the same integration time.

The second field, *Maximum number of ...*, defines the maximum number of elements to be stored. The default value 1000 means that at most 1000 samples will be stored. If the simulation runs over a time interval which needs a larger number of samples, only the last 1000 are stored.

A solution to reduction of the memory request without loss of information is to decimate the signal, i.e. to save data at only every n integration steps. This may be achieved by putting the decimate factor as the second optional parameter in the *Maximum...* field.

If one analyzes the time flow during a simulation, one will notice that it is not equally spaced; in other words the integration method uses a nonuniform sampling. The hypothesis of uniform sampling is required for the use of almost all the signal analysis theorems. The time flow is driven by the differential equation solution method, and its sampling time, as we will see in Section 12.2, depends on the solution itself, i.e. there is a dense sampling in the presence of a discontinuous solution, whereas the integration step is increased if the solution is smoother.

However, the output vector may be forced to have a uniform sampling equal to the third optional parameter in the *Maximum...* field. For example, entering

$$[100,2,0.5]$$

the MATLAB variable will be made by no more than one hundred samples in correspondence of 0,1,2,... seconds, i.e. evaluating the output every 0.5 seconds and skipping one sample each time.

It is important to stress that this parameter influences only the storage process, whereas the integration algorithm goes on with its own sampling. Nevertheless, this option forces the integrator engine to evaluate the solution even at the 'storage' time instants.

Among the *Sinks* there is the *ToFile* block, which allows one to store the time histories on a .mat file.

The file consists of one matrix, whose name may be specified in the second parameter field: each row is the time history of a variable, each column consist of the values assumed by all the variables at a given instant (with respect to the *ToWorkspace* the meanings of rows and columns are interchanged). The time vector is always stored in the first row, hence the time must never be connected to the block.

It is not possible either to specify the storage instant or to give a limit to the maximum number of stored elements.

At the bottom of the scheme shown in Figure 11.3 there is an unconnected structure. It is the *Clock*, i.e. the time generator, connected to a *ToWorkspace* block in order to transfer the time vector to MATLAB.

Even if the *Clock* block is not on the screen, SIMULINK activates the clock generator each time a simulation is started.

The *Clock* is not a real time generator, i.e. you cannot open two time generators. The *Clock* block allows one to 'see' the time, whose flow depends upon the integration algorithm (see page 157).

The *Clock* block opens the description on the input signal generators or, more generally, on the input blocks which are collected in the *Sources* group. There are generators which are able to give differently shaped waves, and blocks which are able to import a signal built in MATLAB. In what follows we focus our attention on the latter, and, more precisely, on the *FromWorkspace* block, which is the opposite of the *ToWorkspace* block. A brief description of all the input and output blocks is presented in Chapter 18.

As mentioned before, the *FromWorkspace* block allows one to use as input a signal designed in MATLAB. The signal must be arranged in a matrix: the first column is the time vector, the second one gives the values assumed by the signal at the corresponding sample instants. Then the first column must contain a monotonically increasing number sequence.

The matrix name is specified in the parameter field.

The *FromWorkspace* is multi-variable: if one uses a matrix with n columns, a generator of $n-1$ input signals is implemented.

The first column of the *FromWorkspace* block must be the time vector. If one wants to use the result of a previous simulation as input for a new one, a MATLAB matrix whose first column is the time vector and whose second consists of signal values must be built. Obviously, the time vector of the first simulation must also have been stored.

The time vector as the first column needs to extrapolate the values assumed by the input at each sampling time correctly. Indeed, SIMULINK calculates the input value at each time instant by means of a table look-up, where the first column is considered to be a 'time column', i.e. the independent variable, and the other columns are 'data columns' to be interpolated.

The simulation runs over the time range specified in the menu *Parameters* of **Simulation**; the *FromWorkspace* values of the inputs out of this range are neglected. On the other hand, if the simulation runs over the range specified by the first column of the input matrix, the inputs are evaluated by means of a linear interpolation of the values assumed in the two nearest time instants; hence the input is constant until the simulation time falls into the input-time range.

The *FromFile* block is the opposite of the *ToFile* block, i.e. the inputs are the data stored in a file.

The data must be arranged as a matrix where, as seen for the *ToFile*, the first row is the time vector, and the other rows consist of signal values. In the multi-variable case, the block *FromFile* follows the same rules as the *FromWorkspace*.

11.5   Some nonlinear blocks

In this section we study some of the *Nonlinear* blocks. They are the fundamental elements in SIMULINK that reproduce a wide class of physical processes (a complete description of the *Nonlinear* blocks is given in Chapter 18).

Among *Nonlinear* there are some blocks whose meaning is self-explanatory: the *Abs* block, whose output is the absolute value of the input; the *Product*, whose output is the product of the n inputs; the *Logical Operator* and the *Relational Operator*, which implement a logic function and a relational operation on the inputs, respectively.

Almost all the SIMULINK blocks will work with multi-variable signals, see Section 13.1.

The *Saturation* block imposes lower and upper bounds on a signal. Within the range of the specified limits, the input signal passes unchanged. Outside these bounds, the signal is clipped to the maximum or minimum bound. In other words

$$y = \begin{cases} U^+ & \text{if} \quad u > U^+ \\ u & \text{if} \quad U^- \leq u \leq U^+ \\ U^- & \text{if} \quad u < U^- \end{cases}$$

where y is the *Saturation* output, u is the input and U^- and U^+ are the lower and upper bounds respectively.

In the presence of multi-variable signals, one may specify the bounds for each signal, or use a couple of bounds for all the signals.

Table 11.1 Truth table of a logical sum with carry

Input1	· Input2	Output1	Output2
0	0	0	0
0	1	1	0
1	0	1	0
1	1	0	1

The *Rate Limiter* allows one to reproduce the saturation on the rate of variation of the actuators. The output changes no faster than the specified limit. Both the rising slew rate and the falling slew rate may be specified.

This block follows the same rules as the *Saturation* in the presence of multi-variable signals.

The *Relay* block reproduces a relay with hysteresis. The output switches between two specified values called 'on' and 'off'. When the relay is 'off', it switches to 'on' as soon as the input exceeds the specified value ON, whereas the relay switches from 'on' to 'off' as soon as the input drops below the OFF value. If one uses the same value for the ON and OFF bounds, one has an ideal relay, i.e. a relay without hysteresis.

This block follows the same rules of the *Saturation* in the presence of multi-variable signals.

The *Sign* block implements the signum function

$$y = \begin{cases} 1 & \text{if} \quad u > 0 \\ 0 & \text{if} \quad u = 0 \\ -1 & \text{if} \quad u < 0 \end{cases}$$

where u is the input and y the output.

In the presence of multi-variable signals the signum function is applied on each signal.

The *Combinatorial Logic* block implements a truth table, and is useful for reproducing a programmable logic array (PLA) or, in general, a logic circuit.

The logical sum with carry of Table 11.1 is reproduced by the *Combinatorial Logic* block entering the following string in the parameter field:

$$[0 \ 0; \ 1 \ 0; \ 1 \ 0; \ 0 \ 1]$$

Each output is associated with a column; the value of the output depends on the input values. The reader will note that the input part of the truth table has not been inserted. SIMULINK assumes that the inputs are arranged as shown, i.e. the inputs are always arranged so that the corresponding binary sequence is in ascending order.

Block inputs different from 0 are attended as a logic 1.

The *Switch* block allows one to implement a controlled switch. The output equals the first input until the control signal (which is the second input) is larger than a threshold; otherwise the output is equal to the third input. This block is useful for all situations in which the topology of the scheme must be modified in correspondence with some events that happen during the simulation.

In the presence of multi-variable signals: if the control signal is scalar, its values specify whether the block output is equal to the first or third input (these signals may have different dimensions). On the other hand, if the control input is multi-variable (in this case all the inputs must have the same dimension) the ith output component is equal to the ith entry of the first input if the ith component of the control input exceeds the corresponding threshold value. The threshold may be a scalar or a vector of the same dimension as the inputs quantities.

The *Look Up Table* block performs a piecewise-linear mapping of the input. Given *a priori* some values of the output which are the mapping of the corresponding input values, in the presence of inputs falling between two of the given values the output is assumed to be the linear interpolation of the corresponding outputs. The block icon displays a graph of input versus output (if the look-up table is modified, the icon changes automatically).

In the presence of multi-variable signals the same piecewise mapping is done on each signal.

The *2D Table Lookup* block performs a piecewise-linear mapping from two input variables to an output.

It works only on scalar signals.

The *Fcn* is the most versatile SIMULINK block. It allows one to implement any input–output function expressed by its mathematical expression. In the parameter field of this block the input must be indicated by the letter u and the following functions may be used:

sin(x)	asin(x)	sinh(x)	exp(x)
cos(x)	acos(x)	cosh(x)	ln(x)
tan(x)	atan(x)	tanh(x)	log10(x)
abs(x)	atan2(x,y)	floor(x)	sqrt(x)
rem(x,y)	hypot(x,y)	power(x,y)	ceil(x)

where we have highlighted the number of arguments of any function. The meaning of this function is the same as that of the homonym MATLAB functions (the MATLAB help command may be used for more information on these functions). This is true except for the following functions: ln(x) natural logarithm of x (like the MATLAB log), power(x,y) x raised to y (like the MATLAB command x^y; however, the *Fcn* also supports the latter expression), hypot(x,y), which gives as output the quantity $z = \sqrt{x^2 + y^2}$. In the *Fcn* expressions one may use the parentheses (), the binary operators + - * / ^, and the unary + - with the usual meanings. Moreover, one may use the relation operators == != > < >= <= with the same meanings as MATLAB operators (except for the *not equal* operator, the symbol by != is used instead of ~=) and the following logical operators

&&	logic AND	like MATLAB &
||	logic OR	like MATLAB |
~	logic NOT	

MATLAB scalar variables may also be used. But one cannot use MATLAB vectors either matrices or elements extracted from these objects (e.g. *Fcn* does not admit such expressions as A(1,2)). As an example, let us consider the following function:

$$y = \begin{cases} 5\cos(u) & \text{if} \quad u > 3 \\ \sin(u*r) & \text{if} \quad 2 < u \le 3 \\ u^2 & \text{if} \quad u \le 2 \end{cases}$$

where r is a MATLAB scalar constant. An implementation via the *Fcn* block is

```
5*cos(u)*(u>3)+sin(r*u)*(u>2 && u<=3)+u^2*(u<=2)
```

The *Fcn* block also may work with more than one input signal. This signal, as shown in Section 13.2, must be multiplexed on a single channel. The input components are labeled u[1],..., u[n]. Note that the input index is surrounded by brackets [ ], instead of the usual MATLAB parentheses (). The *Fcn* parameter field implement a C language expression, which explains all the syntactic peculiarities.

The *Fcn* output is always a scalar variable.

The *MATLAB Fcn* block is an interface between SIMULINK and MATLAB. It applies a specified MATLAB function to the inputs. This allows one to use all the MATLAB functions in a SIMULINK scheme. The main drawback is the loss of computational optimization.

The *S-function* block allows one to include an S-Function in a SIMULINK scheme.

An *S-function* is an ASCII file which implements a generic function. The structure of the S-function is such that SIMULINK is able to optimize the computational load. The S-function may be designed via a SIMULINK scheme (input and output ports of the scheme must be highlighted), via a MATLAB-like structure (see Appendix C), or via a C or Fortran procedure.

11.6 Exercises

Exercise 11.1 Title the blocks of the scheme given in Figure 10.3, and add to it the text: SIMULINK scheme of a mass-spring system.

Exercise 11.2 Build up the SIMULINK scheme of the system

$$\ddot{x} + 0.5\,\dot{x} + 2\,x + x^2 = u$$

and study the response to a unit step.

   Hint. The coefficients of the expression may be seen as a *Gain*.

Exercise 11.3 Build up, in a parametric way, the SIMULINK scheme of

$$m\ddot{y} + c\dot{y} + ky = f$$

and use the following parameter values for the simulation: $m = 2$ kg, $c = 2.5$ kg s^{-1} and $k = 40$ N m^{-1}.

Exercise 11.4 Analyze the response of the system described by the equation $\dot{x} = -x + x^2$, for different initial conditions. In particular, consider the following initial conditions: $0.5, 1, 1.5$.

   Hint. The system is stable with respect to the origin if $|x(0)| < 1$.

Exercise 11.5 Analyze, in the phase plane, the behavior of the system $\ddot{x} + x = 0$ for different initial conditions.

   Hint. Remember that in the phase plane the x variable is on the horizontal axis, and the $\dot{x}$ variable is on the vertical. Use the *XY Graph* block, or transfer the data in MATLAB.

Exercise 11.6 Analyze the response of the system whose input-output relation is described by the following integro-differential equation:

$$\dot{v} + 20\,v = -100\int v\,d\tau \ + u$$

when it is forced by $u = 2\sin(4\,t)$.

   Hint. Transform the system into one described by ordinary differential equations via the change of variables $x_1 = \int v\,d\tau,\ x_2 = v$.

Exercise 11.7 Analyze the behavior of the system

$$\ddot{x} + \left|x^2 - 1\right|\dot{x}^3 + x = \sin\left(\frac{\pi x}{2}\right)$$

with the initial condition $x(0) = 1$, $\dot{x}(0) = 0.5$.

   Hint. Use the *Abs* block.

Exercise 11.8 Analyze the output time history of the system described by means of the following transfer function:

$$f(s) = \frac{1}{1 + \frac{1}{15\omega}s}$$

when it is forced by the input $u = \sin(\omega t) + 0.2\sin(300\,\omega t)$.

Hint. It is interesting to compare the signal spectrum before and after the system by means of the *Spectrum Analyzer* block of the *Extras*. This system is low-pass, hence the high-frequency components of the input are much more attenuated than the low-frequency components.

Exercise 11.9 Force the mass-spring system of Exercise 11.3 by different frequency sine waves. Is the system low-pass? Are you able to estimate the cut-off frequency of the system?

Hint. Remember that, for a low-pass system, the cut-off frequency is the lowest frequency for which the amplitude of the frequency response is attenuated by 3 dB.

Exercise 11.10 The classical predator–prey model, which describes the evolution of two interacting species, is modeled by the Lotka–Volterra equations

$$\dot{N}_1 = aN_1 - bN_1N_2$$
$$\dot{N}_2 = -cN_2 + dN_1N_2$$

where N_1 represents the prey and N_2 the predators, a is the prey growth rate in the absence of predators, c the predators' reduction rate in the absence of prey, whereas the b and d coefficients represent the interaction phenomena. Analyze the behavior of the system when: $a = 3.5$, $b = 0.1$, $c = 4$, $d = 0.06$, $N_1(0) = 100$, $N_2(0) = 20$. Plot the response both with respect to time and in the phase plane (plotting the prey number as a function of the predator number). Modify the initial conditions and analyze the different behaviors.

Exercise 11.11 Build the SIMULINK scheme of the system

$$\dot{x} = -(1 + \sin^2(x))x$$

Hint. Use just one *Integrator* block and just one *Fcn* block.

Exercise 11.12 Compare the response obtained by acquiring a sinusoidal voltage with a peak value of 10 V by means of a digital voltmeter at 4 and 8 bits.

Hint. The resolution of a digital instrument is equal to the peak-to-peak value of the analyzed signal divided by the number of quantization levels. In this exercise the quantization level are 2^4 and 2^8 respectively. Use this relation to find the quantization interval for the *Quantizer* block of the *Nonlinear*.

Exercise 11.13 An easy model of an epidemic progress is that of Kermac and McKendrick:

$$\frac{dx}{dt} = -\beta xy$$
$$\frac{dy}{dt} = \beta xy - \gamma y$$
$$\frac{dz}{dt} = \gamma y$$

where x is the potentially infectable population, y the number of contaminated subjects which are able to transmit the disease and z the number of ill subjects no longer infectable, or dead.

The epidemic extends only if the number of potentially infectable subjects exceeds a given threshold. Otherwise, even in the presence of y_0 infected subjects the epidemic does not extend (i.e. the number of infected subjects falls).

Assume that $\beta = 0.1$, $\gamma = 0.5$ and that at time $t = 0$ there are $y(0) = 10$ infected subjects; determine the maximum number of the potentially infectable population for which the epidemic does not extend.

Exercise 11.14 Modify the scheme of Exercise 11.3 taking into account the further constraint that the spring, over 30 cm, is incompressible.

Hint. Use the *Saturation* block.

Exercise 11.15 The dynamic of a vehicle may be described by the following equation:

$$M\ddot{y} + B\dot{y} + Ky = f$$

where y is the displacement of the vehicle chassis with respect to the equilibrium point, M is the car mass, B the friction coefficient, K is the snubber spring constant and f represents the forces transmitted by the road to the car.

Assume that $M = 100$ kg, $K = 10 \text{ N m}^{-1}$, $B = 1 \text{ N m s}^{-1}$ and analyze the oscillation of the passenger compartment for different speeds assuming that the road has a sinusoidal contour with amplitude 0.1 m and period 10 m.

Hint. The forcing input f is a sinusoidal function whose frequency depends on the ratio between speed and road contour period.

Exercise 11.16 Let us consider the series RLC circuit with the following parameters: $R = 1 \text{ m}\Omega$, $C = 3 \text{ }\mu\text{F}$ and $L = 0.2$ H. Force the system by means of a 1 V sinusoidal voltage at frequencies of 145 Hz, 190 Hz and 300 Hz, and study the voltage response at the capacitor electrodes.

Exercise 11.17 The attitude control of a satellite along a single axis may be done via the 'bang–bang' (VSS) technique. In this case a very easy model is the following:

$$\ddot{\theta} = u(t) \quad \text{where } u(t) = \begin{cases} U & \text{if } \theta \geq 0 \\ -U & \text{if } \theta < 0 \end{cases}$$

where θ is the satellite roll angle. Analyze the behavior of the system, for different values of U, both with respect to time and in the phase plane.

Hint. Use the *Relay* block.

Exercise 11.18 Find the asymptotic stability region of the system

$$\dot{x} = -x + x^3$$

Exercise 11.19 Let us consider a system constituted by the series of a symmetric saturation (with bound equal to 0.5) and the system described by the transfer function

$$G(s) = \frac{5(1+4s)}{(1+s)(1+2s)(1+6s)}$$

Analyze the behavior of the closed loop system in the presence of a negative unit feedback.

Assume a constant input equal to 2 and find the steady-state output value. In this situation is the system stable with respect to small input perturbations?

Hint. Use the *Zero-Pole* block of the *Linear*.

Exercise 11.20 Analyze the behavior of the system described via the block diagram of Figure 11.6 forced by the input signal shown in the same diagram.

Hint. Build the input signal in MATLAB and use the *From Workspace* block. To design the SIMULINK scheme, use the *Transfer Fcn* block of the *Linear*.

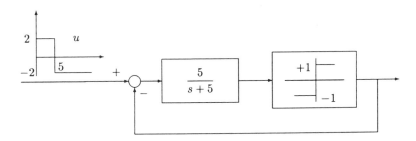

Figure 11.6 Block diagram and input signal time history of Exercise 11.20.

Exercise 11.21 Consider the electric circuit of Figure 11.7. The output tunnel diode characteristic is shown in Figure 11.8. Find the response of the system to $u = 6\sin(20\pi t)$, with the following parameter values: $R = 5$ mΩ, $L = 0.01$ H and $C = 0.05$ F.

Hint. Use the *Look-Up Table* block to model the tunnel diode.

Exercise 11.22 Analyze the response of the system shown in Figure 11.9, subject to the voltage feed $10 + 2\sin(0.1\,t)$ V and in the presence of a light source which irradiates on the phototransistor with a power of $5 + 4\sin(t)$ mW cm^{-2}. The phototransistor characteristic is shown in Figure 11.10.

Hint. Use the *2-D Look-Up Table* block.

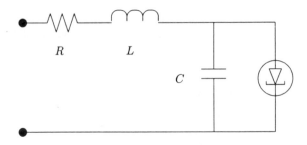

Figure 11.7 Exercise 11.21: scheme of a tunnel diode amplifier.

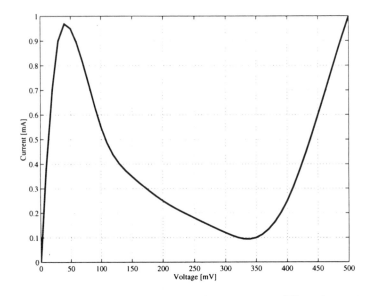

Figure 11.8 Tunnel diode output characteristic of Exercise 11.21.

Exercise 11.23 The dynamics of the spread of corruption at the top (i.e. by politicians) may be described by the following model (Rinaldi, Feichtinger, and Wirl, 1994):

$$\frac{\dot{x}(t)}{x(t)} = \mu^+ \left(\frac{\alpha}{\beta + x(t)} - k \right) - \mu^- \gamma y(t) z(t)$$

$$\frac{\dot{y}(t)}{y(t)} = \varepsilon x(t) - \gamma z(t) - \rho$$

$$\frac{\dot{z}(t)}{z(t)} = \sigma \gamma y(t) - \delta$$

where $x(t)$ is the popularity of politicians, $y(t)$ the (hidden) assets that the cor-

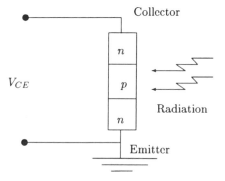

Figure 11.9 Exercise 11.22: phototransistor polarization circuit.

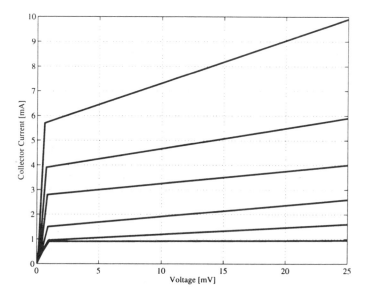

Figure 11.10 Output characteristic of the phototransistor of Exercise 11.22. The curves are labeled in terms of incident power [mW cm^{-2}].

rupt politicians hold, and $z(t)$ the investigation effort. The parameters μ^+ and μ^- express how intensely people react to positive actions and corruption, σ characterizes the immediate reinforcement of the investigative action and δ represents the persistence of investigations, ρ takes into account the discount rate and the marginal propensity for private consumption by politicians, ε measure the politicians' lifestyle (enjoying life rather than accumulating a nestegg).

Italian history from the end of the Second World War (more exactly from the election of 1948) till today may be described by means of the following numerical values:

$$\mu^+ = 1, \quad \mu^- = 10, \quad k = 1, \quad \alpha = 1.5, \quad \beta = 0.5$$
$$\rho = 0.1 \quad \varepsilon = 0.4, \quad \delta = 2, \quad \gamma = 1, \quad \sigma = 2$$

assuming that in 1948 the popularity was half of its maximum $x(0) = 0.5$, low hidden capital $y(0) = 10^{-5}$ and investigative action $z(0) = 10^{-5}$. Analyze the dynamic of the corruption spread.

> **Hint.** Popularity reached its maximum in 1955 and remained constant afterwards, while corruption was practically absent till the early 1970s. The investigation action (and crash of popularity) was predicted in 1989 (note that the 'Mani Pulite' and Di Pietro investigations started in 1990).

Exercise 11.24 A model of the VCON (Voltage Controlled Oscillator Neural) is:

$$\dot{x} = 1.1 + \cos(x) - \cos_+(y)$$
$$\dot{y} = 0.01(1 + \cos(y) - 20\cos_+(y))$$

where the function $\cos_+(y) = \max(\cos(y), 0)$. Analyze its behavior.

> **Hint.** To realize the $\cos_+(y)$ function use the *Switch* or the *Fcn* block.

Exercise 11.25 A pendulum of mass 0.3 kg and length 1 m is hung in a carriage ceiling. The carriage moves in the horizontal plane with a constant acceleration $a = 5 \text{ m s}^{-2}$. Analyze the behavior of the pendulum's free end.

> **Hint.** The pendulum is subject to the vertical acceleration of gravity $g = 9.8 \text{ m s}^{-2}$, and the horizontal force due to carriage motion. The pendulum's motion is due to the force component, which is not balanced by the constraints.

Chapter 12

Simulation in SIMULINK

Many of the systems considered in the previous chapters were linear. Hence their behavior may be studied directly in MATLAB via the `lsim` function. Conversely, the advantage in using SIMULINK instead of MATLAB becomes evident when dealing with nonlinear systems (as with the ship model of Section 11.3).

In fact, the simulation of nonlinear systems in MATLAB is very complex and imposes a huge computational burden: the only MATLAB integration method is the Runge–Kutta algorithm (in the three- and five-point versions, see Section 9.3) and this method is not feasible for some classes of system, e.g. the stiff ones.

On the other hand SIMULINK, having more than one integration algorithm, is able to solve a wider class of problems and at the same time it is able to optimize the computational load.

12.1 Simulation problems

A simple scheme for a DC motor used to move a robot arm is represented in Figure 12.1. If the flux is constant, the mathematical model of the motor is

$$\frac{d}{dt}(J\omega) = ki_a - B\omega$$
$$\frac{d}{dt}(L_a i_a) = -k\omega - R_a i_a + V_a$$

where i_a, V_a, R_a, and L_a are: armature current, input voltage, armature resistance and inductance respectively; ω and J are angular speed and the moment of inertia with respect to the motor axis respectively; B is the friction coefficient; and k is the torque constant.[1] The SIMULINK block diagram is shown in Figure 12.2.

---

[1] With an abuse of notation k indicates both the torque constant k_t and the constant due to the e.m.f. k_v.

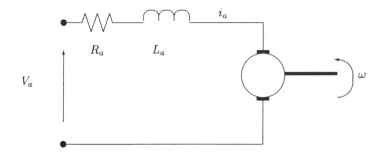

Figure 12.1 Scheme of a DC motor with constant flux.

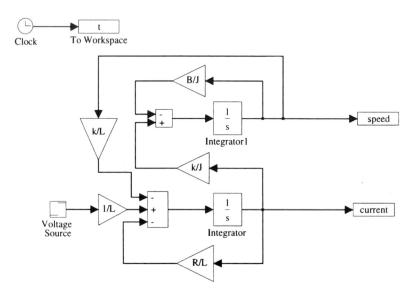

Figure 12.2 SIMULINK scheme of the DC motor.

Let us assume the following numerical values for the parameters:

J	6	kg m^{-2}
L	1	$^-$H
R	0.3	Ω
B	0.01	kg m^2 s^{-1}
k	0.5	

and assume that a step input voltage with 1 V amplitude is applied at the instant $t = 0.01$ s.

The set of differential equations which describe this problem is stiff, i.e. the time constants are very different from each other. In particular, it is easy to recognize that the mechanical part evolves very slowly with respect to the electric one. This is due to the remarkable difference (six orders of magnitude) between the inductance and the moment of inertia values.

The main aim of this example is to show that different systems must be handled by means of different integration algorithms.

We want to analyze the armature current behavior on the time interval $[0, 0.1]$ s. Selecting the option *Parameters* in the **Simulation** menu opens the window shown in Figure 12.3.

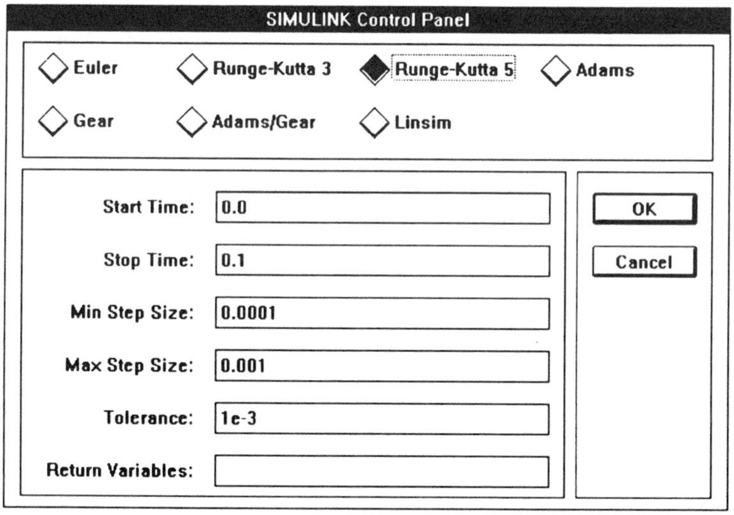

SIMULINK Control Panel			
◇ Euler	◇ Runge-Kutta 3	◆ Runge-Kutta 5	◇ Adams
◇ Gear	◇ Adams/Gear	◇ Linsim	

Start Time:	0.0	OK
Stop Time:	0.1	Cancel
Min Step Size:	0.0001	
Max Step Size:	0.001	
Tolerance:	1e-3	
Return Variables:		

Figure 12.3 SIMULINK integration methods and relative parameters.

The buttons to choose the integration method are at the top of the window (the selected method is highlighted).

Below there are the fields for choosing the simulation initial instant *(Start Time)* and the final one *(Stop Time)*, then the field for defining the minimum integration step size *(Min Step Size)* and the maximum one *(Max Step Size)* (some integration methods, as we will see later, modify the step size according to the function to be integrated), and the *Tolerance* field, which represents a measure of accuracy of the result. The last field *Return Variables* allows one to transfer to MATLAB the time vector, the state and the output histories.
Typing in this field the command

$$[\mathtt{t,x,y}]$$

at the end of the simulation in the MATLAB workspace are the t, x and y variables which are the time vector, the state time histories (see Appendix C) and the output variables (defined using *Outport* blocks).

However, it is preferable to use the *ToWorkspace* block, due to its options.

We now use the default integration method, i.e. Runge–Kutta 5, and set the integration parameters to:

Start time	0.0
Stop time	0.1
Min step size	0.0001
Max step size	0.001
Tolerance	1e-3

Then we start the simulation.

You may notice that the armature current rises indefinitely (see Figure 12.4), i.e. the system seems to be unstable. This behavior is clearly wrong, as you may see by evaluating the eigenvalues of the dynamic matrix, or taking energy considerations into account.

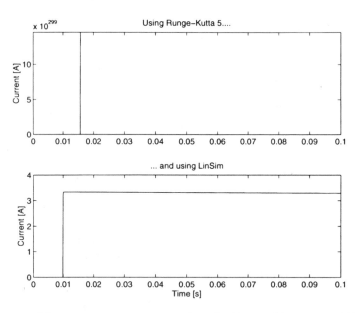

Figure 12.4   Motor armature current using the Runge–Kutta 5 and LinSim algorithms.

However SIMULINK gives a warning message and recommends one to try a different method.

As mentioned before, the Runge–Kutta methods are not feasible in the presence of stiff systems. Hence one must change the integration method. Since our system is linear, we may choose the LinSim algorithm. Using this method (without changing the integration parameters), one obtains the correct result and the simulation runs in less time (see Figure 12.4).

You may notice that, even halving the minimum and maximum integration steps, the Runge–Kutta algorithm gives the wrong results and the simulation requires more time.

12.2   Integration methods

As mentioned before, SIMULINK has six different methods for integrating differential equations.

The numerical solution methods of differential equations are based on the idea of *moving the solution forward starting from the point where it is known*. Let us consider the generic first-order system $\dot{x} = f(x, t)$.

If the solution is known at t_i, then the solution at the next step t_{i+1} may be approximated by

$$x(t_{i+1}) = x(t_i) + h\Delta(t_i, x(t_i); h, f) \tag{12.1}$$

where h is the integration step, i.e. $t_{i+1} - t_i$. The numerical solution assumes that the values of the function in t_{i+1} may be approximated by means of the 'known' solution at t_i and a suitable function Δ which depends on the known solution, on the integration step and on the differential equation itself. The easiest method uses a Taylor series approximation; the classic *Euler* algorithm truncates the series to the first term.

From a theoretical point of view,[2] by decreasing the integration step the numerical solution is more and more accurate. However, this increases the computational burden, and, due to numerical errors (round-off), it is not possible to guarantee the accuracy of the solution. Different solutions to overcome these drawbacks have been proposed in the literature. A solution is to use a multi-step method, i.e. at each step the function is also evaluated in some points interior to the interval, and the value of the solution at t_{i+1} is given by a suitable linear combination of these (*Runge–Kutta* methods).

Moreover, it is desirable that the integration step can adapt itself to the solution behavior (normally known as *variable step methods*). At each integration step the algorithm checks an error bound, i.e. a measure of the solution accuracy (there are different definitions of numeric accuracy). If the error exceeds a given tolerance bound, then the algorithm backtracks, i.e. repeats the integration procedure using a smaller step. Otherwise, the algorithm moves forward using a larger integration step. This procedure allows one to use a thin step only where it is needed, e.g. in the presence of discontinuities; whereas it uses a larger step if the function is smooth.[3] Hence the solution is accurate and, at the same time, the computational burden is reduced.

---

[2] This is true only in the case of measurable functions.

[3] From a theoretical point of view a function is smooth if it is continuously differentiable with all its derivatives. However, in the present context, it is enough that the function is differentiable a 'sufficient' number of times.

The so-called Adams methods solve the equation by integrating a suitable interpolating polynomial of the f function. They are divided into *explicit* methods, where the solution at t_{i+1}, say η_{i+1}, is calculated from the solution η_i at t_i; and *implicit* methods, where the solution is found by solving an implicit equation in η_i and η_{i+1}. The accuracy of the *implicit* method is greater, but the implementation of these methods needs the solution of an implicit equation at each step.

A technique for overcoming this difficulty is the so-called *Predictor–Corrector* procedure. It uses an *explicit* Adams method to find an approximate solution, which is used as a starting point for the *implicit* method. This is iterated at each step until the difference between two consecutive iterations is below a given tolerance value.

No integration method exists which is feasible for all differential equations. Some methods are feasible on some function classes, but they may diverge if the function is of a different type.

Another important parameter to be considered when choosing the appropriate integration method is the computational burden (too small a step dramatically increases the time needed for the simulation).

The integration methods available in SIMULINK are:

- **Euler** is the classic Euler method with variable step. It is feasible for any system, but usually needs a very small step, hence the related computational burden is never justifiable. It should be used only for verifying the results.

- **Runge–Kutta 3** and **Runge–Kutta 5** are methods that are suitable for a wide class of functions. In particular, rk3 solves the equation on the basis of values assumed by the function in three inner points of the integration step, whereas rk5 uses five inner points, and, usually, a larger integration step.

  They are general-purpose methods and hence they cannot reach the performances of more specialized methods. They are to be preferred in the presence of discontinuous functions and for highly nonlinear systems (they do not work with stiff systems).

- **Adams** is a predictor–corrector method which works well with smooth non-stiff systems.

- **Gear** is a predictor–corrector method for stiff systems. It may not work well when there are singularities in the system or when the system is perturbated by rapidly changing inputs.

- **Adams/Gear** SIMULINK chooses between the two predictor–corrector methods on the basis of the stiffness of the system.

- **LinSim** is the method to use in the presence of a linear system. If the system is substantially linear but contains a few nonlinear blocks, linsim also works well.

Once the proper method for the problem under study has been chosen, one must set the parameters congruently.

The *Min Step Size* parameter is the size of the minimum step to use in simulation. In order to obtain good accuracy, this parameter must be set to small values, even if this may produce an increase in the time needed for the simulation.

The *Max Step Size* allows upper bounding of the integration step. Sometimes a simulation produces results which are accurate but are not suitable for producing good plots. In such cases, it is useful to limit the maximum size of the integration step.

The numerical appropriate values for these parameters depends on the system. The *Min Step Size* should be 'small' with respect to the smallest system time-constant (typically $0.01 \div 0.0001 \; \tau_{min}$), while the *Max Step Size* should at the most be equal to the largest one (typically $0.1 \div 0.01 \; \tau_{max}$). (See also page 253.)

All the integration methods reduce the integration step down to the *Min Step Size* value if the error bound is greater than the tolerance value.

Notice that the *Euler* method reduces the integration step, but it does not backtrack on the previous interval.

In SIMULINK schemes algebraic loops should be avoided. Indeed, during the simulation, the loops are detected, signaled and solved via Newton–Raphson iterative algorithms. This increases the time needed for the simulation considerably, or even stops the simulation if SIMULINK is not able to solve the loops in one hundred iterations at most. A way to break algebraic loops without modifying the SIMULINK scheme considerably is to use the *Memory* block of the *Nonlinear* group. This block introduces a delay equal to the current integration step, hence it yields to nonalgebraic loop.

12.3   System with discontinuities

Many physical systems return a solution that is always continuous except in the correspondence of particular events where the solution has discontinuities. For example the trajectory of a ball in the presence of some obstacles: the ball has a regular trajectory until it hits an obstacle; in that time instant its speed and direction of motion are 'instantaneously' changed; after the collision the ball's trajectory is again regular.

For this class of system, the optimal selection of the integration step is a hard problem. Indeed, the use of too small an integration step (in order to handle discontinuities correctly) involves an unjustifiable increase in the computational burden, because it uses small steps even in the presence of smooth solutions.

For these situations, SIMULINK incorporates the *Hit Crossing* block. This block allows one to vary the integration step when the control variable assumes particular values.

To understand the way this block works better, let us consider the following problem.

Let us model the behavior of a unit mass ball dropped from 10 m in the presence of the gravity force and assuming that contact with the ground is an elastic collision. The SIMULINK scheme is shown in Figure 12.5.

The ball's behavior is that of a body subject to a constant acceleration until it hits the ground; at the instant of contact the ball's velocity is inverted (reduced

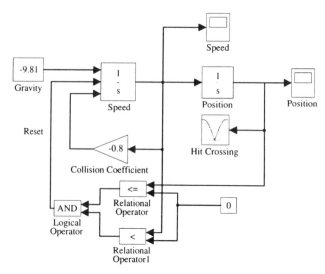

Figure 12.5 Rebounding ball.

by a factor of 0.2), and the ball begins to rise, showing classic parabolic behavior.

The SIMULINK scheme has two integrators in series which allow one to describe the ball's positions knowing its initial position, velocity and the applied forces (Newton law). In particular, the first integrator is a *Reset Integrator* block. The output of the *Reset Integrator* is equal to the integral of the first input if the second input is zero; otherwise (reset condition) the third input value becomes the new 'initial condition' for the integrator.

In the present case, the reset condition is commanded when the ball reaches the ground (because of numerical problem, reaches or 'passes' throw) and when its velocity is negative, i.e. it has fallen. Due to numerical problems the implemented condition is that the ball reaches or 'passes' throw the ground and has a negative velocity.

If both these conditions hold, the first integrator (whose output is the ball's velocity) is set to the opposite value of the one assumed by the ball at the instant of contact (reduced by 20% due to nonideal elastic collision).

When the ball's position is near to zero the *Hit Crossing* block reduces the *Tolerance* parameter from 10^{-3} to 10^{-6}. Hence, around the instant of collision, the integration step is reduced from 10^{-2} to 10^{-4}.

In the presence of problems, where some events change the system topology, classic analysis methods impose evaluation of the system response up to the time instant when the modifying event occurs (impact with the ground, turning a switch, etc.). They then solve the problem in the new configuration starting from the conditions just evaluated.

SIMULINK allows one to simplify the analysis of this class of problems. Using the appropriate logic blocks (*Logical Operator, Relational Operator, Fcn, Switch*) it is possible to draw a single SIMULINK scheme which includes many different topological situations. This allows one to avoid the explicit evaluation of the state at switching instants, and the need for separate schemes for different configurations. The scheme of Figure 12.5 is a lightly modified version of the SIMULINK demo *Reset Integrator and Hit Crossing Block*. In this demo the MATLAB commands for designing a graphic interface (see Appendix B) are also described.

12.4   Time-varying systems

SIMULINK also allows one to model time-varying systems easily. As an example, let us consider the following differential equation:

$$\dot{x} = t \sin \left(x(t) \right)$$

which describes a nonlinear time-varying system. The SIMULINK scheme is depicted in Figure 12.6. It is easy to see that this scheme is similar to those studied up to now. The explicit dependence on time is obtained by considering the time t as a further input.

The way to build the scheme is, practically, that proposed by Lord Kelvin (see Figure 11.5).

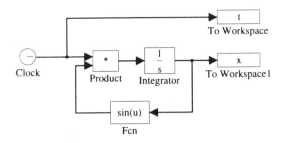

Figure 12.6 Time-varying SIMULINK scheme.

We saw in Section 11.4 that it is possible to build the input signals in MATLAB as a matrix and to import them in SIMULINK via the *FromWorkspace* block. However, this solution may be inaccurate or even incorrect in the presence of 'fast' time-varying input signals and systems which require very small integration steps. Indeed, SIMULINK calculates the input value at each time instant by means of a table look-up, where the first column is considered to be the 'time column', i.e. the independent variable, and the other columns are 'data columns' to be interpolated. If the step size used to define the input matrix is greater than the step used during the simulation (which is selected by the integration method), the input signal will be approximated as a piecewise-linear signal. This may lead to an incorrect solution.

This problem may be avoided by using a smaller step size in defining the input matrix. However, the storage requirements and the computational burden increase. A different solution is that of building the input signal directly into SIMULINK. A series of a *Clock* block with a suitably programmed *Fcn* block allows us to create any signal.

SIMULINK is also able to simulate time-delay systems. Let us consider the system

$$\ddot{x}(t) + x(t) = x(t - 0.5)$$

whose SIMULINK scheme is proposed in Figure 12.7.

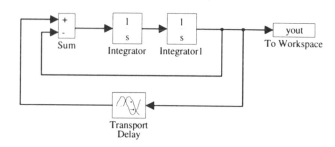

Figure 12.7 Time-delay system.

The output of the *Transport Delay* block is the input signal delayed by τ seconds. The *Transport Delay* stores the input values in order to reconstruct the delayed signal. It is possible to put a negative value into the parameter field *Time Delay* in order to design an anti-casual system. In this case the output values are calculated by linear extrapolation of the last two input values.

The *Variable Transport Delay* block has a second input which is used to indicate, at each time instant, the amount of delay of the input signal (first input signal).

The mathematical model of some physical processes holds until the state variables (or a combination of them) fall within given bounds. When the variables exceed these values the mathematical equations (which are still formally correct) lose any physical meaning. This is what happens for the model of the trajectory of a bullet which holds until the bullet strikes the target, or for biological models where there is no meaning in considering a number of subjects less than zero.

In all these cases the simulation must be interrupted as soon as the solution becomes nonfeasible. This may be achieved by means of the *STOP* block of the *Sinks*.

The *STOP* block interrupts the simulation as soon as its input becomes different from zero. In other words, the simulation is interrupted when the logical condition which commands the *STOP* is verified.

If there is more than one stop condition one may use the same number of *STOP* blocks. A more elegant solution is to consider a logical network which is able to intercept all the stop conditions and to command a single *STOP* block.

12.5 Exercises

Exercise 12.1 Compare the solution of the van der Pol equation

$$\frac{d^2x}{dt^2} - \varepsilon\left(1 - x^2\right)\frac{dx}{dt} + x = 0$$

with the initial conditions $x(0) = 1$ and $\dot{x}(0) = 0$ for $t \in [0,6]$, when $\varepsilon = 0.1$, using the Runge–Kutta 5 integration method and the Adams/Gear method.

> **Hint.** The results cannot be compared immediately since the sample instants chosen by the two algorithms are different. In order to compare the results one should extrapolate the values given by a method at the sample instants of the other method. This operation may be done by SIMULINK using the *From Workspace* block.

Exercise 12.2 Design a SIMULINK scheme whose output is a sine wave with a 'slowly varying' frequency. A model of this system is $y = \sin\left(10\,t + 2\cos(2\,t)\right)$.

> **Hint.** Use the *Clock* and *Fcn* blocks.

Exercise 12.3 An approximated model which describes the evolution of a battle between two armies (Lanchester, 1916) is:

$$\dot{x} = f(t) - \alpha y$$
$$\dot{y} = g(t) - \beta x$$

where x and y are the armies' 'power' (expressed as number of soldiers), α and β are the lethality coefficients (they give a measure of the attack's 'accuracy'); f and g are the reinforcements.

If, in the hypothesis of no reinforcement, an army of 40 000 soldiers and a lethality coefficient of 0.2 fights an army of 10 000 soldiers and a lethality coefficient of 0.9, who wins?

If, by same hypothesis, the second army is made up of 20 000 soldiers, what will the result be?

Analyze the 'battle's' history both with respect to the time and in the phase portrait. Find the number of soldiers in the second army for which the battle's result would be the complete destruction of both armies.

> **Hint.** There is no physical meaning in considering a negative 'power', so use the *STOP* block to interrupt the simulation as soon as an army is destroyed.
>
> In spite of the model's simplicity it has been able to 'reproduce', with incredible accuracy, all the phases of the Iwo Jima (1944) battle between the American and Japanese armies.

Exercise 12.4 The model proposed in the previous exercise, lightly modified, may describe the battle between a regular army and a guerrilla army:

$$\dot{x} = f(t) - \alpha y$$
$$\dot{y} = g(t) - \beta xy$$

With respect to the previous example, in this case we have a nonlinear term. This term allows us to model the limited knowledge of guerrilla spatial dislocation; then the lethality of the regular army attack depends on the number of guerrillas.

Study the results of the battle using the conditions of the previous exercise.

Exercise 12.5 The clock's arms state six o'clock. How long will it be before the hours', minutes' and seconds' arms are aligned again?

> **Hint.** The arms' motion may be represented as a uniform rotational motion. Use the *STOP* blocks and a suitable logic network. Moreover, set the menu *Parameters* fields.

Exercise 12.6 The clock's arms state three o'clock. How long will the arms take to form a square angle?

> **Hint.** Use the *STOP* block and a suitable logic network. Moreover, set the menu *Parameters'* fields.

Exercise 12.7 Analyze the behavior of the system

$$\frac{d^2y}{dt^2} + \frac{4}{t}\frac{dy}{dt} + \frac{2}{t^2}y = 0$$

for $t > 1$ and, starting from the initial condition $y(1) = 1$, $dy/dt(1) = 1$.

Exercise 12.8 Analyze the response of the system of Figure 12.8.

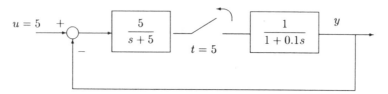

Figure 12.8 Exercise 12.8: switch system.

> **Hint.** Use the *Switch* block. The input to the second block is the first *Transfer Fcn* output before the instant $t = 5$, and zero after.

Exercise 12.9 Evaluate the trajectory of a bullet (with unit mass) shot from 8 m height, with a rear sight of 30°, and with an initial velocity of 12 m s^{-1}. Assume the presence of gravity action only.

> **Hint.** Split the bullet's motion into its horizontal and vertical components. The former is a uniform rectilinear motion. The latter is a uniformly accelerated motion.
>
> The simulation must be stopped as soon as the bullet hits the ground. Use the *XY Graph*.

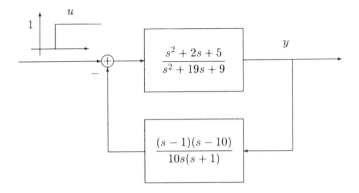

Figure 12.9 Exercise 12.9: closed loop connection.

Exercise 12.10 Analyze the response of the system of Figure 12.9. SIMULINK gives a warning message, why?

Hint. This is a closed loop configuration of two improper systems.

Exercise 12.11 Analyze the behavior of the current flowing in the circuit of Figure 12.10, when the system is forced by $u = 25 + 2.5|\sin(t)|$ V. The nonlinear block represents an electric arc.

The electric arc may be modeled by the equation $v = 10 + 10/i$, where v is the voltage at the arc electrodes and i is the arc current.

Compare the solution with the one obtained using a feed of $u = 12 + 2.5 |\sin(t)|$ V. Assume unitary values for resistance and inductance.

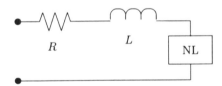

Figure 12.10 Exercise 12.11: electric arc scheme.

Hint. This exercise shows the usefulness of varying step methods. Even in the presence of the same differential equation, different integration steps should be used for different operating points.

Exercise 12.12 There are some classes of problems, termed 'ill-posed', where a small perturbation of initial conditions and/or of some parameters modify the solution considerably.

Compare the behavior of the system

$$\ddot{x} + 0.05\,\dot{x} + x^3 = 7.5\cos(t)$$

starting from $x(0) = 3$ and $\dot{x}(0) = 4$, with the solution obtained from the slightly perturbed initial conditions $x(0) = 3.01$ and $\dot{x}(0) = 4.01$.

Hint. Compare the response on a time range of about ten seconds.

Chapter 13

Multi-variable systems

In all the examples studied up to now, the blocks have one input and one output. In other words, the input and output quantities of each block are scalars.

In this chapter we show how to design multi-variable schemes, where the blocks may be multi-inputs and multi-outputs.

In SIMULINK the multi-variable schemes may be designed both via blocks with more than one input and output port, and via blocks with a single input (output) port corresponding to n-logic inputs (outputs). The latter strategy allows us to minimize the number of connections requested, thus increasing the clarity of the scheme.

13.1   Multi-variable schemes

As an example, let us consider the linearized model of the vertical plane dynamics of an aircraft flying in a horizontal and equilibrium situation. A state-space description is (Huge and Mac Fralen, 1982)

$$\dot{x} = A\,x + B\,u$$
$$y = C x$$

where the control variables are the spoiler angle (measured in tenths of a degree), the forward acceleration (m s^{-2}), and the elevator angle (degrees). The state variables are altitude relative to some datum (m), forward speed (m s^{-1}), pitch angle (degrees), pitch rate (deg s^{-1}), and vertical speed (m s^{-1}).

Assume the following numerical values:

$$A = \begin{pmatrix} 0 & 0 & 1.1320 & 0 & -1.0000 \\ 0 & -0.0538 & -0.1712 & 0 & 0.0705 \\ 0 & 0 & 0 & 1.0000 & 0 \\ 0 & 0.0485 & 0 & -0.8556 & -1.0130 \\ 0 & -0.2909 & 0 & 1.0532 & 0.6859 \end{pmatrix}$$

$$B = \begin{pmatrix} 0 & 0 & 0 \\ -0.1200 & 1.0000 & 0 \\ 0 & 0 & 0 \\ 4.4190 & 0 & -1.6650 \\ 1.5750 & 0 & -0.0732 \end{pmatrix}$$

$$C = \begin{pmatrix} 1 & 0 & 0 & 0 & 0 \\ 0 & 1 & 0 & 0 & 0 \\ 0 & 0 & 1 & 0 & 0 \end{pmatrix}$$

$$D = \begin{pmatrix} 0 & 0 & 0 \\ 0 & 0 & 0 \\ 0 & 0 & 0 \end{pmatrix}$$

In this case, i.e. when the system is described via its A, B, C, D, matrices it is useful to use the same form of description in the SIMULINK scheme.

Among the *Linear* blocks, there is the *State-Space* which allows one to describe a linear time-invarying system by its matrices. Double-clicking on this block opens a dialog box like the one shown in Figure 13.1. The system matrices are inserted in the corresponding parameter fields (as in MATLAB, the rows are separated by ';'). A different solution is to define the matrices in the MATLAB workspace and then to use their names in the *State-Space* fields.

Once the *State-Space* block is set, you may notice an apparent discrepancy: the block shows one input port and one output port, even if the system is multi-input and multi-output (MIMO). In this case the connection ports are multi-variable. In other words, the connection lines which go into and out of the block are 'multi-channel', i.e. more signals are multiplexed on the same line.[1]

The block input (output) is constituted by all the input (output) signals grouped together. This way of working, i.e. grouping more signals in a unique multi-variable channel, even if at first it may appear odd, allows us to work with more clear schemes, and to speed up the drawn process.

Let us return to our example. Force the system by a sine wave. Connect a *Sine Wave* block to the input port, and a *Scope* to the output port. In this case we have made a wrong connection: we have connected a single signal generator to a model having three inputs. If we try to start the simulation, SIMULINK gives the error and stops the simulation.

The problem may be easily corrected by putting the following vector into the first parameter field of the *Sine Wave*:

---

[1] It is just the same procedure as is used in signals transmission. More signals are transmitted 'at the same time' on a single physical channel by means of time or frequency multiplexing.

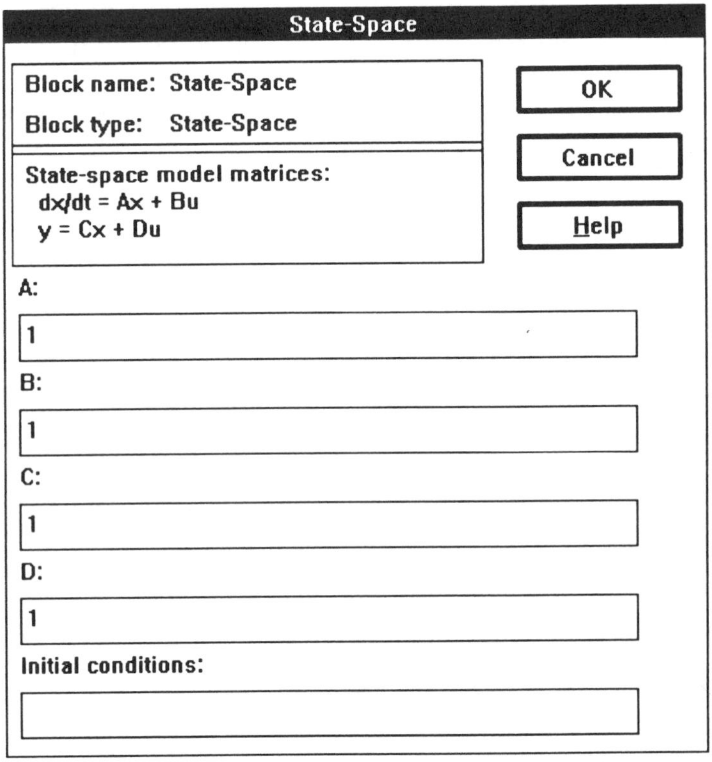

Figure 13.1 *State-Space* dialog box.

[1 1.5 0.5]

This time, the output of the *Sine Wave* is given by three sine waves, with the same frequency and phase but with amplitudes of 1, 1.5 and 0.5, respectively.

Start the simulation again. Three lines move on the *Scope* display. These lines, in accordance with the output matrix C, are the altitude, the forward speed and the pitch angle respectively.

Almost all the SIMULINK blocks[2] are able to match the dimensions of the input signal automatically. In our example, the *Scope* automatically represents the three signals.

To better understand the way SIMULINK works, let us consider the *Integrator* block. We have seen that, in the presence of a scalar input, the output is the time integral of the input signal, starting from given initial conditions. On the

---

[2]The multi-variable system handling has been improved starting from the release 1.3 of SIMULINK and hence most of the following capabilities are not feasible for earlier versions.

other hand if the input is multi-variable, the output is multi-variable, too (with the same dimension as the input), and each output signal is equal to the time integral of the corresponding input signal. Moreover, if the initial condition is a scalar quantity it is used for every signal, whereas if the initial condition is a vector (of the same dimension as the input signal), each signal is integrated starting from the corresponding initial condition.

A lot of SIMULINK blocks work as the *Integrator* (i.e. in the presence of scalar parameters, they automatically match the input dimension). Unfortunately, this is not true for all the SIMULINK blocks: some are intrinsically scalar (e.g. the *Transfer Fcn* block); some others may be only multi-inputs (e.g. the *Fcn* block). There are blocks where vectorial conditions may be used on every parameter field in order to implement *n* different functions (e.g. the *Saturation* block), others where the vectorial conditions are supported only on some parameter fields (e.g. in all the *Discrete* blocks the *Sample Time* parameter must be a scalar), and others which, even working on multi-variable signals, support only scalar parameters (e.g. the *Slider Gain* block).

At a first glance the reader may think that the exceptions number more than the rules; however it is not difficult to understand how to handle multi-variable schemes.

Almost all the SIMULINK blocks are intrinsically multi-variable, i.e. in the presence of multi-variable input signals and using scalar parameters they implement the same function on all the input components, whereas using a vectorial parameter (whose dimension must be equal to that of the input) they implement *n* different functions.

In the presence of multi-variable signals SIMULINK implements the 'scalar expansion', i.e. SIMULINK repeats *n* times the scalar value of the parameter in order to create a vector of the same dimension as the input (output) signal.

In all the blocks which support vector parameters, if only some parameters are different for the signal components, the vector conditions may be imposed only in these parameter fields. For example, let us consider a saturation which works on three multiplexed signals; the lower limit is zero for all the signals whereas the upper ones are 5, 5 and 1 respectively. This block may be simulated via the *Saturation* block entering 0 in the *Lower limit* field, and the vector [5 5 1] in the *Upper limit* field.

A multi-variable SIMULINK scheme appears graphically as a scalar scheme. This may be avoided by using the *Wide Vector Lines* option of the **Style** menu: the multi-variable connections are drawn with thick lines. This option is useful when both multi-variable and scalar connections are present.

13.2   Connections

Often, one needs to connect multi-variable systems with scalar systems (this happens frequently in *group* operation, see the next chapter). To this end, in the

Connections group are the *Mux* and *DeMux* blocks. The meaning of these blocks is explained by their icons: the *Mux* (*DeMux*) has many input ports and one output port (has one input port and many output ports); it allows one to multiplex n scalar signals into a single multi-variable output channel (to split a multi-variable signal into its scalar components).

For example, if some data must be stored via the *To File* block, a solution is to connect each variable to a different *To File* block; in this case n files are opened and the stored data are very redundant: in the first row of each file is stored the same time vector. A more elegant solution is to store data into a single file: connect all the output variables to a *Mux*, and then connect its output to the *To File* block.

On the other hand, if some inputs are created by MATLAB (usually designed on the basis of the same time vector and hence arranged in a matrix with the time vector as the first column), they may be imported into SIMULINK by means of one *From Workspace* block connected to a *Demux* to separate the components.

The *Mux* block also allows one to multiplex a multi-variable channel. If one wants to group two signals on one multi-variable channel, the first one scalar and the latter made up of three grouped signals, the *Mux* parameter field must contain the vector [1 3]. This means that the *Mux* block has two input ports, the first being scalar, the second being multi-variable and of dimension 3.

The same applies for the *Demux* block. If one enters the vector [2 4 1] in the parameter field then the *Demux* has three output ports: the first one is multi-variable and of dimension 2, the second one is also multi-variable and of dimension 4, and the last is scalar (obviously the *Demux* block input must be of dimension 7).

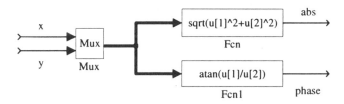

Figure 13.2 Example of *Fcn* block with *Mux*.

In many applications the *Mux* block is used in series with the *Fcn* block. This is the only way to pass more than one variable to the *Fcn* block. As an example, let us consider the problem of converting the position representation of an object from Cartesian coordinates to polar coordinates. A possible solution is proposed in Figure 13.2.

In the scheme the *Wide Vector Lines* option is set; indeed the *Mux* output lines are thick. As seen in Section 11.5, the input components are labeled u[1],..., u[n].

13.3   An example: a planar manipulator

In this section a two-link planar robot scheme is considered in the presence of the
gravity force. This model is rather complex and to draw it the reader must use
most of the tools studied up to now.

The robot dynamics are described by the equation

$$B\left(q\right)\ddot{q}+h\left(q,\dot{q}\right)+g\left(q\right)=k\tau$$

where q is the joint position vector which describes the link configuration (see
Figure 13.3), B is the inertia matrix (depending on the robot configuration), $h\left(q,\dot{q}\right)$
represents the Coriolis and centrifugal force vectors, $g\left(q\right)$ are the torques due to
gravity, the τ are the torques applied at joints, and k is a constant.

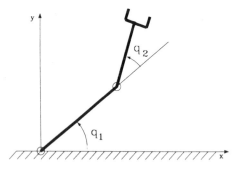

Figure 13.3 Two-link planar manipulator.

In particular we assume that the two links are identical. Each of them has constant section, 1 m length, 50 kg mass, and 10 kg m^2 moment of inertia. Moreover,
we assume that the acceleration of gravity is 9.8 m s^{-2}, and $k = 1$. Then

$$B\left(q\right)=\left(\begin{array}{cc}50\cos\left(q_2\right)+95 & 25\cos\left(q_2\right)+22.5 \\ 25\cos\left(q_2\right)+22.5 & 22.5\end{array}\right)$$

$$h\left(q,\dot{q}\right)=\left(\begin{array}{c}-25\,\dot{q}_2\left(2\dot{q}_1+\dot{q}_2\right)\sin\left(q_2\right) \\ 25\,\dot{q}_1^2\sin\left(q_2\right)\end{array}\right)$$

$$g\left(q\right)=\left(\begin{array}{c}245\cos\left(q_1+q_2\right)+735\cos\left(q_2\right) \\ 245\cos\left(q_1+q_2\right)\end{array}\right)$$

A SIMULINK scheme is shown in Figure 13.4, where a simple position controller
is also proposed.

On the left of the scheme we see the controller defined by the 'shadowed' blocks.
The two blocks on the top left are the position references for the joints. In particular, the reference position is $3\,\pi/4$ for the first joint and $-\pi/3$ for the second
one.

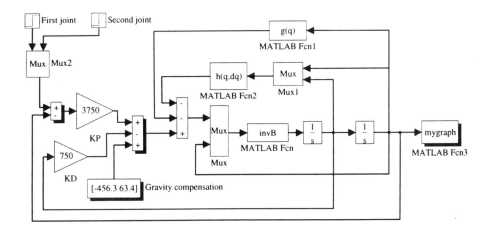

Figure 13.4 SIMULINK scheme of two-link robot.

After these blocks there is a *Mux* which groups the reference inputs into a single channel. Then the position errors feed two standard PD controllers whose proportional and derivative constants are, for both joints, $KP = 3750$ and $KD = 750$ respectively. Moreover, there is a further constant term.[3]

Then there is a *Sum* and a *Mux* whose output is a vector collection of the two outputs of the second *Sum* (the total torques at the joints) and the actual position of the two joints. Hence, in the *Mux* parameter field we have $[2, 2]$, i.e. the output block is a multi-variable signal of dimension 4, determined by grouping two signals of dimension 2. This vector is sent to the *MATLAB Fcn* block which evaluates the joint accelerations $\ddot{q} = B(q)^{-1}(-h - g + \tau)$.

The *MATLAB Fcn* has one input port; then if more variables must be sent to it they must be grouped into a single multi-variable channel. This solution was used because it is fairly complex to evaluate the accelerations due to the computation of the inverse of the B matrix directly in SIMULINK.

Since the input dimension of the *MATLAB Fcn* block is different from that of the output (2, the joints' accelerations), the output dimension must be explicitly defined in the *Output width* parameter field.

Then there is a pair of vectorial integrators, the first one computing the joint angular rates, the second computing the joint angular positions. The first *Integrator* has a zero scalar initial condition, i.e. at the initial time instant the robot is at rest. The condition is the same for the two joints; hence it is useful to insert one scalar value. SIMULINK expands this condition, assuming the same value for all the joints. The initial condition of the second integrator is the vector $[0.7854 \ 0.5236]$; this means that in the initial configuration the first link makes an angle of 45°

---

[3]Required to compensate for the gravity term in the steady-state configuration.

with the ground, and the second one makes an angle of 30° with respect to the first link. In this case, since we are using a vectorial parameter, the dimension of the parameter must be the size of the input.

The *MATLAB Fcn1* and *MATLAB Fcn2* blocks implement the computation of the centrifugal torques and those due to gravity respectively.

In particular, the *MATLAB Fcn2* inputs are the position and the velocity of the joints grouped together, and its outputs are the Coriolis and centrifugal torque components at the two joints.

MATLAB Fcn1 receives as inputs the joint positions and gives as outputs the torques due to gravity. In this case, since the input and output dimensions are the same, one may use the value −1 in the *Output width* field (SIMULINK assumes the output dimension to be that of the input).

These functions have been implemented via *MATLAB Fcn* blocks because of their multi-variable outputs. A different solution is to use a couple of *Fcn* blocks for each function. However, the latter solution reduces the clarity of the scheme. The use of *Fcn* blocks instead of *MATLAB Fcn* blocks allows one to reduce the time required for the simulation drastically. The *MATLAB Fcn* should be used only in the presence of functions that are not supported in the *Fcn* or in the presence of very complex calculations. In the present case, an elegant solution is to use a couple of *Fcn* blocks and then to *group* them together, thus increasing the computational efficiency without reducing the scheme's clarity (for the *group* option see the next chapter).

The plot of the joints' angle behavior does not provide an easy view of the robot's configuration. A more useful representation is obtained via the *MATLAB Fcn3* blocks which call the `function`:

```
function nill=mygraph(q);
% draw the robot behavior
x1=cos(q(1));
y1=sin(q(1));
x2=x1+cos(q(1)+q(2));
y2=y1+sin(q(1)+q(2));
plot([0 x1 x2],[0 y1 y2]); axis([-2 2 0 2])
nill=0;
```

which represents the robot with two segments on the MATLAB graphic window. The *MATLAB Fcn* must have at least one output; hence in the function the fictitious variable `nill` is used.

The use of the MATLAB graphic window, even though it improves the representation quality, slows down the simulation dramatically. This drawback may be reduced by using the more sophisticated graphic commands described in Appendix A.

13.4   Exercises

Exercise 13.1 The three-phase voltage supply consists of three sinusoidal signals of the same amplitude and frequency, each with a phase shift of 120° such that their sum is zero. Design a three-phase generator which supplies voltages with peak values of 220 V and a frequency of 50 Hz. Verify, moreover, that the three-phase voltage sum is identically zero.

> **Hint.** Use one *Sin Wave* block with the *Phase* parameter field set to the phase shift values' vector (the other parameters are scalar quantities). The signal generated is multi-variable; to test the zero-sum connect the generator to a *Sum* block with a single + in its parameter field (the summation symbol appears in the block icon).

Exercise 13.2 Heat diffusion in a 1D stick is described by the following partial differential equation:

$$c_p\rho\frac{\partial T(x,t)}{\partial t} = \lambda\frac{\partial^2 T(x,t)}{\partial x^2}$$

where $T(x,t)$ is the temperature at the abscissa x and time t.

One way to solve a problem described via partial differential equations is to spatially discretize the model and hence study the behavior at given abscissae.

In particular, the discretized model of the stick, taking into account five equally spaced points, assuming that the temperature at the stick's extremes form the system inputs, and assuming all constants equal to 1, is

$$\frac{dT}{dt} = \begin{bmatrix} -2 & 1 & 0 \\ 1 & -2 & 1 \\ 0 & 1 & -2 \end{bmatrix} T + \begin{bmatrix} 1 & 0 \\ 0 & 0 \\ 0 & 1 \end{bmatrix} u$$

where the state variables T_1, T_2, T_3 are the temperature in the three inner points.

Analyze the temperature behavior at the inner stick points assuming that a heat source with a temperature of 10° greater than the equilibrium one is connected to the stick's left end, and that after seven seconds a heat source with a temperature of $-20°$ is connected to the right-hand end.

> **Hint.** Use one *Step Input* block with suitable vector parameters.

Exercise 13.3 Build a SIMULINK scheme which reproduces the trapezoidal signal shown in Figure 13.5. The extreme value p and the slope of the oblique pieces q must be the scheme's inputs.

> **Hint.** Use a *Mux* block to multiply the p, q and the *Clock* signal in an *Fcn* block which implements the suitable logic relation.

Exercise 13.4 The behavior of a three-story building under earthquake conditions may be described, in a first approximation, by means of the model

$$M\ddot{x} + Kx = Bu$$

where the state variables consist of the displacement of the floors with respect to their equilibrium positions, M is the mass matrix, K is the stiffness matrix, B is

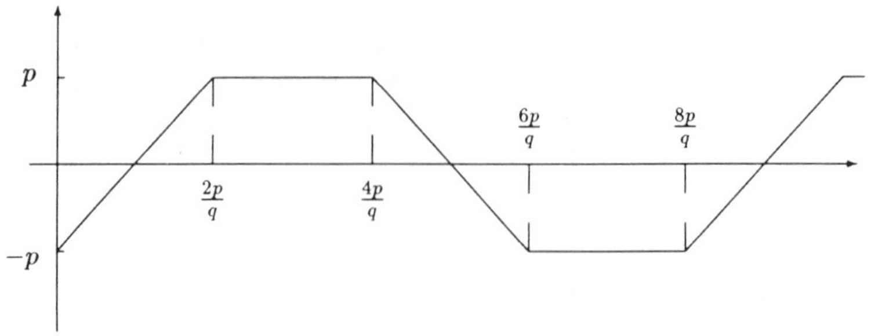

Figure 13.5 Exercise 13.3: trapezoidal wave.

the earthquake's input matrix and u represents the earthquake wave. Assume the following numerical values:

$$K = 10^7 \begin{bmatrix} 1 & -1 & 0 \\ -1 & 2 & -1 \\ 0 & -1 & 2 \end{bmatrix}, \quad M = 10^4 \begin{bmatrix} 1 & 0 & 0 \\ 0 & 1 & 0 \\ 0 & 0 & 2 \end{bmatrix}, \quad B = \begin{bmatrix} 0 \\ 0 \\ 1 \end{bmatrix}$$

The earthquake may be modeled via a sine wave with frequency below 10 rad s^{-1}. Is the building able to resist the earthquake? Moreover, analyze the building's behavior in the presence of an anomalous earthquake with a frequency of 13.3 rad s^{-1}.

> **Hint.** One of the causes of building collapse during earthquakes is that the earthquake wave bring the system to resonance. In our case the proper lower frequency of the building is approximatively 13.3 rad s^{-1}.

Exercise 13.5 Draw the SIMULINK scheme and analyze the behavior of the system

$$\dot{x}_1 = x_2 - x_1 \left(x_1^4 + 2 x_2^2 - 10 \right)$$
$$\dot{x}_2 = -x_1^3 - 3x_2^5 \left(x_1^4 + 2 x_2^2 - 10 \right)$$

with the initial condition $x_1(0) = 1$ and $x_2(0) = 2$.

> **Hint.** Use one *Fcn* block for the term in parentheses.

Exercise 13.6 Draw the SIMULINK scheme for the two-link planar robot

$$M(q)\ddot{q} + h(q, \dot{q}) + f(\dot{q}) = \tau$$

where M is the inertia matrix, h are the torques due to centrifugal forces, f are the frictions at the joints and τ are the command joint torques. Remember that a planar robot has its links in a plane parallel to the ground and hence the gravity is balanced by the mechanical structure instead of, as for the robot in Section 13.3, via the joints torques. Assume the following matrices:

$$M(q) = \begin{pmatrix} 3.316 + 0.234 \cos{(q_2)} & 0.117 + 0.163 \cos{(q_2)} \\ 0.117 + 0.163 \cos{(q_2)} & 0.117 \end{pmatrix}$$

$$h(q, \dot{q}) = \begin{pmatrix} -0.163\,\dot{q}_2\,(2\dot{q}_1 + \dot{q}_2) \sin{(q_2)} \\ 0.163\,\dot{q}_1^2 \sin{(q_2)} \end{pmatrix}$$

$$f(\dot{q}) = \begin{pmatrix} 5.3\,\mathrm{sgn}\,(\dot{q}_1) \\ 1.1\,\mathrm{sgn}\,(\dot{q}_2) \end{pmatrix}$$

and that there are mechanical joints limits, i.e. $|q_1|_{MAX} = 2.3$ rad and $|q_2|_{MAX} = 2.15$ rad.

Exercise 13.7 The linearized model of the lateral axis dynamics of an aircraft is (Harvey and Stein, 1978)

$$\dot{x} = Ax + Bu$$

where the state variables are

$$x = \begin{bmatrix} p_s \\ r_s \\ \beta \\ \varphi \\ \delta_r \\ \delta_d \end{bmatrix} = \begin{bmatrix} \text{stability axis roll rate} \\ \text{stability axis yaw rate} \\ \text{sideslip angle} \\ \text{bank angle} \\ \text{rudder deflection} \\ \text{aileron deflection} \end{bmatrix}$$

and the inputs are

$$u = \begin{bmatrix} \delta_{rc} \\ \delta_{ac} \end{bmatrix} = \begin{bmatrix} \text{rudder command} \\ \text{aileron command} \end{bmatrix}$$

Assume the following numerical values:

$$A = \begin{pmatrix} -0.746 & 0.387 & -12.9 & 0 & 0.952 & 6.05 \\ 0.024 & -0.174 & 4.31 & 0 & -1.76 & -0.416 \\ 0.006 & -0.999 & -0.0587 & -0.0369 & 0.0092 & -0.0012 \\ 1 & 0 & 0 & 0 & 0 & 0 \\ 0 & 0 & 0 & 0 & -10 & 0 \\ 0 & 0 & 0 & 0 & 0 & -5 \end{pmatrix}$$

$$B = \begin{pmatrix} 0 & 0 \\ 0 & 0 \\ 0 & 0 \\ 0 & 0 \\ 20 & 0 \\ 0 & 10 \end{pmatrix}$$

$$C = \begin{pmatrix} 1 & 0 & 0 & 0 & 0 & 0 \\ 0 & 1 & 0 & 0 & 0 & 0 \end{pmatrix}$$

and analyze the plane's behavior when the rudder deflection has a unit step perturbation in the presence of the feedback control law $u = Kx$, with

$$K = \begin{pmatrix} -0.13 & -0.88 & 1.58 & 0.03 & 0.68 & -0.03 \\ 0.52 & 0.42 & -2.83 & 0.02 & -0.01 & 0.86 \end{pmatrix}$$

Hint. To implement the feedback control law the whole state is needed. This may be recovered via an augmented output matrix $\tilde{C}$ which includes the actual output variables and the state. The two sets of variables may be separated via a *Demux* block with the string [2,6] in the parameter field, i.e. the eight input signals of the *Demux* are separated into two multi-variable channels: the first one of dimension 2 (system outputs) and the second one of dimension 6 (system states).

Use the *Matrix Gain* block of the *Linear* for the state feedback.

Chapter 14

Group operation

The readability and clarity of a block scheme becomes highly deteriorated when the blocks' numbers increase. However, almost always, a complex scheme may be split into simpler subschemes.

To split a complex scheme into its lower-level components increases the scheme's clarity and, moreover, allows improvement of the analysis and debugging phases. Indeed, each component may be tested separately, and only in a second step is the whole scheme is tested.

There is no unique criterion for finding the 'best' decomposition of a scheme: sometimes it is useful to use a two-level description, sometimes it is better to use a deeper hierarchical description.

The choices depend on the designer's experience. However, the best result is a trade-off between scheme clarity and a limited number of sublevels (in a scheme with too many levels one may lose global sense).

In this chapter we analyze the SIMULINK capacity to handle modular structures.

14.1   Using the group option

One of the advantages of a modular scheme is its ease of modification. For example, the design of a controller for a generic plant may be performed as shown in Figure 14.1, where the highlighted blocks are *Plant* block and the *Controller* block.

This structure allows one to check the plant model first, and then to test different controllers (each of them is a single block, hence changing the controller simply means changing the *Controller* block).

Moreover if a further element, for example an *Observer*, should be added, the scheme is easily modified as shown in Figure 14.2. Obviously, the design and the test of each block is done separately.

Splitting a complex scheme into its low-level components allows to improve the

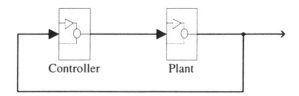

Figure 14.1 Plant-controller scheme.

design, the debugging, and the maintenance of the scheme itself.
Schemes as such those of Figure 14.1 and Figure 14.2 are usually multi-variable; hence the block connections are multi-variable channels.

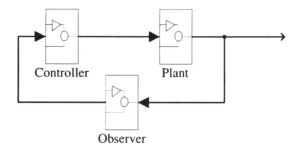

Figure 14.2 Introduction of a further block, the observer.

The way SIMULINK works is slightly different from that explained before, but the final result is the same.

SIMULINK is able to group a set of elementary blocks (a 'simple' scheme) in a superblock called a *group*. This *group* becomes very similar to SIMULINK blocks, and it may be used to draw more complex schemes.

A scheme obtained via *group* blocks may be further grouped in order to build up a super*group*, and so on.[1]

Let us elaborate with an example: the ship model of Figure 11.3. Our aim is to build a 'ship block', i.e. a block which has the same input–output behavior as the ship. Then that portion of the ship scheme after the input block up to the output y should be grouped. To this end, the **selection via bounding box** of Section 11.1 must be used, i.e. we click on the screen and drag the mouse to include in the dashed box all the elements to be grouped.

---

[1]The methods described are known as *top-down* (the first one) and *bottom-up* (the latter one). The first one is to be preferred for analysis purposes, whereas the second one is preferable for design and to synthesize phases.

Select the part of the scheme to be grouped and use the *Group* option of the **Options** menu.

The same result may be obtained via the hot-key CTRL-G.

Once we have done this, a block such as that of Figure 14.3 appears. SIMULINK labels the block *Subsystem*. In order to avoid confusion, we immediately change the block title to a more mnemonic one, such as ship. In the presence of more *grouped* blocks, the blocks may be distinguished only by their titles; hence the *Hidden* option should never be used.

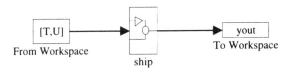

Figure 14.3 Ship block.

This group is like a SIMULINK block, it may be used, for example, to design a transverse ship vibration controller.

In the scheme of Figure 14.3 one may note that all the connections with non-grouped blocks are retained.

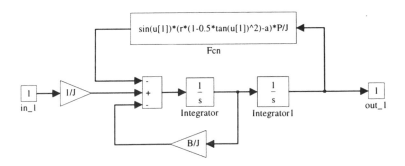

Figure 14.4 '*Ungroup*' of the ship block.

To better understand how SIMULINK works it is useful to 'explode' the grouped block. Double-clicking on the ship block opens the window of Figure 14.4. One may recognize the ship scheme, but there are some new blocks: *in_1* and *out_1*.

These blocks are the input and output of the ship block: the *in_1* is the input (*Inport*) and *out_1* is the output (*Outport*).

The *Group* operation may, in the presence of very complex schemes, modify the topology of the scheme considerably, even if its input–output relations still hold. This is due to the SIMULINK internal representation of the group. In these cases, also the *Ungroup* option has the same drawback. It is useful to save the scheme, with a different name, before performing the *group* operation.

However, very complex schemes should be avoided. If the scheme complexity rises it is convenient to increase the hierarchy by assembling some 'sub-*groups*'.

If the *group* has more inputs and/or outputs, each port is labeled with a number. The input (output) port numbering must start from 1 and must be strictly sequential.

SIMULINK automatically numbers the ports so that the first input is on top of the *group* block. However, this order may be modified by numbering the ports differently.

If the group-block ports are not specified, SIMULINK inserts them automatically at the bounding-box crossing connection (as we saw in the previous example).

However one may override SIMULINK, putting the input and output ports (by means of the *Connections* blocks) explicitly before the *Group* operation.

The *Ungroup* option is able to ungroup a block, i.e. placing the block in the main parent diagram.

The same result may be obtained via the hot-key CTRL-U.

There is no limit to the number of input and output ports. However a huge number of ports and their related connection lines reduce the scheme's clarity. In such cases it is better to multiplex the input (output) on a multi-variable channel and then to use a single port.

14.2   Exercises

Exercise 14.1 The biological competitive exclusion principle says that if two or more like species (i.e. they have the same food requirements) are in the same habitat, the limited resources yield to the extinction of all the species except for the fitter one.

This principle is described via the Volterra model

$$\dot{x}_1 = [\beta_1 - \gamma_1 F(x)]\, x_1$$
$$\dot{x}_2 = [\beta_2 - \gamma_2 F(x)]\, x_2$$
$$\dot{x}_3 = [\beta_3 - \gamma_3 F(x)]\, x_3$$

where

$$F(x) = \sum_{i=1}^{3} \alpha_i x_i.$$

All the constants are given positive numbers, β_i is the natural growth rate of the *i*th specie, and γ_i is the sensitivity of the *i*th specie to the whole population. Show that, no matter how fixed the α_i, β_i, and γ_i parameters and the initial conditions are, only one specie survives and the others are extinct after the transient.

Hint. Consider the $F(x)$ function as a *group*.

Exercise 14.2 Modify the scheme of Figure 13.4 using the *Fcn* block to evaluate the torques due to gravity $g(q)$ and to Coriolis forces $h(q, \dot{q})$. Use the *group* option to preserve the scheme's clarity.

Compare the performances, in terms of the time needed for the simulation, of this scheme with that of Section 13.3 (where the torque calculus is implemented via *MATLAB Fcn* blocks).

> **Hint.** Do not use the MATLAB graphic window when comparing the performance of the schemes.

Exercise 14.3 Usually, the force produced by a spring is assumed to be proportional to the displacement. This is a very simplified model. A more sophisticated model is the so-called *hard spring* (Duffing, 1908): the force increases more than proportional to the displacement.

Analyze the behavior of a mass-spring system perturbed by $u = A\cos(0.1\,t)$ when the spring is modeled, respectively, as:

$$\begin{array}{cc} \text{linear spring} & \text{hard spring} \\ k\,y & k\,y + k_2\,y^3 \end{array}$$

where y is the displacement with respect to the equilibrium position. Assume unit mass, $k = 0.3\ \mathrm{N\,m^{-1}}$, $k_2 = 0.02\ \mathrm{N\,m^{-3}}$, and a perturbation magnitude A equal to 0.1 N. Next, compare the behavior of the two models when A is equal to 2 N.

> **Hint.** Build up the spring models as two *groups* and substitute each of them in the unique system scheme.
>
> In the presence of small displacements the behavior of the two systems is almost the same, whereas when the displacements increase, the behavior is different.

Exercise 14.4 Let us consider the system of Figure 14.5. A gun is set on a platform 0.85 m in height and fires a bullet of mass 2 kg with velocity 50 m s^{-1} and rear sight of 40°. A pendulum of 10 kg mass, 2 m height and with a friction coefficient of 0.002 N m s^{-1}, is 501.8 m from the gun. At the firing time instant the angle between the pendulum and the vertical is 90° .

Figure 14.5 Exercise 14.4: gun with moving target.

Assuming that the acceleration of gravity is 9.8 m s^{-2}, verify whether the bullet hits the pendulum.

Hint. Define two *group* blocks, one for the bullet and the other for the pendulum. The pendulum is hit if a time instant exists such that the two objects' coordinates coincide (i.e. the difference between the coordinates is less than ε).

Set the **Simulation** menu parameters in an appropriate way.

The pendulum equation is

$$M\,l\,\ddot{\theta} + M\,g\sin(\theta) + M\,k\,\dot{\theta} = 0$$

Chapter 15

Discrete-time systems

Until now we have analyzed only systems described by differential equations, but there are also systems that can be described by difference equations. They derive either from intrinsically discrete-time processes, or are approximations of continuous-time processes.

Discrete-time systems may be simulated directly in MATLAB by means of `for`-like structures; however this solution is not computationally efficient (see Section 8.1) and does not provide the global insight of a SIMULINK scheme. Moreover, the MATLAB simulation of high-dimension discrete-time systems, especially in the presence of elements with different sample rates and/or nonsynchronous sample instants is quite complex.

On the other hand, as we will see in this chapter, SIMULINK automatically solves all the synchronization problems and simplifies the handling of hybrid systems considerably (i.e. systems with both continuous-time and discrete-time elements), as will be shown in Chapter 16.

15.1   An example: the repopulation of a lake

A simple model of the life cycle of a fish species was proposed by Leslie (1945). He assumed as state variables the number $x_i(T)$ of females of age i in year T. The main phenomena related to the life cycle are: aging (described by the survival coefficients s_i, which take into account the interaction of the fish with predators and the natural deaths) and the reproduction (described by the fertility coefficients f_i):

$$
\begin{aligned}
x_1(T+1) &= s_0 \sum_{i=1}^{5} f_i x_i(T) \\
x_2(T+1) &= s_1 x_1(T) \\
x_3(T+1) &= s_2 x_2(T)
\end{aligned}
\tag{15.1}
$$

$$x_4(T+1) = s_3 x_3(T)$$
$$x_5(T+1) = s_4 x_4(T)$$

The first equation represents the number of females born in the year T, whereas the last equations model the aging phenomenon.

This process is intrinsically discrete-time, since the number of fish varies only at some fixed moment (in particular, the number of fish varies appreciably only during the reproduction period, and it is almost constant between two reproduction periods).

It is important to stress that this is a very crude approximation of the fishes' life cycle; indeed the mathematical model does not take into account food and predator variation.

Usually the f_i and s_i coefficients are highly time-varying depending on the varying habitat conditions and on the number of fish itself.

Let us consider the life cycle inside a lake to be repopulated, and focus our attention on studying whether the repopulation operation reaches its goal, i.e. if the new fish species fit the lake habitat.

In our example we assume that the coefficients are time-invariant, that the females are not fertile until their second year, that the percentage of survival in the first and fourth years is reduced, and that the number of fish living beyond their fifth year is negligible. These hypotheses may be summarized by the following coefficients:

$$s_0 = 0.01 \qquad\qquad f_1 = 0$$
$$s_1 = 0.5 \qquad\qquad f_2 = 55$$
$$s_2 = 0.9 \qquad\qquad f_3 = 69.5$$
$$s_3 = 0.9 \qquad\qquad f_4 = 90$$
$$s_4 = 0.5 \qquad\qquad f_5 = 23.5$$

Moreover we assume that a newly hatched fish colony (with one hundred females) is introduced in year 0, and that there are no further introductions over the next years.

The discrete-time blocks are collected in the *Discrete* group. In all the discrete-time blocks there is the *Sample time* field which defines the sample step. To solve the problem we use the *Dis. State-space* block and set the parameter *Sample time* to 1.

The time behavior of the four-year-old females shows some oscillations during the first twenty years before reaching a steady-state value. This means that the repopulation of the lake was successful. From a mathematical point of view this means that the system has a pole with unit modulus.

The asymptotic stability of a linear discrete-time system, unlike the linear continuous-time system case, where the asymptotic stability is guaranteed if all the poles have a negative real part, is guaranteed if all the poles are inside the unit disk.

A situation of nonasymptotic stability is the boundary between the the asymptotic stability and the instability regions: a small perturbation may modify the system behavior dramatically. Indeed, if the survival coefficients or the fertility coefficients are slightly perturbed the fish population will either undergo a demographic explosion (the unstable system) or it will be extinguished (asymptotically stable system).

The real world is not so 'sensitive', and this proves that the 'control' developed by Nature, the ecosystem, is very efficient (and that our model is too approximate). Sometimes, elements with different sampling rates must be implemented. SIMULINK is powerful in these situations: it is enough to insert in each block its own sampling rate; all the synchronization work will be done automatically by SIMULINK.

In the same way, SIMULINK automatically handles blocks with different sampling instants. To this end, the *offset* of each block, i.e. the delay of the block's sample instant with respect to the 'nominal' sample instant, is entered as a second optional parameter in the *Sample time* field.

The behavior of two discrete-time systems is shown in Figure 15.1. These systems have different sampling periods: the first block has a sample period of 1 s, whereas the second one has a sample period of 0.5 s with a delay offset of 0.3 s with respect to the first block (this is obtained by setting the second block's parameter field to [0.5 0.3]).

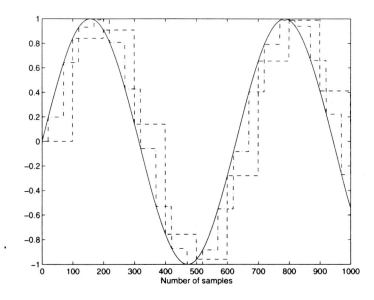

Figure 15.1 Signal sampled with two different sampling rates and sample instants.

The simulation scheme is shown in Figure 15.2. The *Zero-Order Hold* block reproduces an ideal sample and hold.

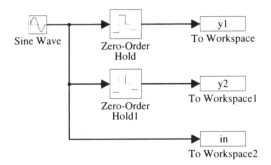

Figure 15.2 SIMULINK scheme of two sample and hold with different sampling rates and different sampling instants.

The *Digital Clock* block allows optimization of the computational burden in the presence of discrete-time multi-rate systems.

A scalar must be entered in the sample time field, even in the presence of multi-variable signals. All the signals elaborated by a single block are sampled with the same sample time (and at the same sample instant).

The *Sine Wave* generator has the parameter field *Sample Time* which allows us to generate a signal similar to a sinusoid through a sample and hold. The algorithm implemented is numerically more robust and efficient for the simulation of discrete-time systems.

15.2   Exercises

Exercise 15.1 The Newton method for evaluation of the minimum of a function reduces the problem to one of finding the limit of a series. Then this method may be seen as a discrete-time system.

Verify that the minimum of the function $f = e^x + x^3 \sin x$ is the steady-state solution of the following series:

$$x_{k+1} = -\frac{f'(x_k)}{f''(x_k)} + x_k$$

Hint. The 'memoryless' blocks are the same both for the discrete-time systems and the continuous-time ones.

Exercise 15.2 Study the behavior of the difference equation

$$y(k + 2) + y(k) = 0$$

with initial condition $y(0) = 1$ and $y(1) = 0$.

Exercise 15.3 Solve this problem, formulated by Leonardo da Pisa in 1202: 'Assume that there are a couple of rabbits, a male and a female, and that a rabbit is fertile after a month, and that each couple of rabbits generates a new rabbit couple (a male and a female) each month; how many rabbits are there after one year?'

> **Hint.** Leonardo da Pisa, who is known by the name of Fibonacci, introduced the Fibonacci series with this problem.

Exercise 15.4 A way of evaluating the square root of a number is by means of successive approximations. Verify that the steady-state solution of the system

$$x(k + 1) = x(k) + \alpha - x(k)^2$$

for $\alpha = 1/4$ and $\alpha = 1/3$ is the square root of α, whatever the initial condition, provided that $0 \leq x(0) \leq 1$. This method holds only for a positive number α smaller than 1.

Exercise 15.5 The growth model of a species in a habitat with limited resources is

$$x(k + 1) = x(k) + a \left[1 - \frac{1}{c} x(k) \right] x(k) + u(k)$$

where the bracketed term represents the instantaneous growth rate and depends on the actual number of animals; in particular c is the maximum number of animals which may co-exist. Analyze the influence of the parameters a and c on the species' evolution, assuming that the forcing term $u(k)$ is identically zero.

Exercise 15.6 The dynamic of the student population inside a five-year course at an Italian University may be described by means of the following model:

$$
\begin{aligned}
x_2(k + 1) &= \alpha_1 u(k) \\
x_3(k + 1) &= \alpha_2 x_2(k) \\
x_4(k + 1) &= \alpha_3 x_3(k) \\
x_5(k + 1) &= \alpha_4 x_4(k) + (1 - \beta) x_5(k)
\end{aligned}
$$

where the $x_i(k)$ are the number of students in the ith year course at year k, $u(k)$ is the number of students admitted at year k, $(1 - \alpha_i)$ is the rate of resignation from the ith to the $(i+1)$th year course, and β is the rate of graduation. Analyze the behavior of the total number of students assuming that the number of students admitted each year is a Gaussian random variable with mean 130 and variance 0.1, zero initial conditions and the following numerical values:

$$\alpha_1 = 0.91; \quad \alpha_2 = 0.73; \quad \alpha_3 = 0.80; \quad \alpha_4 = 0.98; \quad \beta = 0.15$$

> **Hint.** The numerical values are those of the Faculty of Engineering of the University of Benevento (Italy).

Exercise 15.7 As mentioned before, the *Zero-Order Hold* block is an ideal sample and hold. Build an ideal sample and hold without using this block.

Hint. Use the *Discrete Transfer Fcn* block.

Exercise 15.8 The restore process of capital via six-monthly installments may be described by the equation

$$y_k = \left(1 + \frac{i}{2}\right) y_{k-1} - u_k$$

where y is the amount to pay, i the annual interest rate and u the six-monthly installment.

After how long will a one million pound loan be paid off via 80 000 pound six-monthly installments with an annual interest rate of 12%?

    Hint. In order to put the model in the usual state-space form make the following change of variables: $x_k = y_{k-1}$. Stop the simulation as soon as the loan is paid off (use the *STOP* block).

Exercise 15.9 Analyze the behavior of the system shown in Figure 15.3 forced by the periodic input signal shown in the same diagram.

    Hint. Use the *Repeating Sequence* block to generate the input signal. The output of the *Repeating Sequence* repeats, regularly over time, an arbitrary signal specified in the parameter fields. In the presence of discontinuities in the signal, the discontinuous points must be doubled in the sequence: at one time with the left-hand value and then with the right-hand value.

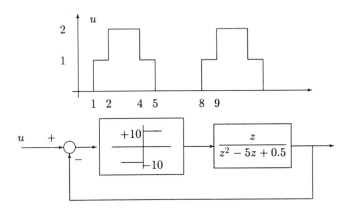

Figure 15.3 Exercise 15.9: system scheme with input time history.

Exercise 15.10 Design the SIMULINK scheme of the following Cremona difference equations

$$x_1(k+1) = x_1(k)\cos(a) - \left(x_2(k) - x_1(k)^2\right)\sin(a)$$

$$x_2(k+1) = x_1(k)\sin(a) + \left(x_2(k) - x_1(k)^2\right)\cos(a)$$

and analyze its behavior assuming that $a = 1.328$ and the following four initial conditions:

$$(0.1, 0.1) \quad (0.2, 0.2) \quad (0.5, 0.5) \quad (0.6, 0.6)$$

Represent all the solutions on the same phase plane and use the ' . ' option for the `plot`.

Exercise 15.11 The following model (Streeter–Phelps, 1925) allows one to analyze river water quality taking into account dissolved oxygen concentration (DO) with respect to the oxygen quantity needed by biological processes (BOD). Study the behavior of

$$
\begin{aligned}
DO_{k+1} &= 1.335 + 0.8142\ DO_k - 0.00001\ BOD_k \\
BOD_{k+1} &= 0.7455\ BOD_k + 0.1471\ DO_{k-3}
\end{aligned}
$$

Hint. Use some *Unit Delay* blocks to realize DO_{k-3}.

Exercise 15.12 Analyze the behavior of the system

$$ky_{k+1} = [1 + (k-1)\,y_{k-1}]$$

with zero initial conditions and for $k > 3$.

Hint. Use *Clock* block to produce the k values and 'delay' it appropriately.

Exercise 15.13 Difference equations, too, have ill-posed problems. Compare the solution of the predator–prey model

$$
\begin{aligned}
x_1(k+1) &= ax_1(k)\,(1 - x_1(k)) - x_1(k)x_2(k) \\
x_2(k+1) &= \frac{1}{b}x_1(k)x_2(k)
\end{aligned}
$$

starting from the initial condition $x_1(0) = 0.2$ and $x_2(0) = 0.2$ when the parameter values are $a = 3.43$ and $b = 0.31$, and when one parameter is slightly perturbed: $a = 3.6545$.

Hint. Take into account at least 5000 points; compare the responses on the phase plane using the ' . ' option for the `plot` command.

Exercise 15.14 Design an automaton whose input is a sequence of $\{0, 1\}$ and which is able to recognize the string $\{1, 1, 1\}$.

Hint. Every finite state machine may be described via a combinatorial network and some delay elements.

Build up the combinatorial network via the *Combinatorial Logic* block and the delay elements by means of *Unit Delay* (in this way one obtains a synchronous machine). Design the input sequences in MATLAB.

Exercise 15.15 Simulate a machine whose inputs are coins of 5, 10, 25 and 50 cents and that is able to warn if it was inserted at least 1 $.

Hint. This is a finite state machine; see the previous exercise hint.

The *Combinatorial Logic* block is able to handle only the {0,1} values, then a block which is able to codify the coin-type information (this is four-level information) into a {0,1} string must be placed before it.

Build the code block via a *group*.

Chapter 16

Hybrid systems

In real world situations it is very unusual to find purely discrete-time systems, or systems which are completely analog (these are more and more uncommon due to the diffusion of digital technologies). On the other hand *hybrid* systems are usually found, i.e. systems where both continuous- and discrete-time elements are present.

Very often, a microprocessor controller (a digital system) must be designed for a generic plant (usually an analog one). To this end, two substantially different 'worlds' must be interconnected.

From system theory, it is known that a possible solution is to convert the analog part into a discrete-time system via sample data system techniques and then to work with purely discrete-time systems. This solution, besides being computationally hard, may be difficult to achieve, in particular in the presence of highly nonlinear systems.

A different solution consists of neglecting the sample time with respect to the analog time constant, i.e. one assumes that the digit system is so fast that it may be considered instantaneous with respect to the analog system's dynamics. However, this solution may also not always be applicable.

Even in this situation SIMULINK allows one to solve all the problems automatically and allows easy simulation and study of hybrid systems.

Before analyzing how SIMULINK works, it is useful to detail down one important concept. The computer, besides being a discrete-time system, works on quantized quantities; hence the quantization process should also be considered in the plant–computer interface. In what follows this phenomenon is neglected. However, if the quantization process is relevant it may be easily simulated via the *Quantizer* block of the *Nonlinear*.

16.1 Hybrid schemes

If a digital controller has been designed for the ship model, in order to analyze its performances in SIMULINK it is enough to draw it and to connect it to the ship scheme. All the synchronization problems are automatically solved by SIMULINK during the simulation.

In this case, a 'Hold block' should be placed after the digital controller and a 'Sample block' before it. However, these blocks (which must be present in the actual implementation) are not included in the SIMULINK scheme because SIMULINK automatically implements their functions.

SIMULINK is able to distinguish continuous-time elements from discrete-time ones. The first elements evolve on the basis of the chosen differential equation integration method, i.e. the instants at which they are evaluated are 'determined' by the integration algorithm. On the other hand, for discrete-time elements, the instants at which the blocks must be evaluated are the sampling instants; hence SIMULINK is forced to evaluate the system's response at those instants.

The *Sample Time Colors* option of the **Style** menu associates with each block (and with its output connection lines) a color on the basis of its sample time. In particular,

Color	Meaning
Black	continuous-time
Magenta	constant with respect to time
Yellow	hybrid
Red	shortest sample time
Green	second shortest sample time
Blue	third shortest sample time
Cyan	fourth shortest sample time

The 'hybrid' term indicates that it is not possible to uniquely associate a 'sample time' to the block: *group* where there are both continuous- and discrete-time parts, *Mux* where signals with different sample times are multiplexed, etc.

The Cyan color is used for blocks with a sample time equal to, or larger than, the fourth shortest sample time.

Some blocks (such as *Gain*) have no color of their own; they assume the color of the immediately preceding block.

If a new block is added, the *Update Diagram* option of the **Style** menu must be used in order to display it with its correct color.

To return to the true blocks' color, it is enough to deselect the *Sample Time Colors* option.

Before going on, it is important to expend some words on the SIMULINK way of working. Each SIMULINK block may be interrogated (or, more exactly, may be 'called') in order to obtain the values of: the block outputs, the state derivative (only for continuous-time systems), the next step state (only for discrete-time systems), the next sample instant (only for discrete-time systems).[1]

In the presence of continuous-time blocks only the integration method determines, step-by-step, the next instant at which to evaluate the response and to update the state.

In the presence of discrete-time blocks only SIMULINK reads the information for the next sample instant from the blocks and evaluates the system at that time instant. If in the system there are blocks with different sampling rates and/or different sample instants, SIMULINK reads the values of the next sample time from all the blocks, sorts them in ascending order and evaluates the whole system at the very next sample instant. Then SIMULINK evaluates the next sample instant of those blocks whose sample instant is the actual one, again sorts all the sample instants in order to find the next one, and so on. It is important to stress that at each sample instant SIMULINK evaluates the output of all the blocks. In other words, SIMULINK assumes that after each discrete-time block there is a 'hold' device.

In the presence of hybrid systems, SIMULINK reads the values of the discrete-time sample instants and sorts them, then receives the sample instant evaluated by the integration algorithm and evaluates the system at the nearest time instant.

To better understand the way in which SIMULINK proceeds let us consider the following example: design a pulse generator, i.e. a block whose output is a sequence of rectangular pulses of given amplitude, frequency and duty-cycle.

A possible SIMULINK scheme is shown in Figure 16.1. The scheme is obtained by a nonlinear continuous-time function which allows one to set the pulse parameters (ht is the amplitude, start is the switch-on position in the period Ts, on is the pulse length).

Moreover, in the scheme there are two other discrete-time blocks. They have no influence on the block output, but they force SIMULINK to evaluate the system at the switching instants. Indeed, these discrete-time blocks have a sampling period equal to the pulse frequency and sample instants coincident with the switch-on and switch-off instants respectively.

If one deletes the discrete-time blocks, the pulse generator output seems to be the same. But if one watches more carefully, one notices that the switching instants are slightly different from the actual ones. Deleting the discrete-time blocks has removed the constraint to update the system at switching instants. Now the system is updated at the time instant determined by the integration algorithm, which may not coincide with the switching one.

What we have proposed in Figure 16.1 is the scheme of the *Pulse Generator* implemented in the 1.2 SIMULINK release; in the versions following, it was implemented differently.

At the end of this chapter it is useful to say something about the aliasing phenomenon. In the real world the sampler is preceded by an anti-aliasing filter. In SIMULINK schemes since the sampler is not needed, the filter is not present usually.

The aliasing phenomenon depends on the ratio between the sample time and the faster system dynamics. SIMULINK intrinsically assumes that the sample time has been correctly chosen, i.e. the aliasing phenomenon is avoided (notice that, usually, the aliasing phenomenon is due to nonmodelled high-frequency dynamics, which are obviously not

---

[1]A more exhaustive treatment is given in the S-Function Appendix.

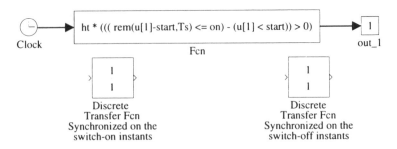

Figure 16.1 Pulse generator.

present in simulation schemes).

However, if one needs to use a sample time which introduces aliasing phenomena, the anti-aliasing filter may be simulated by means of a filter (analog or digital) between the plant outputs and the controller inputs.

16.2   Exercises

In the following exercises we use s to indicate the continuous-time transfer function complex variable, and z for the discrete-time one.

Exercise 16.1 Analyze the output of the *Unit Delay* block with sample time equal to 1 s, when the block is forced by a sine wave with 1 rad s^{-1} frequency and amplitude equal to 1.

> **Hint.** Use the ' . ' option for the `plot`; the result is very nice, but it is not a sine wave.

Exercise 16.2 Analyze the response of a *Zero-Order Hold* with a sampling rate of 1 kHz. In particular, analyze the response of the block to sine wave inputs with unit amplitude and different frequencies: 100 Hz, 300 Hz, 500 Hz, 900 Hz, 1000 Hz, 1100 Hz.

> **Hint.** This exercise emphasizes some sampling problems. Is there any difference between the responses to the 500 Hz and 1000 Hz inputs?
>
> The response to the 100 Hz, 900 Hz and 1100 Hz inputs are 'similar', why?

Exercise 16.3 Analyze the response of the system shown in Figure 16.2 when it is forced via a sine wave with unit amplitude and frequency of 1 rad s^{-1}. Assume that the discrete-time block has a sample time of 0.05 s.

> **Hint.** In the scheme of Figure 16.2 the left-hand block is a continuous-time system, the middle one is memoryless, and the right-hand one is a discrete-time system.

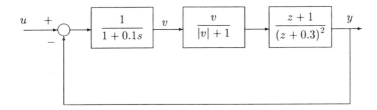

Figure 16.2 Exercise 16.3.

Exercise 16.4 Analyze the behavior of the system shown in Figure 16.3. Assume that the sample time of the discrete part is equal to 0.1 s, and that the inputs are constant and equal to $u_1 = 10$, and $u_2 = 2$.

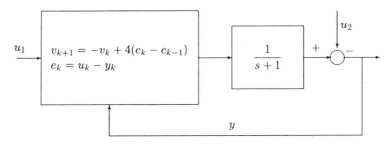

Figure 16.3 Exercise 16.4.

Chapter 17

Advanced topics

In this chapter we discuss some more SIMULINK sophisticated tools and options.

17.1 System analysis

SIMULINK schemes are stored in files with an extension .m (as with any MATLAB file), and they are called *S-functions*.

An analysis of these files by means of a text editor shows that they have quite a sophisticated structure. It is very hard to write the *S-function* 'by hand' instead of drawing the related scheme (however, it is explained in the manual how to write an *S-function* directly).

The *S-function* may even be used directly in MATLAB. Indeed, there are some MATLAB tools which may be applied to the *S-function* and which allow one to evaluate some useful information about system properties.

It is important to stress that in what we discuss in the following the system inputs and outputs are solely those connected to *Inport* and *Outport* blocks respectively. This means that a signal generator is not seen as an input, but rather as an element of the system, so that the presence of a *Sink* block does not imply that this variable is handled as an output.

Hence, when one uses the SIMULINK analysis routines the signal generators present in the scheme must be replaced by *Inport* blocks, and *Outport* blocks must be attached to all *Sink* elements.

17.1.1 Integration

SIMULINK integration routines may be called from the MATLAB prompt with the following syntax:

```
[t,x,y]=linsim('model',tfinal,x0,options,ut,P1,P2,...);
```

which gives as output the time t, the state x, and the outputs y of the system described by the *S-function* model.m, integrated from time zero to time tfinal, with the vector of initial conditions being x0. The value x0 replaces the initial condition parameters possibly present in the blocks of the *model* scheme.

It is possible to insert a two-element vector as tfinal; in this case the first value is the simulation start time and the second is the final time.

The optional parameter options includes some integration method parameters; ut is the external input signal; and P1,P2, ... are variables that are passed directly to the *S-function model* (these may amount at the most to ten).

If a parameter is omitted (i.e. [] is inserted), the default value is assumed.

A more detailed description of the integration routines and their parameters is given on page 157.

The elements of the vector options represent:

	Meaning	Default value
options(1)	tolerance	1e-3
options(2)	minimum step size	tfinal/2000
options(3)	maximum step size	tfinal/50
options(4)	used to automatically select one from the Adams and Gear methods	0 (no)
options(5)	display parameters	0 (no)
options(6)	plot the outputs	0 (no)

The plot of the output variable is obtained every time the function is called without output arguments.

The external input signal ut, may be expressed either via a table or via a string. In the first case the first column of the table must be the time vector (it must contain monotonically increasing numbers), whereas the other columns (as many as the system inputs) are the input signals' time histories. Values for the inputs are linearly interpolated based on the time vector.

If a string is inserted, SIMULINK evaluates the related expression at each time step. For example, entering

$$ut = 't*sin(3*t+2)'$$

means that the system is forced by a sine wave whose amplitude is increased proportionate to time. The letter t represents the current time instant. All MATLAB expressions are supported, and it is possible to refer to MATLAB variables.

The integration routines (the same as those of Figure 12.3) are:

linsim	method for linear systems
rk23	Runge–Kutta method on three points
rk45	Runge–Kutta method on five points
gear	predictor–corrector Gear method for stiff systems
adams	predictor–corrector Adams method
euler	Euler method

The syntax of these routines is the same as that of the linsim routine.

The gear and adams methods also use the fourth elements of the vector options. If this parameter is set to 1 SIMULINK automatically chooses between the two predictor–corrector methods in accordance with the stiff nature of the system.

The reader should have noticed some similarities between the syntax of these routines and that of the MATLAB functions working on functionals, i.e. quad, ode23 and ode45 (see Section 9.3). The main difference is in the structure of the functional used. The functions presented in this chapter work only on the *S-function*.

The *Sink* elements present in the scheme, even if they are not considered as outputs, work correctly. For example, if in a scheme there is a *To Workspace* block, at the end of the simulation in the MATLAB workspace there is the corresponding vector.

If in the scheme there are *Sink* blocks which use the MATLAB graphic window, during the simulation the graphic windows are opened to plot the corresponding quantities. However, it is not possible to access to *Scope* blocks.

17.1.2   Linearization

In many situations it is useful to have the linearized model of a nonlinear system. The SIMULINK function linmod allows one to extract the linear state-space model. This command works on a SIMULINK model, i.e.: *S-function*, and has as output the linearized model around an operating point.

Its syntax is

$$[\text{A,B,C,D}]=\text{linmod}('model',\text{x},\text{u});$$

where x and u are the state and the input vectors which characterize the operating point.

If these parameter are omitted, the linearization is done around an operating point with the state variables set to zero and the input set to zero.

For example, let us consider the ship model represented in Figure 14.4. Store it in the file mship. The instruction

$$[\text{A,B,C,D}]=\text{linmod}('mship')$$

returns the matrices of the linearized model around the state zero and, in the absence of inputs:

$$A = \begin{pmatrix} 0 & 1 \\ -0.153 & -0.079 \end{pmatrix}$$

$$B = \begin{pmatrix} 0 \\ 1.613 \cdot 10^{-6} \end{pmatrix}$$

$$C = \begin{pmatrix} 1 & 0 \end{pmatrix}$$

$$D = 0$$

The linearized model has one input and one output which correspond to the *Inport* and *Outport* blocks respectively. The association between state variables and scheme elements may be recovered by means of the instruction

$$\texttt{[sys, x0, str]=ship([],[],[],0)}$$

which returns the following string in the vector `str`:

/ship/Integrator
/ship/Integrator1

In other words, the first state variable is related to the `Integrator` block, whereas the second one is related to the block titled `Integartor1`.
For further details on the correspondence between the scheme elements and the state vector variables see Appendix C.

Note that the instruction

$$\texttt{[A,B,C,D]=linmod('ship')}$$

where `ship` is the name of the file where the scheme of Figure 11.3 is stored, returns the matrix A as before, but it returns empty matrices for B, C and D. Indeed, in this case there are no *Inport* or *Outport* blocks in the scheme, thus SIMULINK assumes that there are no inputs or outputs. (Remember that *Source* blocks such as *From Workspace* are not considered as inputs, so *Sink* block such as *To Workspace* are not considered as outputs.)
The function `linmod` perturbs the states around the operating point to determine the rate of change in the state derivatives and outputs (Jacobians).

`linmod` has a third optional scalar parameter (called `pert`) which is the perturbation amplitude (the default value is `1e-5`). Moreover, `linmod` has other options: it is possible to indicate the amplitude of the perturbation of each state and output; and other parameter to evaluate the linearized model better. For further details see the manual.

`dlinmod` is the discrete-time version of `linmod`. It produces the discrete-time state-space model with a given sampling time of a continuous-time, discrete-time, multi-rate, hybrid linear and nonlinear system. Its syntax is like the `linmod` syntax:

$$\texttt{[A,B,C,D]=dlinmod('}\textit{model}\texttt{',ts,x,u);}$$

where `ts` is the sample time.
The function `dlinmod` allows one to convert a discrete-time system into a new discrete-time system with different sample times, or to have a continuous-time approximation of a discrete-time system (set `ts=0`) and vice versa.

In the presence of linear, discrete- and continuous-time systems, and in the presence of multi-rate elements, `dlinmod` produces the correct model if

- **ts** is an integer multiple of all sample times of the system;

- the system is stable

If these conditions do not hold, one must verify the **dlinmod** results.

linmod2 is an advanced version of **linmod**. This routine takes longer to run than does **linmod** but it produces more accurate results.

The algorithm used in **linmod2** tries to balance the trade-off which occurs between round-off error (caused by small perturbation levels, which lead to errors associated with finite precision mathematics) and truncation error (caused by large perturbation levels, which invalidate the piecewise approximation). Moreover, **linmod2** produces a warning message in the presence of singularities.

linmod may be also used to get the state-space model of a linear system. In this case, because the truncation problem does not exist, it is better to use a high-perturbation level in order to reduce the round-off errors. Moreover, in this case the operating point values are not influential.

In the presence of a time-varying system (i.e. in the scheme there is a direct dependence upon the time), one may use the optional second entry of the parameter **pert** to set the time instant around which to evaluate the linearized model.

The linearization procedure applied to schemes with *Derivative* or *Transport Delay* blocks may produce incorrect results. In those cases it is useful to replace the blocks by suitable transfer functions or by means of the blocks which appear typing the command **extrlin** from the MATLAB prompt.

17.1.3   Equilibrium points

The linearized model has a practical usefulness only if it is evaluated around an equilibrium point. SIMULINK is able to calculate the system equilibrium points by means of the function **trim**.

Its syntax is

$$[\text{x,u,y}]=\text{trim}('model',\text{x0,u0,y0});$$

the output is the equilibrium point closer to the vectors **x0,u0,y0**: i.e. **trim** attempts to find values for inputs and states (and then evaluates the output values) that set the state derivative to zero and minimize the following functional:

$$\min \ \max \{|x - x_0|, |u - u_0|, |y - y_0|\}$$

If the **x0,u0,y0** parameters are omitted, **trim** tries to find an equilibrium point to minimize the maximum of $\{|x|, |u|, |y|\}$.

The function **trim** reduces the equilibrium points search to a constrained optimization problem.

trim has further parameters which increase its flexibility. One may include in the minimizing functional only some states, input and output components, fix the derivatives to nonzero values and some other options. More details may be found in SIMULINK and *Optimization Toolbox* manuals.

The `trim2`, `trim3`, `trim4` functions determine the equilibrium point on the basis of different algorithms. They do not use a *minmax* algorithm, as does `trim`, but rather a least square one. However, they need the *Optimization Toolbox*.

17.2 Using Mask

In Section 14.1 it was shown how to design a SIMULINK-like block by means of the *group* option. Unfortunately these 'blocks' have a few drawbacks. First, they do not have a self-explanatory icon; their icon is always the same hence they can be distinguished only by means of their name. There is no information on the block properties, i.e. there is no *help*. Moreover, it is quite difficult to obtain a parameterized *group*. Indeed, the only way to obtain a parameterized *group* is to use global variables. However, this solution does not hold in the presence of more copies of the same *group* in one scheme.

These drawbacks may be overcome by masking the *group* by means of the **Mask** option of the **Options** menu.

Select a *group* and use the *Mask* option. A dialog box such as that of Figure 17.1 is opened.

At the top of the window is some information on the block type, whereas at the bottom are five fields which allow modification of the block interface, and some 'structural' (see page 135) and graphic properties.

In particular, the fields are

- *New block type:* defines the block type.
- *Dialog strings separated by |:* allows definition of the parameter fields of the new block dialog box. More precisely, the first entry (each entry is separated by |) is the dialog box title; the remaining entries are the labels of the dialog box parameter fields (six at the most).
- *Initialization commands:* allows definition of the initialization commands and the correspondence between the dialog box parameters and the underlying block variables.
  The strings of the dialog box are accessed by the notation `@1`,`@2`, and so on. These variables are local, hence they cannot produce any side effects. This field supports any MATLAB expression.
- *Drawing commands:* allows one to draw a customized block icon. At present three drawing commands are supported: `dpoly` draws a transfer function of the numerator and denominator, `droots` draws a zero-pole-gain transfer function and `plot` is similar to the MATLAB command. Unfortunately the `plot` command does not support the color and type line attributes. Any other expression is interpreted as text. A custom icon is shown in Figure 17.2.
- *Help string:* allows one to include more details on block functionalities, on the implemented algorithm and on the block's limits.
  This string is the text which appears on clicking the *HELP* button.

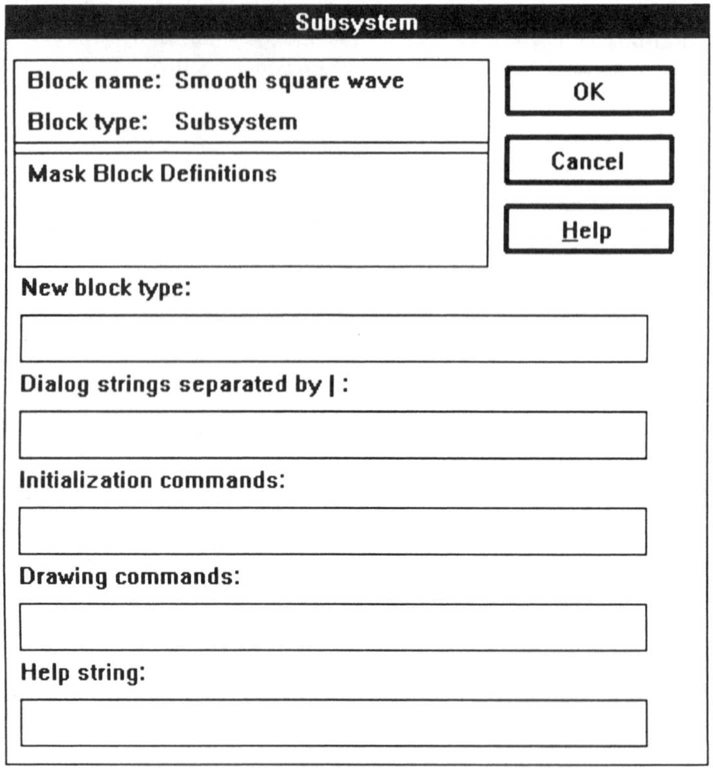

Subsystem	
Block name: Smooth square wave **Block type:** Subsystem	OK
Mask Block Definitions	Cancel
	Help

New block type:

Dialog strings separated by | :

Initialization commands:

Drawing commands:

Help string:

Figure 17.1 *Mask* dialog box.

Satellite

Figure 17.2 Icon realized by means of the plot command.

All text elements, such as *Help*, may be formatted using a C-like syntax. In particular: \n enter a newline, \t enter a tab.

The *Help* window is dynamically resized to include all the inserted text. On the other hand, the dimensions of the dialog box are fixed, hence it may happen that part of the text inserted into the *Dialog strings...* field does not appear in the dialog box.

A simple example of **Mask** is shown in the following.

We build a new block whose output is a square wave with smooth rising and

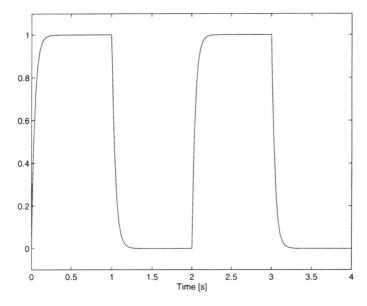

Figure 17.3 Smooth square wave.

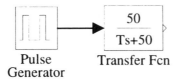

Pulse
Generator   Transfer Fcn

Figure 17.4 SIMULINK scheme of 'Smooth square wave' block.

falling edges (see Figure 17.3).

- Draw the SIMULINK scheme of the smooth square wave (see Figure 17.4).
  Set the *Pulse Generator* parameter fields with the following variables: T for
  the pulse period, T/2 for the pulse width, A for the pulse height and 0 for the
  pulse start time. Moreover, set the *Transfer function* parameters to: 50 for
  the numerator and [T 50] for the denominator.
- Select all the scheme blocks and use the *group* option. Change the *group*
  name to Smooth square wave.
- Select the *group* and *Mask* it. Fill the *Mask* dialog box fields as follows:

  - *New block type*:   Smooth Square Wave

  - *Dialog strings...*:   Produces a smooth square wave | Square wave
    period (sec) | Amplitude

- *Initialization...*:   `T=@1; A=@2; x=0:0.1:1; y=1-exp(-10*x);`
`z=exp(-10*x);`
- *Drawing...*:   `plot(x,y,x+1,z,x+2,y,x+3,z)`
- *Help string*:   `Produces a smooth square wave.\n The output`
`is the one of a first order filter forced by a square wave.`
`\n\n\t The filter is tuned on the square wave \t period.`

The icon block, its dialog box, and its help are shown in Figure 17.5.

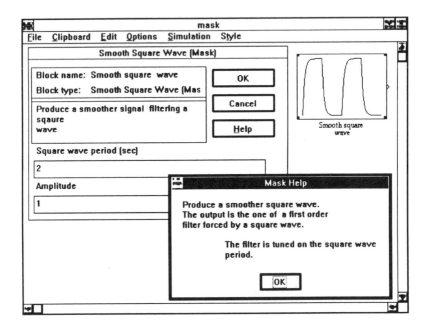

Figure 17.5 Icon block, dialog box, help window.

It is possible to associate to the double click on a *masked* block a call to a MATLAB function, instead of opening the block dialog box. To this end insert the MATLAB function inside an `eval` command[1] in the *Dialog string separated by* field.

For example, inserting the string

$$\text{eval('load data\_sat')}$$

MATLAB loads the file `data_sat.mat` each time the block is double clicked. The block icon may also be drawn by means of the MATLAB function `iconedit`, which allows one to draw it using the mouse.

---

[1] See Section 8.2.

17.3 Customizing SIMULINK

SIMULINK splits functional blocks into groups on the basis of their properties. This placement is the most logical one, but it is sometimes not the most efficient. The SIMULINK user often adopts the same set of blocks (e.g. the *Sum*, the *Integrator*, the *Scope*) and to select them (s)he must open many windows.

One way of avoiding this drawback is to put all the most commonly used blocks in one group. Put these blocks in a 'scheme' and save it with a suitable name (e.g. `my_simu`), and use this name to call the SIMULINK environment by the MATLAB prompt.

A collection of the most commonly used blocks is in the *Extras* group, to call it type `blocklib` from the MATLAB prompt.

17.4 Exercises

Exercise 17.1 Evaluate the linearized model of the ship (see Section 11.3) and compute the maximum values of the inputs which do not invalidate the piecewise-linear approximation.

Exercise 17.2 Draw the SIMULINK scheme of two cascade tanks with sections A equal to 40 m^2 and 60 m^2, respectively, both of height 20 m, and both with a hole of section $a = 50$ cm on the bottom. Assume that the fluid level dynamic in a tank is described by the equation:

$$A\dot{x} = -\sqrt{2gx} + u$$

where $g = 9.8$m s^{-2} is the acceleration of gravity, x is the fluid height in the tank, and u is the volumetric feed flow.

> **Hint.** Build one 'tank' block, and use two *Mask* copies to define the two tanks with different sections.

Exercise 17.3 Draw a SIMULINK block which is able to generate: a constant, a ramp, a parabola, a cubic or a signal given by a linear combination of these.

> **Hint.** Define a *Mask*. All the parameters related to the same input type should be inserted as one vector. Indeed, it is not possible to handle ten different parameter fields.

Exercise 17.4 Analyze the behavior of a ball on a billiard table of dimensions 1.2 m × 2.5 m given the ball's initial position and velocity. Assume the impact with the cushions is elastic (with an elasticity factor of 0.9) and model the ball–cloth interaction as pure friction with a damping factor $\alpha = 0.05$.

> **Hint.** Design a *Mask* which receives the ball's initial position and velocity (magnitude and orientation) and evaluates the motion components with respect to the billiard table's edges.

The dynamics with respect to these directions are independent, hence they may be simulated by means of two copies of the same *Mask* (pay attention to the different billiard table dimensions).

Exercise 17.5 Find the equilibrium point of the system

$$\dot{x}_1 = -x_1 x_2 + x_2$$
$$\dot{x}_2 = x_1 - x_1^2 x_2 + 0.5 x_2 + u$$
$$y = x_1 - x_1^2 x_2$$

for the different values of the input $u \in [0, 1, 3, 9]$. Find the equilibrium point closer to: $u = 9$ and $y = 3$. Which equilibrium points are stable?

Hint. Use the `trim` command to evaluate the equilibrium points. The stability may be checked using the `linmod` command to evaluate the linearized model (around the analyzed equilibrium point) and then to check the eigenvalues.

Exercise 17.6 Consider the time-varying system $\dot{x} = A(t)x$ (Vinogradov, 1952), where

$$A(t) = \begin{pmatrix} -1 - 9\cos^2 6t + 6\sin 12t & 12\cos^2 6t + 4.5sin12t \\ -12\sin^2 6t + 4.5\sin 12t & -1 - 9\sin^2 6t - 6sin12t \end{pmatrix}$$

Evaluate the eigenvalues of the 'frozen time' system, i.e. evaluate the eigenvalues in each time instant handling t as a parameter. Analyze the system behavior with the initial condition $x(0) = [1;\ 2]$.

Hint. Use the `linmod` command and evaluate the linearized model at different time instants.

Check that the system eigenvalues are constant (i.e. they do not depend on time) and have negative real part. Nevertheless, the state is unbounded, as shown by the simulation.

Exercise 17.7 Consider the time-varying system $\dot{x} = A(t)x$ (Wu, 1974), where

$$A(t) = \begin{pmatrix} -\frac{11}{2} + \frac{15}{2}\sin 12t & \frac{15}{2}\cos 12t \\ \frac{15}{2}\cos 12t & -\frac{11}{2} - \frac{15}{2}\sin 12t \end{pmatrix}$$

Evaluate the eigenvalues of the 'frozen time' system, i.e. evaluate the eigenvalues at each time instant handling t as a parameter. Analyze the system behavior with the initial condition $x(0) = [1;\ 2]$.

Hint. Use the `linmod` instruction and evaluate the linearized model for different time instants.

Check that the system eigenvalues are constant (i.e. they do not depend on time) and that there is an eigenvalue with positive real part. However, the system state variables are bounded, as shown by simulations.

Chapter 18

Blocks library

The SIMULINK blocks are described in this chapter. The object of this chapter is not an exhaustive description of each block (for this purpose the reader may refer to the manual), but, rather, a short description of their properties, along with some useful hints for their use.

In the following, all the blocks (unless explicitly stated) are assumed to be able to work with multi-variable signals and to perform *scalar expansion*, i.e in the presence of multi-variable signals if scalar quantities are used in the block parameter fields, they are expanded in order to implement the function on all the input signals.

18.1 Sources

This group includes signal generator blocks and signal input blocks (see Section 11.4 for further details).

Clock

Clock: Provides the time vector.
Notes: Opened during a simulation, displays the time continuously during which the simulation evolves.

It is important to stress that *Clock* is not a time 'generator'; it is only able to display the simulation time. Connection to a *ToWorkspace* block allows one to transfer the time vector in MATLAB.

Signal
Generator

Signal Gen.: Produces one of four different waveforms:

- Sine wave;
- Square wave;
- Sawtooth wave;
- Random noise with uniform distribution.

Parameters: Peak value, frequency (in rad s^{-1}). (The *range* fields define the maximum value of the relative bars.)
Notes: The waveform properties may be dynamically changed during a simulation. This block is intrinsically scalar.

Sine Wave

Sine Wave: Generates a sine wave.
Parameters: Amplitude, frequency (in rad s^{-1}), phase (in rad), sample time (only for discrete-time systems).
Notes: If in the last field there is a nonzero value, SIMULINK uses a different algorithm which is numerically more robust for the discrete-time code. Using a two-element vector in the *Sample Time* field, the second entry is the *offset* of the sampling instant. All the sine waves generated have the same sample rate and the same offset (the *Sample Time* field is intrinsically scalar).

From
Workspace

From Workspace: Reads data from a MATLAB matrix.
Parameters: Matrix name.
Notes: The matrix must consist of at least two columns, the first of which must be the time vector (it must contain a monotonically increasing number sequence).

If an output value is needed at some time between two given values, the output is linearly interpolated between the two time values bracketing the required time. N.B. The start and end simulation instants are specified in the *Simulation* menu. If one of them is out of the first-column range, the input data are evaluated by linear extrapolation.

untitled.mat ▷

From File

From File: Reads data from a file.
Parameters: File name.
Notes: The data must be arranged as a matrix. Each column consists of the values assumed by the n inputs at a given time (the first element of the column). Then the first row is the time vector (compared to the *From Workspace*, the meaning of the rows and columns is reversed). The way to extrapolate data is the same as in the *From Workspace* block.

Step Input

Step Input: Generates a step function.
Parameters: Switching time, initial and final values.
Notes: The switching time can also be negative and the initial value can be larger than the final one.

Random
Number

Random Number: Generates a zero mean random sequence with normal Gaussian distribution.
Parameters: Starting 'seed' values.
Notes: Different random sequences are generated using different seed values. You may use a vector of seeds.

In the presence of discrete-time systems, due to numerical problems, it is preferable to use the *Band-Limited White Noise* block.

Constant

Constant: Generates a constant value.
Parameters: Constant value.

Repeating
Sequence

Repeating Sequence: Generates a signal replicating a given waveform.
Parameters: Time vector and related function values vector (of the same dimensions).
Notes: When the simulation time falls within the time vector range the output is evaluated as for the *FromWorkspace*. Otherwise, the function is extended by means of a periodic replication of the basic waveform. The signal generated is periodic with period equal to the amplitude of the time vector.
This block is intrinsically scalar.

Pulse
Generator

Pulse Generator: Generates a sequence of pulses at regular intervals.
Parameters: Period, pulse width, pulse height, pulse start time.
Notes: Vectorial parameters may be used only in the *Pulse height* field in order to generate n pulses with different amplitudes.

Digital Clock

Digital Clock: Provides the time for discrete-time systems.
Parameters: Sample time.
Notes: Allows one to generate a more efficient code for multi-rate discrete-time

systems.

It is not able, as is *Clock*, to display the simulation time in progress.

Band-Limited
White Noise

Band-Limited White Noise: Generates a band-limited white noise sequence.
Parameters: Spectral power, sampling time, 'seed'.
Notes: This block should be used for discrete-time systems.

Chirp Signal

Chirp Signal: Generates a signal whose frequency increases at a linear rate with time.
Parameters: Initial frequency (in Hz), target time (in seconds), frequency at target time (in Hz).
Notes: This block may be used for the spectral analysis of nonlinear systems. The initial frequency can be larger than the target frequency, but after the target time the frequency starts increasing.
The block is intrinsically scalar.

18.2   Sinks

The output blocks are collected in this group (see Section 11.4 for further details).

Scope

Scope: Displays signals during the simulation with an oscilloscope-like graphic.

Parameters: Horizontal (time) and vertical axis range.
Notes: The vertical axis is zero centered and the horizontal one scrolls at each range.

> yout
To Workspace

To Workspace: Stores data in a MATLAB matrix.
Parameters: Matrix name, maximum number of samples to export.
Notes: Each matrix column represents a different variable. The data are transferred to MATLAB only at the end of the simulation.
If the simulation needs a larger number of steps than the *Maximum* ... one, the block saves the last n values only, where n is the specified maximum number of samples.
It is possible to include a second optional parameter in the *Maximum number of rows (timesteps)* field which allows one to save data only every n integration steps. Usually, the simulation time vector is not uniformly spaced (see page 157); the third optional parameter of the *Maximum* ... field allows one to specify a fixed sampling time at which to collect data.
For example, if you insert

$$[100, 3, 0.4]$$

the output matrix consists of one hundred rows (the number of columns is equal to the block input dimensions) whose values are stored each $T = 3*0.4 = 1.2$ seconds, i.e. at $T_0 = 0$, $T_1 = 1*3*0.4 = 1.2$, $T_2 = 2*3*0.4 = 2.4$, $T_3 = 3*3*0.4 = 3.6$, ..., $T_k = k*3*0.4$ seconds.

untitled1.mat
To File

To File: Stores data into a file.
Parameters: File name, matrix name.
Notes: The data are arranged as a matrix: the first row is the time vector, the others are the output variables (compared to the *To Workspace* the meaning of the rows and columns is reversed).

The time vector is always stored as the first row (it does not need an explicit connection with the *Clock*).

Graph

Graph: Plots data on the MATLAB graphic window.
Parameters: Time axes range, lower and upper bounds of vertical axes, color and type of each line.
Notes: If the simulation time exceeds the specified time range, the plot restarts from the left edge.
It is possible to connect more *Graphs* to the same scheme; each of them corresponds to a different figure.

Auto-Scale
Graph

Auto-Scale Graph: Plots data on the MATLAB graphic window and automatically resizes the axes.
Parameters: Initial amplitude of the time axis, initial lower and upper bound of the vertical axis, number of points to store, type and color of each line.
Notes: The axes are automatically scaled in order to plot the whole of the signals' time histories.

XY Graph

XY Graph: Plots the first input data vs the second on the MATLAB graphic window.

Parameters: Lower and upper bounds of the axes.
Notes: This block is intrinsically scalar.

STOP

Stop Simulation

STOP: Stops the simulation as soon as its input is different from zero.
Notes: In the presence of multi-variable input signals the simulation is stopped as soon as one of the input components is different from zero.

Hit Crossing

Hit Crossing: Allows one to reduce the simulation steps in the neighborhood of the time instants when the crossing value is assumed by the input.
Parameters: Crossing values, tolerance value around the crossing value.
Notes: This block may be used to improve the simulations accuracy of systems containing discontinuities.
The *Tol* field is intrinsically scalar.

18.3   Discrete

All the blocks in this group have the *Sample Time* field. Even if the blocks are multi-variable, this field is intrinsically scalar, i.e. every component of the input signal is sampled at the same sample time.

The second optional parameter of this field represents the *offset* of the sample instant: in particular it is the lead time (if the offset value is negative) or the delay (if the offset value is positive) with respect to the other blocks' sample instants. (See Chapter 15 for further details.)

$$\boxed{\quad 1/z \quad}$$

Unit Delay

Unit Delay: Delays the input by one sample period.
Parameters: Initial value (value assumed in the first simulation period when the block output is undefined), sample time.

$$\boxed{\dfrac{(z\text{-}1)}{z(z\text{-}0.5)}}$$

Discrete
Zero-Pole

Dis. Zero-Pole: Implements a discrete-time transfer function in zero-pole form.
Parameters: Zeros, poles, gain, sample time.
Notes: For more details on transfer functions in zero-pole form see Section 19.1.3. This block is intrinsically scalar.

$$\boxed{\begin{array}{l} x(n\text{+}1)\text{=}Ax(n)\text{+}Bu(n) \\ y(n)\text{=}Cx(n)\text{+}Du(n) \end{array}}$$

Discrete State-Space

Dis. State-space: Implements a discrete-time system in state-space form.
Parameters: System matrices, initial conditions, sample time.

$$\boxed{\dfrac{1}{1\text{+}2z^{-1}}}$$

Filter

Filter: Implements IIR and FIR filters.
Parameters: Filter numerator and denominator, sample time.
Notes: The parameter fields' vector elements are treated as the coefficients of the numerator and denominator polynomials sorted according to ascending powers of z^{-1}.
This block is intrinsically scalar.

Discrete
Transfer Fcn

Dis. Transfer Fcn: Implements a discrete-time transfer function.
Parameters: Transfer function numerator and denominator, sample time.
Notes: For details on discrete-time transfer function notation, see Section 19.1.2. The parameter fields' vector elements are treated as the coefficients of the numerator and denominator polynomials in descending powers of z.
This block is intrinsically scalar.

Zero-Order
Hold

Zero-Order Hold: Implements a sample-and-hold function.
Parameters: Sample time.
Notes: This block provides a mechanism for discretizing signals. It samples the input signal at the sample instant and holds this value as the output until the next sample instant.

First-Order
Hold

First-Order Hold: Implements a first-order sample-and-hold latch.
Parameters: Sample time.
Notes: The output is delayed by one sample interval since its values are obtained by linear interpolation between two successive samples.

Discrete-Time
Integrator

Discrete-time Integrator: Implements a discrete-time integrator.
Parameters: Initial conditions, sample time.
Notes: It implements the Z-domain transfer function

$$Y = \frac{T_s}{z - 1} U$$

where T_s is the sample time value.

Discrete-Time
Limited Integrator

Discrete-Time Limited Integrator: Implements an integrator in the discrete-time domain, whose output levels do not exceed the specified levels.
Parameters: Lower and upper bounds, initial conditions, sample time.

18.4 Linear

This group includes the more common linear blocks both in the time and Laplace domains.

Sum Sum

Sum: Sum inputs.
Parameters: Input signs.
Notes: The number of signs represents the number of input ports of the block.

If all the input signals are multi-variable (with the same dimensions) the corresponding components are summed. If there are both multi-variable and scalar inputs, the scalar quantities are summed to all the multi-variable signal components. In the presence of one input signal the block evaluates the sum of the input components (the summation symbol Σ appears in the block icon).

Gain

Gain: Multiplies the inputs by a given constant.
Parameters: Gain values.
Notes: The block displays the text as it appears in the dialog box. You may use this capability to title the block. If the *Gain* data is too long to be displayed, the string `-k-` is displayed. To avoid this the block should be resized.
If a variable is used in the *Gain* field, its contents can be displayed by enclosing the variable within parentheses.
In the presence of a vectorial parameter we must enclose the whole vector within parentheses.

Derivative

Derivative: Evaluates the time derivative of the input signals.
Notes: At each step the derivative is evaluated on the basis of the incremental ratio with respect to the previous step values; in the presence of a large time step the results may be inaccurate.

Transfer Fcn

Transfer Fcn: Implements a transfer function.
Parameters: Transfer function numerator and denominator.
Notes: The parameter fields' vector elements are treated as the coefficients of the numerator and denominator polynomials in descending powers of s.
For more details on transfer function implementation in MATLAB see Section 19.1.2. This block is intrinsically scalar.

Integrator

Integrator: Integrates the input signals.
Parameters: Initial conditions.

State-Space

State-Space: Implements a linear-time invariant system in state-space form.
Parameters: System matrices, initial conditions.

Zero-Pole

Zero-Pole: Implements a transfer function in zero-pole form.
Parameters: Zeros, poles, and gain.
Notes: For more details about transfer function in zero-pole form see Section 19.1.3.
This block is intrinsically scalar.

Matrix
Gain

Matrix Gain: The output is the product of the specified matrix by the input
vector.
Parameters: Gain matrix.
Notes: The input vector dimension must be equal to the number of matrix columns.
The output is a vector of length equal to the gain matrix rows.

Inner
Product

Inner Product: Evaluates the *dot product* of input vectors.

Slider
Gain

Slider Gain: Interactively varying gain.
Parameters: Gain value.
Notes: In the presence of multi-variable inputs each signal is amplified by the same factor (the parameter field is intrinsically scalar).

18.5 Nonlinear

Some of the more common nonlinear functions used to simulate dynamic systems are collected in this group (see Section 11.5 for further details).

Abs

Abs: Evaluates the absolute value of the inputs.

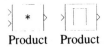

Product   Product

Product: Multiplies its inputs.
Parameters: Number of inputs.
Notes: If all the inputs are multi-variable (with the same dimensions) the output

is multi-variable and is the product of the corresponding components. If there are both multi-variable and scalar inputs the latter multiplies all the multi-variable components. If there is one input signal the *Product* multiplies the signal components (the product symbol $\prod$ appears in the block icon).

Fcn

Fcn: Generic function.
Parameters: Evaluating expression.
Notes: The inputs must be referenced as u[i]. MATLAB variables used must be scalar.
This block is able to implement only SISO (single input single output) and MISO (multi input single output) functions.
See pages 144 and 170 for a complete description of the *Fcn* block capabilities.

Dead Zone

Dead Zone: The output is zero for all the input values inside the Dead Zone; outside this region the output is equal to the input value minus the length of the Dead Zone.
Parameters: Begin and end of the Dead Zone.
Notes: More specifically, if the upper and lower bounds of the Dead Zone are D_z^+ and D_z^-, respectively, the block output y satisfies the following relation:

$$y = \begin{cases} u - D_z^- & \text{if} & u < D_z^- \\ 0 & \text{if} & D_z^- \le u \le D_z^+ \\ u - D_z^+ & \text{if} & u > D_z^+ \end{cases}$$

where u is the block input.

Backlash

Backlash: Models the behavior of a system with *prong*.

Parameters: Dead Zone width, input and output initial values.
Notes: The output is constant until the input is inside the Dead Zone.

Saturation

Saturation: Emulates a saturation, i.e. limits the output signal amplitude.
Parameters: Lower and upper output values.

Switch

Switch: The output is equal to the first input until the command signal (the second input) is greater than or equal to the threshold value; then the output is equal to the third input.
Parameters: Threshold value.
Notes: In the presence of multi-variable inputs, if the command signal is scalar the output coincides with the first or third input (which may have different dimensions). If the command signal is multi-variable (in this case all the inputs must have the same dimension), the ith component of the output is equal to the corresponding one of the first or third input signals on the basis of the relation between the ith component of the command signal and the threshold value (which may be scalar or multi-variable with the same dimension as the inputs).

Look-Up
Table

Look-Up Table: Performs a piecewise-linear mapping of the input.
Parameters: Input and output value vectors (of the same dimension).
Notes: The input vector must be sorted in strictly ascending order. Outputs for input values falling outside the range of the input vector are linearly extrapolated from the nearest two points.
The output vector may be expressed both as a vector and by means of a MATLAB expression.
The *Look-Up Table* icon displays a graph of input vs output; if one changes the underlying law the block icon is modified dynamically.

In the presence of multi-variable signals the same function is performed on all the input components.

2-D Look-Up
Table

2-D Look-Up Table: Performs a linear interpolation on a bidimensional table.
Parameters: X and Y indices' values and related output values.
Notes: The X and Y indices' vector must be in monotonically ascending order. This block is intrinsically scalar.

Rate Limiter

Rate Limiter: Limits the rate of change of the input signals.
Parameters: Maximum values of the rising and falling slew rate.

Relay

Relay: Simulates a relay.
Parameters: Threshold and output values for the *on* and *off* states.
Notes: The output switches between two specified values. When the relay is on, it remains on until the input drops below the off input threshold; the inverse when the relay is off.
The input value for the on and for the off may be equal.

MATLAB Fcn

MATLAB Fcn: Allows one to call a MATLAB function in a SIMULINK scheme.
Parameters: Function name, output dimension (−1 means that the output has the

same dimension as the input signal).
Notes: SIMULINK is not able to check whether the dimension of the MATLAB function output matches that specified in the block.

Quantizer

Quantizer: Simulates a quantizer.
Parameters: Quantization interval.

Coulombic
Friction

Coulombic Friction: Emulates Coulomb friction.
Parameters: Initial offset value, rate of linear friction.
Notes: The friction for zero speed is equal to the offset value; for speed values greater than zero the friction is equal to the offset value plus a term that is proportional to the speed.
In the presence of multi-variable signals the same function is implemented on each input (the parameter fields are intrinsically scalar).

AND
Logical
Operator

Logical Operator: The output is 1 if the logic operation on the inputs returns true; otherwise the output is 0.
Parameters: Logic operator, number of input ports.
Notes: The logic true value is associated with any value different from 0 (which represents the false value). The supported operators are summarized in the HELP window.
In the presence of multi-variable input signals (of the same dimensions) the operation is evaluated on the corresponding components of the inputs. If a signal is scalar the operation is performed between the scalar value and each multi-variable signal component.

If there is a single input signal the operation is performed on the input components (except for the NOT operator, in which case all the input components are complemented).

Relational
Operator

Relational Operator: The output is 1 if the underlying relation holds true; otherwise the output is 0.
Parameters: Relational operator.
Notes: The first input represents the left-hand term of the relation. The operators supported are summarized in the HELP window.
In the presence of multi-variable input signals the relation is evaluated on the corresponding components. If an input is scalar its value is related to all the multi-variable signal components.

Combinatorial
Logic

Combinatorial Logic: Implements a combinatorial network.
Parameters: Truth table.
Notes: Only the output part of the truth table must be entered; the inputs are implicitly assumed ordered so that the corresponding binary sequence is in ascending order.
The number of inputs must be equal to $\log_2$ of the truth table number of rows.

Reset
Integrator

Reset Integrator: If the second input is zero, then it integrates the first input, otherwise the block state is assumed equal to the third input.
Parameters: Initial conditions.
Notes: In the presence of multi-variable input signals (with the same dimension) the reset condition is evaluated for each component of the input signals. If the reset input (the second one) is scalar, its value is assumed for all the input components.

Transport
Delay

Transport Delay: Delays the inputs by a given amount of time.
Parameters: Amount of time delay, initial value (assumed when the output is indeterminate).
Notes: In the presence of a negative value in the *Time delay* field the block emulates a lead time system and the output values are evaluated by linear extrapolation of the last two input samples.

Variable
Transport Delay

Variable Transport Delay: Introduces a variable time delay: the second input is the amount by which to delay the first input.
Parameters: Initial delay value, number of samples to store.

Limited
Integrator

Limited Integrator: Integrator with saturation.
Parameters: Lower and upper bounds, initial condition.

Memory

Memory: Applies a one-integration-step sample-and-hold to the inputs.
Parameters: Initial conditions.
Notes: This block is useful for solving algebraic loops.

Sign

Sign: Implements the signum function.

18.6   Connections

This group includes the blocks needed to connect scalar to multi-variable systems and to specify the inputs and the outputs of SIMULINK schemes.

Inport

Inport: Defines an input port (used in *group* operation or to create an *S-function*) and allows one to specify the operating point around which to analyze the system.
Parameters: Port number.
Notes: The ports must be numbered sequentially in ascending order, starting with 1. The first port is the highest block input.

Outport

Output: Defines an output port (used in *group* operation or to create an *S-function*) and allows one to specify the operating point around which to analyze the system.
Parameters: Port number.
Notes: The ports must be numbered sequentially in ascending order, starting with 1. The first port is the highest block output.

Mux

Mux: Multiplexes several input signals into a multi-variable signal.
Parameters: Inputs' number.

Notes: It is possible to multiplex multi-variable signals. The command [2 4 1] allows one to group three inputs into a signal: the first one is a multi-variable signal with dimension 2, the second is a multi-variable signal with dimension 4, and the last is a scalar signal.

Demux

Demux: Separates a multi-variable signal into scalar ones.
Parameters: Outputs' number.
Notes: It is possible to separate a multi-variable signal into several multi-variable signals of given dimension. The command [2 4 1] splits the input signal (of dimension 7) into three signals of dimension 2, 4, and 1 respectively.

18.7   Extras

This group is composed of some SIMULINK blocks introduced with the 1.2 version and is not completely documented in the Manual. These blocks are very specialized and hence we confine ourselves to their short description. The interested reader may find more details in the *HELP* text or *un-mask* the block and analyze the underlying implementation. All the blocks presented in this section are shown in Figure 18.1.

Some of these blocks need some additional toolboxes such as: *Signal Processing, Control, Robust Control, Neural Network.*

18.7.1   Conversion

Polar to Cartesian: Converts from polar coordinates to Cartesian coordinates.
Cartesian to polar: Converts from Cartesian coordinates to polar coordinates.
Spherical to Cartesian: Converts from spherical coordinates to Cartesian coordinates.
Cartesian to spherical: Converts from Cartesian coordinates to spherical coordinates.

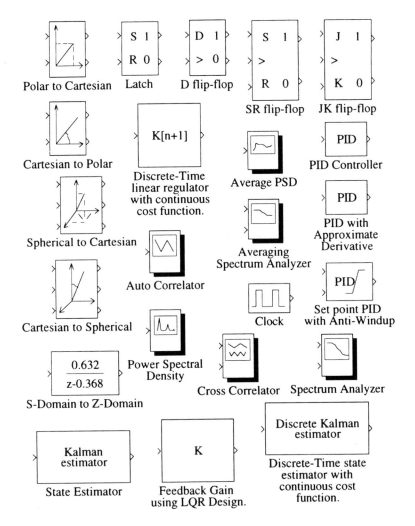

Figure 18.1 Some blocks included in the *Extras* group.

18.7.2   Flip-flop

In the flip-flop descriptions we will use the symbols shown in Figure 18.1.

Latch: The output (1) is one if the set signal (S) is one and remains one until the reset signal (R) is zero.

D flip-flop: If the synchronization signal (>) is one the output (1) is equal to the (D) input; when the synchronization signal is zero the output holds its value.

SR flip-flop: If the synchronization signal (>) is one and the set signal (S) is one then the output (1) is one. If the reset signal (R) is one the output (1) is zero. The output does not switch if the synchronization signal is zero.

JK flip-flop: If the synchronization signal (>) is one, the output (1) is zero and the signal J is one then the output switches; otherwise if the synchronization signal (>) is one, the output (1) is one and the signal K is one then output switches.

Clock: Generates a signal optimized to be the synchronization input of a logic network.

18.7.3 PID controllers

PID Controller: Represents a PID controller; it is able to give a control law constituted by a proportional action, an integral action and a derivative action.

PID with Approximate Derivative: Represents a PID controller whose derivative term is implemented via a high-pass transfer function.

Set point PID with Anti-Windup: Allows one to avoid the *windup* phenomenon due to the use of a PID controller in the presence of limited actuator signals.

The windup phenomenon appears when the actuators saturate: the integrator charges as long as the saturation holds; then there should be a high control action to decharge the integrator. This phenomenon may be overcome eliminating the integrator action as soon as the actuator saturates.

18.7.4 Analyzers

Power spectral density: Displays the power spectral density of the input.

Spectrum analyzer: Analyzes the frequency response of a system and plots the spectrum.

Cross correlator: Displays the cross-correlation between the two input signals.

18.7.5 Controllers

These blocks require the *Control System Toolbox.*

State Estimator: State observer designed by means of the steady-state solution of the LQE continuous-time equation (steady-state Kalman filter).

Feedback Gain using LQR Design: Optimal state feedback gain evaluated by means of the steady-state solution of the associated algebraic Riccati equation.

Discrete-time state estimator with continuous cost function: Discrete-time state observer for a continuous-time plant designed by means of the steady-state solution of the LQE discrete-time equation (steady-state discrete-time Kalman filter).

Discrete-time linear regulator with continuous cost function: Optimal discrete-time state regulator for continuous-time plant evaluated by means of the steady-state solution of the corresponding discrete-time Riccati equation.

S-Domain to Z-Domain: Converts the continuous-time transfer function coefficients into the corresponding coefficients of a discrete-time transfer function using a ZOH approximation.

18.7.6   Filters

A wide set of digital and analog filters are collected in this group. They require the *Signal Processing Toolbox*.

18.7.7   System ID

Some of the most commonly used algorithms for system identification are collected in this group: AR, ARX, OE, ARMAX, BJ, PEM. They require the *System Identification Toolbox*.

Part III

Control System Toolbox

Chapter 19

Models for LTI systems

In this chapter we describe how to operate with the commands of the Control System Toolbox (from now on indicated by CONTROL) dealing with: the possible representations of linear time invariant (LTI) systems and the transformations among them, the conversion between continuous- and discrete-time models, the controllability and observability properties, the model order reduction and the series and parallel and feedback connections.

The representation and the manipulation of single input single output (SISO) and single input multi-output (SIMO) systems will be straightforward, whereas the use of the CONTROL commands for multi-input systems will be more complicated since MATLAB does not operate with 3D matrices.[1]

19.1 LTI system representations

An LTI system can be represented in different forms: in the time domain by means of first-order differential equations (the so-called input–state–output or state-space form), and in the complex domain by means of the system transfer function. In particular, this function can be written as the ratio of two polynomials, by means of their roots, i.e. using zeros and poles, or in terms of partial fractions, i.e. using residues and poles.

In this section, dealing with continuous-time systems, we show how these representations can be defined in a 'MATLAB form' using the CONTROL commands; similar considerations hold for discrete-time systems.

---

[1]On the possibility of defining 3D matrices see the command `eval` in Section 8.3.

19.1.1   State-space models

Any LTI system can be represented by using a set of first-order differential equations in the vector form

$$\dot{x} = A\,x + B\,u \tag{19.1}$$
$$y = C\,x + D\,u \tag{19.2}$$

where x is the n_x-dimensional state vector, u is the n_u-dimensional input vector and y is the n_y-dimensional output vector.

As is well known, the matrix is the basic element of MATLAB. Therefore, 'to think' in MATLAB of a system which is represented in the input–state–output form is equivalent to assigning the four matrices A, B, C, D, defined in (19.1)–(19.2).

For instance, considering a two-integrator cascade system (see Figure 19.1), which can be represented in the state-space form

$$\dot{x} = \begin{pmatrix} 0 & 0 \\ 1 & 0 \end{pmatrix} x + \begin{pmatrix} 1 \\ 0 \end{pmatrix} u$$
$$y = \begin{pmatrix} 0 & 1 \end{pmatrix} x$$

the following MATLAB instructions:

```
A=[0 0;1 0];
B=[1 0]';
C=[0 1];
D=0;
```

assign the matrices of a possible state-space representation.

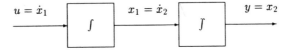

Figure 19.1 Scheme of two series connected integrators.

19.1.2   Transfer matrix models

A different representation for an LTI system can be obtained by using the Laplace transform, which for a continuous-time signal $f(t)$ is defined as

$$F(s) = \mathcal{L}(f(t)) = \int_0^\infty f(t)e^{-st}dt$$

Using the Laplace transform derivative rule, from (19.1) and (19.2) we achieve, after some algebra,

$$Y(s) = G(s)U(s) = (C(sI - A)^{-1}B + D)U(s)$$

where $G(s)$ indicates the so-called transfer matrix (or transfer function for SISO systems).

It is quite simple to assign a transfer function in MATLAB. In fact, for SISO LTI systems $G(s)$ is given by the ratio between two polynomials: the denominator, whose degree is n_x, and the numerator, whose degree is less than or equal to n_x (depending on whether the system is proper or strictly proper respectively). It is then sufficient to assign two vectors given by the coefficients of the numerator and denominator polynomials and sorted in decreasing powers of the variable s. [2]

As an example, if we wish to represent the following transfer function:

$$G(s) = \frac{s+2}{s^2 + 2s + 7}$$

we have to assign

$$\text{num}=[1\ 2];$$
$$\text{den}=[1\ 2\ 7];$$

For SIMO systems the numerator of $G(s)$ is a vector of polynomials, rather than a single polynomial, each one related to a different output. In this case a matrix must replace the vector **num**; in particular the number of rows of this matrix is the number of outputs of the system, whereas the number of columns must be the maximum among the row lengths.

For instance, to represent the transfer matrix

$$G(s) = \frac{\left(\begin{array}{c} s+2 \\ s^3 + 3s^2 + 1 \end{array} \right)}{s^3 + 5s + 1}$$

we must define the matrices

$$\text{NUM}=[0\ 0\ 1\ 2; 1\ 3\ 0\ 1];$$
$$\text{den}=[1\ 0\ 5\ 1];$$

where the polynomials are assumed to be ordered in decreasing powers of the complex variable s and some zeros have been inserted for vector consistency (when an s power term does not appear in the polynomials).

19.1.3   Zeros and poles models

The transfer matrix of an LTI system can also be represented by highlighting the roots of the numerator (zeros) and those of the denominator (poles) polynomials.

---

[2]For polynomial operations in MATLAB see Chapter 6.

For instance, for SISO systems, the transfer function can be written in the following form:

$$G(s) = k\frac{(s - z_1)(s - z_2)\cdots(s - z_{n_z})}{(s - p_1)(s - p_2)\cdots(s - p_{n_p})} \tag{19.3}$$

where k is the gain constant and z and p are the roots (possibly complex) vectors of the numerator and the denominator polynomials (subscripts indicate the vector components). By the way, it is important to recall that MATLAB uses row vectors to indicate the coefficients of a polynomial and column vectors to indicate the roots (see the commands `roots` and `poly` in Chapter 6).

On the basis of the above considerations, the MATLAB representation of the transfer function

$$G(s) = 5\frac{(s - 3)(s + 1)}{(s + 2)(s - 4)}$$

with reference to (19.3), can be achieved by the instructions

```
k=5;
zer=[3;-1];
pol=[-2 4]';
```

When assigning the vectors `zer` and `pol`, some care must be given to the signs to be used: in these vectors the roots of the corresponding polynomials with their signs must be given, i.e. the z_i and p_i shown in (19.3).

Let us consider another example: the representation of the transfer function

$$G(s) = \frac{1}{(s + 1 - j)(s + 1 + j)}$$

Since $G(s)$ has no zeros, it seems sufficient to define the vectors

```
k=1;
pol=[-1+j;-1-j];
```

As will be shown in Section 19.3, in order to obtain the representation conversion from the transfer function form to a different one, the vector `zer` cannot remain undefined. Therefore, in this case, i.e. when no finite zeros are present, since the transfer function has zeros at infinite, the vector `zer` can be assigned as

```
zer=inf;
```

As already discussed, in the SIMO case the numerator of $G(s)$ is a matrix. It is important to stress that the roots of the numerator polynomial related to the ith output must be entered in the ith *column* of the MATLAB matrix ZER. (Recall that in the transfer matrix representation of a LTI system the ith *row* of the matrix

NUM has to contain the coefficients of the numerator polynomial related to the ith output.)

Let us consider as an example the system with one input and three outputs whose transfer matrix is

$$G(s) = \frac{\begin{pmatrix} 1 \\ s \\ 3(s+4)(s+6) \end{pmatrix}}{(s+1)^2(s+2)}$$

Its MATLAB zeros–poles representation can be achieved with the assignments

$$\text{ZER=[inf 0 -4;inf inf -6];}$$
$$\text{pol=[-1 -1 -2]';}$$
$$\text{k=[1 1 3]';}$$

where we have also considered the infinite zeros in order to complete the matrix ZER which must have as many *columns* as outputs of the system considered. In particular, the second column of the matrix ZER is the vector $(0 \quad \infty)$ since the transfer function between the input and the second output has zeros at 0 and ∞.

19.1.4 Residues and poles models

A different representation of the transfer function of a system can be achieved by rewriting (19.3) in terms of partial fractions, i.e.

$$G(s) = \frac{r_1}{s - p_1} + \frac{r_2}{s - p_2} + \cdots + \frac{r_{n_z}}{s - p_{n_z}} + k(s) \qquad (19.4)$$

where we have assumed that the denominator has no multiple roots. This form can be defined by means of a column vector pol which contains the system poles, a column vector res which contains the residues and a row vector k which contains the coefficients of the polynomial which defines the improper part of the system (this vector must not be defined if the number of zeros is smaller than or equal to the number of poles).

When the transfer function has multiple poles, some terms such as the following will appear in (19.4)

$$\frac{r_i}{s - p_i} + \frac{r_{i+1}}{(s - p_i)^2} + \cdots + \frac{r_{i+m-1}}{(s - p_i)^m}$$

where m is the multiplicity of the pole p_i.

The system representation will be the same, except for the vector pol where the pole p_i will appear m times. Considering, for instance, the transfer function

$$\begin{aligned} G(s) &= \frac{s+2}{s(s+1)^2(s^2 - 2s + 10)} \\ &= \frac{0.4}{s} - \frac{0.3125}{s+1} - \frac{0.125}{(s+1)^2} - \frac{0.0437 \pm 0.025j}{s - 1 \pm 2j} \end{aligned}$$

its MATLAB representation can be obtained with the assignments

```
res=[0.4 -0.3125 -0.125 -0.0437-0.025j -0.0437+0.025j]';
pol=[0 -1 -1 1-2j 1+2j]';
k=[];
```

or, by using the `residue` command (see Chapter 6), by the instruction

```
[res,pol,k]=residue([1 2],conv([1 2 1 0],[1 -2 5]));
```

which assigns the vectors containing the residues, the poles and the improper part of the system respectively, by using the numerator and denominator polynomials of the considered transfer function as arguments of the `residue` command.

Note that the MATLAB command `residue` can also be used in an inverse way, i.e. to obtain the transfer function from knowledge of the residues, poles and the improper part of the system.

The representation with residues and poles may be used in order to obtain the analytical expression of the time response of an LTI system (see Chapter 20 for details).

19.2   Conversions among representations

Once the system has been represented in one of the above forms, it is quite simple to obtain conversion to the other forms by the CONTROL commands. In particular, the commands that are useful for conversions among the representations of an LTI system are

`canon`	transformation in canonical form
`ss2ss`*	state transformation
`ss2tf`*	from state-space to transfer function
`ss2zp`*	from state-space to zeros–poles
`tf2ss`*	from transfer function to state-space
`tf2zp`*	from transfer function to zeros–poles
`zp2ss`*	from zeros–poles to state-space
`zp2tf`*	from zeros–poles to transfer function

where the asterisk indicates that the command can also be used for discrete-time systems.

In what follows these commands will be presented in different subsections, each of them devoted to the commands available for conversion from a given representation.

19.2.1 From the state-space

Given an input–state–output representation, it can often be useful or desirable to obtain a different state-space form. This can be achieved by introducing a state transformation.

Let us consider the electrical circuit shown in Figure 19.2.

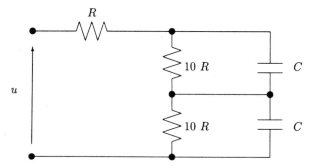

Figure 19.2 Resistances and capacitors bridge.

A possible input–state–output representation for this circuit is

$$\dot{x}_1 = -1.1\,x_1 - x_2 + u$$
$$\dot{x}_2 = -x_1 - 1.1\,x_2 + u$$
$$y = x_1 + x_2$$

where we have assumed $RC = 1$ s, the state variables are the voltages on the two capacitors and the output is their sum. Suppose that a representation where one state variable coincides with the output is requested. To this end the following state transformation:

$$\begin{pmatrix} z_1 \\ z_2 \end{pmatrix} = \begin{pmatrix} 1 & 0 \\ 1 & 1 \end{pmatrix} \begin{pmatrix} x_1 \\ x_2 \end{pmatrix}$$

can be introduced by means of the instructions

```
A=[-1.1 -1;-1 -1.1]; B=[1;1]; C=[1 1];
T=[1 0;1 1]; [At,Bt,Ct,Dt]=ss2ss(A,B,C,0,T);
```

where the last command gives the matrices of the transformed model

$$\dot{z}_1 = -1.1\,z_1 - z_2 + u$$
$$\dot{z}_2 = -2.1\,z_2 + 2\,u$$
$$y = z_2$$

More generally, the command ss2ss allows one to achieve a state transformation such as

$$z = Tx$$

providing the following matrices as output:

$$A_t = TAT^{-1}, \quad B_t = TB, \quad C_t = CT^{-1}, \quad D_t = D$$

The command `canon` can be considered a special case of the instruction `ss2ss`. This command allows one to obtain two different state-space representations starting from an input–state–output model. The first transformed form can be used for diagonalizable systems and is characterized by a transformed dynamic matrix A_t in the so-called modal form, i.e. a block diagonal matrix whose 1×1 blocks are the real eigenvalues of the dynamic matrix A and whose 2×2 blocks have the real parts of the complex eigenvalues of A on the principal diagonal and the imaginary parts on the other diagonal. For instance, with reference to the last example, the instruction

$$[At,Bt,Ct,Dt]=canon(A,B,C,D);$$

gives the matrices

$$A_t = \begin{pmatrix} -0.1 & 0 \\ 0 & -2.1 \end{pmatrix}, \qquad B_t = \begin{pmatrix} 0 \\ -1.41 \end{pmatrix}$$

$$C_t = \begin{pmatrix} 0 & -1.41 \end{pmatrix}, \qquad D_t = 0$$

(Note that the transformed dynamic matrix is diagonal and has the eigenvalues on the principal diagonal.)

We now show what happens if the eigenvalues are not all real. Let us consider a dynamic matrix whose real eigenvalues are -1 and -30 and whose complex ones are $-5 \pm 10j$. The instruction

$$[At,Bt,Ct,Dt]=canon(A,B,C,D);$$

gives transformed matrices such that the dynamic matrix has the following form:

$$A_t = \begin{pmatrix} -30 & 0 & 0 & 0 \\ 0 & -1 & 0 & 0 \\ 0 & 0 & -5 & 10 \\ 0 & 0 & -10 & -5 \end{pmatrix}$$

Actually, the `canon` command also makes provision for a fifth argument, a string which defines the type of canonical form and whose default value is 'modal'. A further way of employing the command `canon` is by using the argument 'companion' instead of 'modal' so that the matrix pair (A_t, B_t) is achieved in the controllability canonical form, i.e.

$$A_t = \begin{pmatrix} 0 & 0 & 0 & \cdots & -a_n \\ 1 & 0 & 0 & \cdots & -a_{n-1} \\ 0 & 1 & 0 & \cdots & \vdots \\ \vdots & \vdots & \vdots & \ddots & -a_2 \\ 0 & \cdots & \cdots & 1 & -a_1 \end{pmatrix}, \qquad B_t = \begin{pmatrix} 1 & 0 & 0 & \cdots & 0 \end{pmatrix}^T$$

where the a_is are the coefficients of the characteristic polynomial

$$|sI - A| = s^n + a_1 s^{n-1} + a_2 s^{n-2} + \cdots + a_{n-1} s + a_n$$

Usually the simplest way to achieve a model for an LTI system is to operate in the state-space domain. Nevertheless, the frequency domain techniques for both system analysis and controller design argue for the use of the transfer function or the zeros–poles representations. Therefore, it is interesting to show how to obtain a transfer matrix representation from a state-space one.

Let us consider the electrical circuit shown in Figure 19.3, where u_1 and u_2 are input voltages.

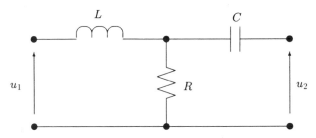

Figure 19.3 RLC circuit.

Assuming as state variables the inductor current and the capacitor voltage, respectively, a possible input–state–output model for the circuit, obtained by applying the Kirchhoff's rules, is

$$
\begin{aligned}
\dot{x}_1 &= -\frac{1}{L}x_2 + \frac{1}{L}u_1 - \frac{1}{L}u_2 \\
\dot{x}_2 &= \frac{1}{C}x_1 - \frac{1}{RC}x_2 - \frac{1}{RC}u_2 \\
y_1 &= x_1 \\
y_2 &= x_2
\end{aligned}
\tag{19.5}
$$

As is evident, this system is MIMO. Then, in order to use the CONTROL commands, the information about the input considered must always be specified.

Choosing $C = 1$ mF, $R = 1$ kΩ and $L = 100$ mH as circuit parameters, the matrices of the model can be assigned by writing

```
A=[0 -10;1000 -1]; B=[10 -10;0 -1];
C=eye(2); D=zeros(2);
```

A transfer matrix model of this circuit w.r.t. the first input and assuming the second one to be zero (a short circuit) is achievable with the instruction

```
[NUM,den]=ss2tf(A,B,C,D,1);
```

which gives the matrices

$$NUM = \begin{pmatrix} 0 & 10 & 10 \\ 0 & 0 & 10000 \end{pmatrix}, \qquad den = \begin{pmatrix} 1 & 1 & 10000 \end{pmatrix}$$

In particular, the last argument in the previous instruction is the index of the input considered and can then be omitted in the single input system case. The matrix NUM, which becomes a vector in the single output case, has as many rows as the outputs of the system and as many columns as the elements of the denominator polynomial, i.e. the number of states plus one.

The tranformation from a state-space representation to a zeros–poles form can be achieved by the instruction

$$[ZER,pol,k]=ss2zp(A,B,C,D,1);$$

which, applied to the previous example, gives the vectors

$$ZER = \begin{pmatrix} -1 & \infty \end{pmatrix}$$
$$pol = \begin{pmatrix} -0.5 + 100j & -0.5 - 100j \end{pmatrix}$$
$$k = \begin{pmatrix} 10 & 10000 \end{pmatrix}$$

Note that in this case ZER is a vector rather than a matrix since the number of its rows must coincide with the maximum number of zeros among all the outputs, and in the example considered only the transfer function related to the first output has a zero in -1 (the second transfer function has no finite zeros and, consequently, the term Inf appears).

19.2.2   From the transfer matrix

Conversion from the transfer matrix representation to the state-space form is achieved by the command tf2ss. In particular, from the matrix NUM and the vector den defined in the previous section, the instruction

$$[A,B,C,D]=tf2ss(NUM,den);$$

gives the matrices

$$A = \begin{pmatrix} -1 & -10000 \\ 1 & 0 \end{pmatrix}, \qquad B = \begin{pmatrix} 1 \\ 0 \end{pmatrix}$$
$$C = \begin{pmatrix} 10 & 10 \\ 0 & 10000 \end{pmatrix}, \qquad D = \begin{pmatrix} 0 \\ 0 \end{pmatrix}$$

which do not coincide with the matrices of the model in (19.5) with $u_2 = 0$. This is due to the fact that the command `tf2ss` provides the matrices in control canonical form , i.e. in the form

$$A_t = \begin{pmatrix} -a_1 & -a_2 & \cdots & -a_{n-1} & -a_n \\ 1 & 0 & \cdots & 0 & 0 \\ 0 & 1 & \cdots & 0 & 0 \\ \vdots & \vdots & \ddots & \vdots & \vdots \\ 0 & 0 & \cdots & 1 & 0 \end{pmatrix}$$

$$B_t = \begin{pmatrix} 1 & 0 & 0 & \cdots & 0 \end{pmatrix}^T$$

(19.6)

It is important to stress that the command `tf2ss` handles only single input systems and, consequently, the matrices B and D will always be column vectors.

Conversion from the transfer function representation to the zeros–poles representation is achieved by means of the instruction

$$[ZER,pol,k]=tf2zp(NUM,den);$$

which, in the example considered, gives the same vectors as were obtained in the previous section by the command `ss2zp`.

19.2.3   From the zeros and poles

Conversion from the zeros–poles representation to the others is quite simple. In particular, to obtain the representations in state-space and transfer matrix forms, the instructions to be used are

$$[A,B,C,D]=zp2ss(ZER,pol,k);$$
$$[NUM,den]=zp2tf(ZER,pol,k);$$

respectively, where it is important to stress that the number of columns of the matrix NUM must coincide with the length of the vector den.

19.3   Continuous to discrete conversion

The commands available for continuous to discrete conversion are

c2d	conversion from continuous to discrete
c2dm	conversion from continuous to discrete with method
c2dt	conversion from continuous to discrete with delay

Given a system in the continuous-time state-space form, a discrete-time version for it is the so-called sampled data system. In particular, considering the system

$$\dot{x} = Ax + Bu$$
$$y = Cx + Du$$

assuming a constant sampling period T_c and assuming that the control signal $u(t)$ is constant during each period, the sampled data version of the system is

$$x(n+1) = e^{AT_c}\, x(n) + A^{-1}\left(e^{AT_c} - I\right) B\, u(n) \tag{19.7}$$
$$y(n) = C\, x(n) + D\, u(n) \tag{19.8}$$

As will be discussed later in this section, the sampling period is a crucial parameter in continuous to discrete conversion.

To demonstrate this dependence, let us consider the following first-order SISO system:

$$\dot{x} = -x + 3\, u \tag{19.9}$$

The time step response of this system, which can easily be calculated, is

$$x(t) = 3(1 - e^{-t})\, u(t) \tag{19.10}$$

where we have assumed a zero initial condition. A sampled data system with sampling period $T_c = 1$ s can be obtained by the instructions

```
Tc=1;  [ad,bd]=c2d(-1,3,Tc);
```

Note that the first and second arguments of the command c2d are the dynamic and input matrices of the continuous-time model, respectively, whereas the third argument is the sampling period. The matrices C and D do not depend on the conversion and therefore must not be indicated.

Using (19.10), the step response of the system (19.9), from zero till 4 s, can be computed by writing

```
t=0:0.1:4;  xc=3*(1-exp(-t));
```

On the other hand, the time response of the discretized system can be achieved by the instructions

```
xd(1)=0;
for i=1:4/Tc
xd(i+1)=ad*xd(i)+bd;
end
```

It is important to stress that this response could also be obtained simply by using the command dlsim (see Section 24.4) or a SIMULINK scheme.

The representation of the two responses is obtained by the instruction

$$\text{plot(t,xc,0:Tc:4,xd,'*');}$$

which gives the plot shown in Figure 19.4. From a graphical comparison it follows that, as expected, the discrete signal values are 'exact' (coincide with the continuous values) at multiple time instants of the sampling period.
Note that a piecewise-continuous version of the discrete signal can be obtained with the instructions

$$\text{[tc,xdc]=stairs(0:Tc:4,xd);}$$
$$\text{plot(t,xc,tc,xdc)}$$

In general, from the Shannon theorem it follows that, given a sampled signal, the corresponding continuous signal can be achieved by the equation

$$f(t) = \sum_{k=-\infty}^{+\infty} f(kT_c) \frac{sin\omega_c(t-kT_c)/2}{\omega_c(t-kT_c)/2}$$

where $\omega_c = 2\pi T_c$. To apply this relation, all the signal samples, past and future, must be available at time t and thus it is not possible to use it in *on-line* computations. Consequently, in order to achieve a continuous-time signal from the discrete-time samples, other methods are implemented (see the description of the command zoh below).

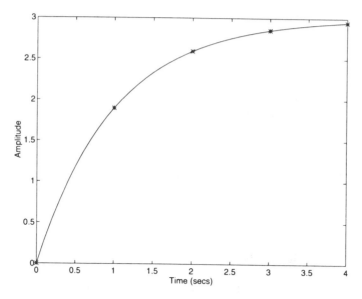

Figure 19.4 Continuous (solid line) and sampled (star points) signals.

A further instruction which provides continuous to discrete conversion is c2dm. To describe how this command operates let us recall that from the definition of the Laplace transform of a continuous-time signal $f(t)$, i.e.

$$F(s) = \mathcal{L}(f(t)) = \int_0^\infty f(t)e^{-st}dt$$

and from the definition of the $\mathcal{Z}$-transform of the corresponding discrete-time signal $\phi(n) = f(nT_c)$, i.e.

$$\Phi(z) = \mathcal{Z}(\phi(n)) = \sum_{n=0}^{\infty} \phi(n)z^{-n}$$

one can obtain the following well-known relation between the complex discrete variable z and the continuous one s:

$$z = e^{sT_c} \tag{19.11}$$

Since the application of (19.11) to compute continuous to discrete conversion could be too difficult, the command c2dm uses some approximated expressions achievable from (19.11). In particular, the instructions

```
[Ad,Bd,Cd,Dd]=c2dm(A,B,C,D,T,'method');
[numd,dend]=c2dm(num,den,T,'method');
```

provide the conversion using a suitable method and starting from a state-space or a transfer matrix representation, respectively, of the system. Note that, differently from the previously described command, in this case one must also specify the output matrices C and D since these can change due to the particular method adopted.

The conversion method that must be specified when using the command c2dm can be chosen from the following: zoh, foh, tustin, prewarp and matched. The option zoh makes the command c2dm coincide with c2d. To show the features of this conversion method consider a series between a sampler and a so-called zero order hold (ZOH), i.e. a system which holds its input for a fixed time interval (the sampling period). This kind of system, which is represented in Figure 19.5, is of practical interest. In fact, the controller input is usually a continuous-time signal, but the widely used digital controllers operate on discrete-time signals and must provide (the input of the process) a continuous-time signal again as output.

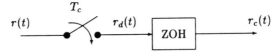

Figure 19.5 Sampler and zero order hold.

An input signal $r(t)$ is sampled with a fixed period T_c in Figure 19.5, providing the signal $r_d(t)$ which becomes the input of the ZOH, whose output is the continuous-time signal $r_c(t)$. To resume we can write

$$r_d(t) \;=\; \sum_{n=1}^{\infty} r(t)\delta(t - nT_c)$$

$$r_c(t) \;=\; \sum_{n=1}^{\infty} r(nT_c)[1(t - (n-1)T_c) - 1(t - nT_c)]$$

where $\delta(t)$ is the Dirac function and $1(t)$ indicates the Heaviside or unitary step function.

By applying the Laplace transform, we can obtain the transfer function of a ZOH from these expressions:

$$G_{ZOH}(s) = \frac{1 - e^{-sT_c}}{s}$$

Introducing $s = j\omega$, and after some algebra the following armonic response is obtained:

$$G_{ZOH}(j\omega) = T_c \frac{\sin(\omega T_c/2)}{\omega T_c/2} e^{-j\omega T_c/2}$$

which allows one to write the frequency relation between the signals $r(t)$ and $r_c(t)$ as

$$R_c(j\omega) = \frac{\sin(\omega T_c/2)}{\omega T_c/2} e^{-j\omega T_c/2} \sum_{n=-\infty}^{+\infty} R\left(j\omega + jn\frac{2\pi}{T_c}\right) \tag{19.12}$$

where we have also used the well-known expression

$$R_d(j\omega) = \frac{1}{T_c} \sum_{n=-\infty}^{+\infty} R\left(j\omega + jn\frac{2\pi}{T_c}\right)$$

Note that the 'gain' T_c of the ZOH is compensated by the 'gain' $1/T_c$ of the sampler. Introducing the hypothesis

$$\frac{\omega T_c}{2} \ll 1 \tag{19.13}$$

expression (19.12) can be rewritten as

$$R_c(j\omega) = e^{-j\omega T_c/2} \sum_{n=-\infty}^{+\infty} R\left(j\omega + jn\frac{2\pi}{T_c}\right)$$

Then, if the sampling period is sufficiently small (see (19.13)), the sampling and holding operations are equivalent to a phase shift, i.e. a time delay in the time domain.

The error introduced by approximating the continuous signal $r(t)$ with the piecewise-continuous one $r_c(t)$ verifies the relation

$$|r(t) - r_c(t)|_{ZOH} \leq T_c \max_t \left|\frac{dr(t)}{dt}\right|$$

The result achieved when the option **foh** is used in the command **c2dm** can still be interpreted by considering Figure 19.5. In particular, the ZOH must be replaced by a first order hold (FOH) so that the output $r_c(t)$ becomes a piecewise-linear signal instead of a piecewise-constant one. More specifically, $\forall t \in [nT_c, nT_c + T_c)$, we can write

$$r_c(t) = r(nT_c) + \frac{t - nT_c}{T_c} (r(nT_c) - r(nT_c - T_c))$$

The output of an FOH is then a linear continuous, but not differentiable, signal which coincides with the input at multiple time instants of the sampling period. Following the same procedure as for the case of a ZOH, the transfer function of an FOH can be written as

$$G_{FOH}(s) = \frac{1 + sT_c}{T_c} \left(\frac{1 - e^{-sT_c}}{s} \right)^2$$

which gives

$$R_c(j\omega) = \sqrt{1 + \omega^2 T_c^2} \left(\frac{\sin \frac{\omega T_c}{2}}{\frac{\omega T_c}{2}} \right)^2 e^{j\left(-\omega T_c + \tan^{-1}(\omega T_c) \right)} \sum_{n=-\infty}^{+\infty} R\left(j\omega + jn\frac{2\pi}{T_c} \right) \quad (19.14)$$

which can be compared with (19.12). In the FOH case the error between $r(t)$ and $r_c(t)$ satisfies the condition

$$|r(t) - r_c(t)|_{FOH} \leq T_c^2 \max_t \left| \frac{d^2 r(t)}{dt^2} \right|$$

Note that this transformation is not invertible, i.e. the corresponding discrete to continuous conversion does not exist.

The option `tustin` assumes the following bilinear transformation:

$$s = \frac{2}{T_c} \frac{z - 1}{z + 1} \quad (19.15)$$

The transformation (19.15) is achieved from the trapezoidal approximation of the derivative. This transformation allows one to map the left half-plane of the 'continuous' complex plane (the asymptotic stability region for continuous-time systems) onto the unitary circle of the 'discrete' complex plane (the asymptotic stability region for discrete-time systems), i.e. stable continuous-time systems correspond to stable discrete-time transformed systems and vice versa.

The option `prewarp` is based on a transformation that is similar to (19.15) which allows a perfect frequency matching between continuous and discrete frequency responses at a desired critical frequency. In other words, it can be used when we desire that the two frequency responses be equal at a given frequency $\bar\omega$. Assume the transformation (19.15) and, say, $G(s)$ a continuous transfer function and $H(z)$ the corresponding discrete one, i.e.

$$H(z) = G(s)|_{s = \frac{2}{T_c} \frac{z-1}{z+1}}$$

The frequency response of the continuous system at a frequency $\bar\omega$ is given by $G(j\bar\omega)$, whereas, at the same frequency, the frequency response of the discrete system is given by $H(e^{j\bar\omega T_c})$, which in general is different from $G(j\bar\omega)$. Noting that

$$\frac{2}{T_c} \frac{e^{j\omega T_c} - 1}{e^{j\omega T_c} + 1} = \frac{2j}{T_c} \tan\left(\frac{\omega T_c}{2} \right)$$

it is straightforward to show that the value of the continuous frequency response at $\bar\omega$ will equal the value of the discrete frequency response at the frequency $\hat\omega$, which satisfies the following relation:

$$\bar\omega = \frac{2}{T_c} \tan\left(\frac{\hat\omega T_c}{2} \right)$$

A transformation which eliminates this warping at a desired frequency $\bar{\omega}$ is then

$$s = \frac{\bar{\omega}}{\tan(\bar{\omega}T_c/2)} \frac{z-1}{z+1} \tag{19.16}$$

In fact, substituing $s = j\bar{\omega}$ and $z = e^{j\bar{\omega}T_c}$ in (19.16), we obtain

$$H(e^{j\bar{\omega}T_c}) = G(j\bar{\omega})$$

as was required. The option **prewarp** assumes a transformation such as (19.16). It should be noted that this transformation does not eliminate the warping at frequencies different from the one specified.

The option **matched** uses the following zeros–poles equivalences: a pole in $s = -a$ becomes a pole in $z = e^{-aT}$, a zero in $s = -b$ becomes a zero in $z = e^{-bT}$, a zero in $s = -\infty$ becomes a zero in $z = -1$. Moreover, the equality condition for the dc gains is imposed, i.e. the continuous transfer function at 0 equals the discrete one at 1.

It is not possible to provide general criteria for the choice of the continuous to discrete conversion method to be adopted. Nevertheless, some suggestions can be given. For first-order systems the options **zoh** and **matched** give good results w.r.t. the frequency responses equivalence. For more complex systems the options **foh** and **tustin** should be preferred, whereas the option **prewarp** should be used when a correction of the bilinear transformation achievable with the option **tustin** at a desired frequency is needed.

Inverse transformations, i.e. conversion from discrete-time to continuous-time systems, can be obtained by the commands **d2c** and **d2cm** (see Section 24.1). The command **c2dt** provides continuous to discrete transformation in the presence of a time delayed input of the continuous system.

To conclude this section some considerations about the selection of the sampling period are in order; since no rigorous general rules can be pointed out, only some empirical considerations will be given. Firstly, for a feedback controlled system the choice of the sampling period must be made with reference to the closed loop system; this is because, usually, the processes to be controlled are low–pass systems and the bandwidth of the closed loop system is larger than that of the open loop system. To select T_c it is convenient to refer to a nondimensional parameter, say N. In the case of real poles, N can be defined as

$$N = \frac{t_r}{T_c}$$

where t_r is the rise time of the system; in this case N should be chosen larger than 4. For transfer functions with complex poles, N can be defined as

$$N = \frac{2\pi}{\omega_n T_c \sqrt{1 - \zeta^2}}$$

where ω_n and ζ are the natural frequency and the damping ratio; in this case the condition $N > 10$ should be verified.

Among the other empirical rules for the sampling period selection, of interest are those which consider the phase margin deterioration due to the phase shift introduced by the ZOH or, considering the signal sampling problems, those which determine T_c on the basis of an upper limit in the reconstruction error introduced by the holding operation.

19.4   Controllability and observability

The CONTROL commands for the controllability and observability properties are:

ctrb	controllability matrix
ctrbf	controllable blocks form
gram	controllability and observability gramians
obsv	observability matrix
obsvf	observable blocks form

Given a system

$$\dot{x} = Ax + Bu$$

the pair (A, B) is said to be controllable if, for any x_0 and x_1 and for any time instants t_0 and t_1 with $t_0 < t_1$, there exists a control u which brings the state from the state x_0 at the time instant t_0 to the state x_1 at the time instant t_1. This definition does not provide a formal procedure to determine whether or not a system is controllable. To this end the following equivalent definition is more useful: the pair (A, B) is said to be controllable if

$$rank \left(\begin{array}{cccc} B & AB & \ldots & A^{n_x-1}B \end{array} \right) = n_x \tag{19.17}$$

where n_x is the system order.

The CONTROL commands provide simple ways of verifying the controllability of a system. Let us consider the following dynamic model of a dc motor with constant flux:

$$\left(\begin{array}{c} \dot{x}_1 \\ \dot{x}_2 \end{array} \right) = \left(\begin{array}{cc} -\dfrac{R}{L} & -\dfrac{k_v}{L} \\ \dfrac{k_t}{J} & -\dfrac{B}{J} \end{array} \right) \left(\begin{array}{c} x_1 \\ x_2 \end{array} \right) + \left(\begin{array}{cc} \dfrac{1}{L} & 0 \\ 0 & -\dfrac{1}{J} \end{array} \right) \left(\begin{array}{c} u \\ T_L \end{array} \right)$$

$$y = \left(\begin{array}{cc} 0 & 1 \end{array} \right) \left(\begin{array}{c} x_1 \\ x_2 \end{array} \right)$$

where the first state variable is the armature current, the second is the rotor speed and the other parameters have the following meanings:

R armature resistance;
L armature inductance;
J moment of inertia;
B friction coefficient;
k_t torque constant;
k_v e.m.f. constant;
u armature voltage;
T_L load torque.

Assuming zero load torque and unitary parameters, except for the armature inductance $L = 1$ mH, we obtain the model

$$\begin{pmatrix} \dot{x}_1 \\ \dot{x}_2 \end{pmatrix} = \begin{pmatrix} -10^3 & -10^3 \\ 1 & -1 \end{pmatrix} \begin{pmatrix} x_1 \\ x_2 \end{pmatrix} + \begin{pmatrix} 10^3 \\ 0 \end{pmatrix} u \qquad (19.18)$$

$$y = \begin{pmatrix} 0 & 1 \end{pmatrix} \begin{pmatrix} x_1 \\ x_2 \end{pmatrix} \qquad (19.19)$$

Assigning the matrices A and B of this system, the instruction

$$\text{Mcont=ctrb(A,B);}$$

gives the matrix

$$M_{cont} = \begin{pmatrix} 10^3 & 10^6 \\ 0 & 10^3 \end{pmatrix}$$

which is the controllability matrix, defined in (19.17), of the system (19.18)–(19.19). The system is then controllable as it can be verified, on the basis of the previous considerations, from the MATLAB instruction rank(Mcont), which provides the rank of the controllability matrix Mcont.

Noncontrollable systems can be divided into their controllable and noncontrollable parts (standard controllable Kalman form). In other words, if a pair of matrices is noncontrollable, a transformation matrix T will exist such that the transforméd matrices

$$A_t = TAT^{-1}, \qquad B_t = TB$$

have the following block matrix forms:

$$A_t = \begin{pmatrix} A_{nc} & 0 \\ A_{21} & A_c \end{pmatrix}, \qquad B_t = \begin{pmatrix} 0 \\ B_c \end{pmatrix}$$

with nonzero A_{nc} and where the pair (A_c, B_c) is controllable. Note that only the matrices A_{nc} and A_c will definitely be square matrices.

To show how this form can be achieved by the CONTROL commands let us consider the electrical circuit shown in Figure 19.6. A possible state-space model for this circuit is

$$\dot{x}_1 = -\frac{R_1 + R_3}{C_1 R_1 R_3} x_1 \frac{1}{C_1 R_3} x_2 + \frac{1}{R_1 C_1} u \qquad (19.20)$$

$$\dot{x}_2 = \frac{1}{C_2 R_3} x_1 - \frac{R_2 + R_3}{C_2 R_2 R_3} x_2 + \frac{1}{R_2 C_2} u \qquad (19.21)$$

$$y = x_1 - x_2 \qquad (19.22)$$

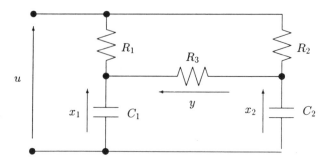

Figure 19.6 Resistances and capacitors bridge.

Assume that all the resistances are equal to 1 kΩ and that both the capacitors are equal to 1 mF. This system is noncontrollable as can be deduced from the fact that the command `ctrb` gives the matrix

$$M_{cont} = \begin{pmatrix} 1 & -1 \\ 1 & -1 \end{pmatrix}$$

which has unitary rank. Defining the circuit matrices in MATLAB,

$$A=[-2 \ 1;1 \ -2]; \ B=[1 \ 1]'; \ C=[1 \ -1];$$

the instruction

$$[At,Bt,Ct,T,k]=ctrbf(A,B,C);$$

gives the matrices of the transformed system

$$\begin{pmatrix} \dot{z}_1 \\ \dot{z}_2 \end{pmatrix} = \begin{pmatrix} -3 & 0 \\ 0 & -1 \end{pmatrix} \begin{pmatrix} z_1 \\ z_2 \end{pmatrix} + \begin{pmatrix} 0 \\ 1.4 \end{pmatrix} u$$

$$y = \begin{pmatrix} -1.4 & 0 \end{pmatrix} \begin{pmatrix} z_1 \\ z_2 \end{pmatrix}$$

the transformation matrix

$$T = 0.707 \begin{pmatrix} -1 & 1 \\ 1 & 1 \end{pmatrix}$$

and the vector k, whose elements are the ranks of the blocks of the matrix At. Note that the output variables T and k can also be omitted.

From the transformed system it is apparent that only z_2 is controllable, which means that it is not possible to control the difference $x_1 - x_2$, which will be zero at steady state, as should be expected because of the so-called equilibrium of the bridge, i.e. $R_1 C_1 = R_2 C_2$.

We can now consider the CONTROL commands dealing with the observability property. A pair of matrices (A, C) is said to be observable if the observability matrix defined as

$$M_{obs} = \begin{pmatrix} C \\ CA \\ \vdots \\ CA^{n-1} \end{pmatrix}$$

has full rank.

The observability and controllability properties are linked by the duality principle: a pair (A, C) is observable if and only if the pair (A^T, C^T) is controllable.

Also, for nonobservable systems it is always possible to define a transformation which highlights the observable and the nonobservable parts of the system (the observability Kalman form), i.e. such that

$$A_t = \begin{pmatrix} A_{no} & A_{12} \\ 0 & A_o \end{pmatrix}, \quad C_t = \begin{pmatrix} 0 & C_o \end{pmatrix} \tag{19.23}$$

Using the CONTROL commands we can verify that the system (19.20)–(19.21)–(19.22), with the previously defined parameters, is also nonobservable. In particular, the instruction

```
Mobs=obsv(A,C);
```

gives the matrix

$$M_{obs} = \begin{pmatrix} 1 & -1 \\ -3 & 3 \end{pmatrix}$$

which is not full rank. By means of the command

```
[At,Bt,Ct,T,k]=obsvf(A,B,C);
```

we obtain the Kalman observability form.

A different way of checking the controllability and observability of a pair of matrices consists of using the so-called controllability and observability gramians which are defined as

$$G_{cont} = \int_0^T e^{At} B B^T e^{A^T t} dt \tag{19.24}$$

and

$$G_{obs} = \int_0^T e^{A^T t} C^T C e^{At} dt \tag{19.25}$$

respectively. The instructions gram(A,B) and gram(A',C') allow one to compute these gramians for $\tau = \infty$. It can be proved that a pair (A, B) (a pair (A, C)) is controllable (observable) if the corresponding gramian is full rank.

258 Models for LTI systems

19.5 Commands for other properties

In this section we describe how to use the following commands:

covar	covariance matrices to white noise
damp	damping ratios and natural frequencies
dcgain	DC gain
esort	eigenvalues sorted by real part
tzero	transmission zeros

19.5.1 Natural frequencies and damping ratios

Firstly, we recall that given a pair of complex poles $\alpha \pm j\beta$ the natural frequency and the damping ratio are defined as

$$\omega_n = \sqrt{\alpha^2 + \beta^2}, \quad \zeta = -\frac{\alpha}{\sqrt{\alpha^2 + \beta^2}}$$

i.e. the modulus and the opposite of the cosine of the phase of the complex number $\alpha + j\beta$, respectively.

Let us consider the dynamic system characterized by the following matrices:

$$A = \begin{pmatrix} -4 & 3 \\ -3 & -4 \end{pmatrix}, \qquad B = \begin{pmatrix} 0 \\ 1 \end{pmatrix}$$

$$C = \begin{pmatrix} 0 & 1 \end{pmatrix}, \qquad D = 0$$

which correspond to the transfer function

$$G(s) = \frac{s - 5}{s^2 + 8s + 25}$$

By means of the instruction

$$[\text{wn,zita}] = \text{damp(A)};$$

we obtain the vectors $\omega_n = (5,5)^T$ and $\zeta = (0.8, 0.8)^T$, i.e. the column vectors wn and zita containing the natural frequencies and the damping ratios of the considered system.

By using the command eig(A) we can calculate the eigenvalues of the matrix A, which are $-4 \pm 3j$. The instruction

$$[\text{wn,zita}] = \text{damp([-4-3j;-4+3j])};$$

gives the result previously obtained. Moreover, defining the vector of the coefficients of the characteristic polynomial of A by the command poly(A), by means of the instruction

$$[\text{wn,zita}]=\text{damp}([1\ 8\ 25]);$$

we again obtain the same results $\omega_n = (5,5)^T$ and $\zeta = (0.8, 0.8)^T$.

To resume, if, instead of a matrix as the argument of the damp command we use a column vector, the command operates by assuming this to be a vector containing the eigenvalues of a matrix, whereas if we use a row vector, the command operates by assuming it to be a vector containing the coefficients of the characteristic polynomial.

The command damp can be used with systems of any order and gives the natural frequencies and the damping ratios which correspond to the different pairs of complex eigenvalues of the considered matrix. In the case of real eigenvalues, as with the definitions given at the beginning of this section, the command gives as natural frequency the absolute value of the eigenvalue and as damping ratio ± 1, depending on the negative or positive sign of the eigenvalue.

19.5.2 DC gain

Let us consider the transfer function

$$G(s) = \frac{s-2}{s^2 + 2s + 2}$$

which can be defined in MATLAB by means of the vectors

$$\text{num}=[1\ -2]; \ \text{den}=[1\ 2\ 2];$$

Writing

$$\text{dcgain(num,den)}$$

we achieve the result -1, which is the dc gain of the defined transfer function, i.e. $\lim_{s\to 0} G(s)$. Note that the command dcgain gives 0 and Inf in the case where the transfer function has zeros or poles, respectively, at the origin.

This instruction can also be used for multi-output systems (with transfer matrix) and also in the form

$$\text{k=dcgain(A,B,C,D)};$$

i.e. when the input–state–output representation is given.

19.5.3   Eigenvalues sorting

Consider the vector

$$p = (\;-1 - 3j \quad 4j \quad -1 + 3j \quad -3.4 \quad 5.2 \quad -4j \;)$$

and use the command `esort` in the following way:

$$[\texttt{pord,ind}]\texttt{=esort(p)};$$

The first vector obtained is

$$p_{ord} = (\; 5.2 \quad 4j \quad -4j \quad -1 + 3j \quad -1 - 3j \quad -3.4 \;)$$

which contains the elements of the vector p, sorted according to the decreasing values of their real parts. The second output variable is the vector

$$ind = (\; 5 \quad 2 \quad 6 \quad 3 \quad 1 \quad 4 \;)$$

which contains the indices of the original positions on the vector p of the elements of the vector `pord`. The second output variable `ind` may be omitted. This function can be used for any polynomial, but is useful for sorting the eigenvalues of a matrix.

Note that the command `esort` is different from the command `sort` (see p. 37), which sorts the elements of the vector according to their increasing absolute values. For instance, by using the command `sort(p)` with the previously defined vector p the following result is obtained:

$$(\;-1 - 3j \quad -1 + 3j \quad -3.4 \quad -4j \quad 4j \quad 5.2 \;)$$

19.5.4   Covariance matrices

Given an LTI system, the instructions

$$[\texttt{P,Q}]\texttt{=covar(A,B,C,D,W)};$$
$$\texttt{P=covar(num,den,W)};$$

return the covariance output and state matrices

$$P = E[yy^T], \qquad Q = E[xx^T]$$

where E indicates the mean operator when the system is forced by a Gaussian white noise characterized by a covariance matrix

$$E[w(t)w(\tau)^T] = W\delta(t - \tau)$$

where δ is the Dirac function.

19.5.5 Transmission zeros

Firstly, let us recall the definition of the transmission zeros. Given a rational matrix $G(s)$ whose rank is r (we can think of a transfer matrix of a MIMO system) it is always possible to transform it in a pseudo-diagonal block form (the transfer matrix may be not square) such as

$$M(s) = diag\left\{\frac{\varepsilon_1(s)}{\psi_1(s)}, \frac{\varepsilon_2(s)}{\psi_2(s)}, \ldots, \frac{\varepsilon_r(s)}{\psi_r(s)}, 0, 0, \ldots, 0\right\}$$

which is called the Smith–McMillan form, where in all polynomials the coefficient of the term with the maximum power of s is one, (ε_i, ψ_i) have no common factors and any polynomial ε_{i-1} is divisible by ε_i. The transmission zeros are defined as the roots of the polynomials ε_is.

The definition is then a generalization of the definition of the zeros in the SISO case and it preserves the physical meaning related to the situation where the output is zero without the input and the state being zero. In other words a complex number λ is a transmission zero if for some input vector u_0 and some initial state x_0 the following equations hold:

$$\begin{aligned}\dot{x}(t) &= Ax(t) + Bu_0e^{\lambda t}, & x(0) = x_0 \\ 0 &= Cx(t) + Du_0e^{\lambda t}\end{aligned}$$

With one of the instructions

```
trzer=tzero(A,B,C,D);
trzer=tzero(NUM,DEN);
[pol,trzer]=pzmap(A,B,C,D);
[pol,trzer]=pzmap(NUM,DEN);
```

one can obtain the transmission zeros of the defined system.

19.6 Model order reduction

The commands that are useful for the order reduction of an LTI system are

balreal	balanced realization
minreal	minimal realization and poles–zeros cancelation
modred	model order reduction
ssdelete*	deletes inputs, states and outputs from the model
ssselect*	subsystem from a higher-order system

where the asterisk indicates that the instruction can also be used for discrete-time systems.

The possibility of obtaining a reduced model from a given dynamic model is very useful for systems characterized by a 'hard' dynamic separation among the state behaviors. In this case, if, for instance, only the evolution of the 'fast' state variables is of interest, one can introduce a model reduction by neglecting the 'slow' state variables which will be considered as constants and as equal to their initial values. On the other hand, if we are only interested in studying the system behavior on the 'slow' timescale we could assume the 'fast' state variables to be constants and equal to their steady-state values.

As an example, let us consider the system

$$
\begin{pmatrix} \dot{x}_1 \\ \dot{x}_2 \\ \dot{x}_3 \end{pmatrix} = \begin{pmatrix} -8 & -100 & 0 \\ 100 & -8 & 0 \\ 0 & 0 & -1 \end{pmatrix} \begin{pmatrix} x_1 \\ x_2 \\ x_3 \end{pmatrix} + \begin{pmatrix} -1 \\ -0.2 \\ 0.0002 \end{pmatrix} u \tag{19.26}
$$

$$
y = \begin{pmatrix} -2 & 20 & 580 \end{pmatrix} \begin{pmatrix} x_1 \\ x_2 \\ x_3 \end{pmatrix} \tag{19.27}
$$

As is apparent, the third state variable is dynamically separated from the first two. (The step time response of the whole system is shown in part *(a)* of Figure 19.7.) Since the evolution of the third state variable is slower than that of the others, a reduced model which neglects the 'slow' dynamics can be obtained by means of the instruction

```
[Ar,Br,Cr,Dr]=ssdelete(A,B,C,0,[],[],3);
```

which, assuming a zero initial state, gives the matrices of the reduced order system

$$
\begin{pmatrix} \dot{x}_1 \\ \dot{x}_2 \end{pmatrix} = \begin{pmatrix} -8 & -100 \\ 100 & -8 \end{pmatrix} \begin{pmatrix} x_1 \\ x_2 \end{pmatrix} + \begin{pmatrix} -1 \\ -0.2 \end{pmatrix} u
$$

$$
y_f = \begin{pmatrix} -2 & 20 \end{pmatrix} \begin{pmatrix} x_1 \\ x_2 \end{pmatrix}
$$

The step time response of the 'fast' subsystem is shown in part *(b)* of Figure 19.7. In general, given a system of the form

$$
\begin{aligned}
\dot{x} &= A\,x + B\,u \\
y &= C\,x + D\,u
\end{aligned}
$$

it is possible to reduce the number of inputs, states and/or outputs by the instruction

```
[Ar,Br,Cr,Dr]=ssdelete(A,B,C,D,indin,indout,indst);
```

which provides the matrices of the reduced model obtained by neglecting the inputs, the outputs and the states whose indices are specified by the vectors indin, indout and indst respectively. Note that ssdelete can also be used without specifying the vector indst, whereas indin and indout must always be indicated (if one

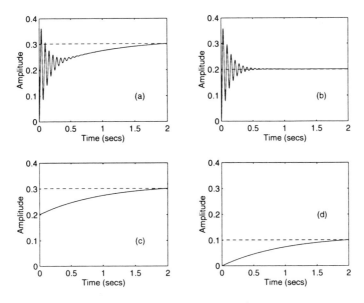

Figure 19.7 Step responses: *(a)* whole system, *(b)* 'fast' subsystem, *(c)* 'slow' subsystem obtained with modred, *(d)* 'slow' subsystem obtained with ssdelete.

desires to eliminate only some states, in place of the vectors indin and indout one must substitute two empty vectors, i.e. [], as in the previous example).

Quite similar to the instruction ssdelete is the instruction ssselect which, with the same possible parameters, allows one to select the inputs, the outputs and the states which must be preserved; in other words, in the vectors indin, indout and indst one must specify the indices related to the inputs, the outputs and the states that must be retained.

Going back to the example presented at the beginning of this section, it is important to stress that it is wrong to think that it is possible to obtain the model which characterizes the 'slow' dynamics of the system by the command ssdelete. In fact, in this case the 'fast' variable will be assumed zero instead of steady state as it should be. By considering the system (19.26)–(19.27), the 'slow' reduced model can be obtained by writing the instruction

$$\texttt{[Ar,Br,Cr,Dr]=modred(A,B,C,0,[1,2]);}$$

where the next to last argument is, as in the ssdelete case, the matrix D of the original system, and the last argument consists of the indices of the state variables that must be eliminated (differently from the ssdelete instruction, one cannot specify any inputs or outputs to be eliminated). In such a way we obtain the matrices of the 'slow' reduced system

$$\dot{x}_3 \;=\; -x_3 + 0.0002\,u$$

$$y_s = \cdot\, 580\, x_3 + 0.2\, u$$

The step response of the 'slow' subsystem is shown in part *(c)* of Figure 19.7, whereas the output of the system obtained by the following instruction is plotted in part *(d)*:

$$[\mathtt{As,Bs,Cs,Ds}]=\mathtt{ssdelete(A,B,C,0,[],[],[1,2])};$$

To stress the operations of the command **modred** we consider a generalization of the model reduction problem with the constraint that the steady-state input-output behavior remains unchanged. To this end, consider the system

$$\begin{pmatrix} \dot{x}_1 \\ \dot{x}_2 \end{pmatrix} = \begin{pmatrix} A_{11} & A_{12} \\ A_{21} & A_{22} \end{pmatrix} \begin{pmatrix} x_1 \\ x_2 \end{pmatrix} + \begin{pmatrix} B_1 \\ B_2 \end{pmatrix} u$$

$$y = \begin{pmatrix} C_1 & C_2 \end{pmatrix} \begin{pmatrix} x_1 \\ x_2 \end{pmatrix} + Du$$

Let us assume that the 'fast' variables are x_1; under this hypothesis we can impose that these variables are at steady state, i.e. $\dot{x}_1 = 0$, and then solve the corresponding equation w.r.t. x_1. Then we can substitute the x_1 value computed in the dynamic equation of x_2 so that the following reduced model is obtained:

$$\dot{x}_2 = \left(A_{22} - A_{21} A_{11}^{-1} A_{12}\right) x_2 + \left(B_2 - A_{21} A_{11}^{-1} B_1\right) u$$

$$y = \left(C_2 - C_1 A_{11}^{-1} A_{12}\right) x_2 + \left(D - C_1 A_{11}^{-1} B_1\right) u$$

which preserves the steady-state input–output behavior. This kind of reduction is achieved by means of the command **modred**.

A further way of obtaining reduced order models consists of eliminating the states which influence the controllability and observability of the whole system 'less'. To obtain this result the commands **balreal** and **modred** can be used simultaneously.

The command **balreal** allows one to obtain the so-called balanced representation of a system in the state-space, i.e. a representation such that the controllability and observability gramians coincide and are diagonal.

In particular, the instruction

$$[\mathtt{At,Bt,Ct,g,T}]=\mathtt{balreal(A,B,C)};$$

gives the matrices in the previously described form, a vector **g** which contains the elements of the gramians diagonals and the transformation matrix adopted.

The usefulness of such a representation is due to the fact that, since the elements of the vector **g** reflect both the controllability and the observability of each state variable, one can eliminate the states characterized by a 'small' **g(i)** since they have a small influence on the input–output behavior of the system. This kind of reduction can be obtained, for instance, after applying the command **balreal**, with the instructions

$$\mathtt{indst=find(g<g(1)/10)};$$
$$[\mathtt{Ar,Br,Cr,Dr}]=\mathtt{modred(At,Bt,Ct,D,indst)};$$

The command `minreal` provides a reduced minimal representation of a system and can be used in the following forms:

$$[\text{Am,Bm,Cm,D}]=\text{minreal}(\text{A,B,C,D,tol});$$
$$[\text{numm,denm}]=\text{minreal}(\text{num,den,tol});$$
$$[\text{zerm,polm}]=\text{minreal}(\text{zer,pol,tol});$$

where the last argument `tol`, which is the tolerance for the zeros and poles cancelation, i.e. the maximum allowed error between the pole and the zero that must be canceled, can be omitted. Note that using the command with transfer function models, `num` and `den` must be row vectors, whereas using zeros–poles models, `zer` and `pol` must be column vectors.

19.7   System connections

Before describing the CONTROL commands that are useful for system connections, it is important to stress that in most cases the same results can be obtained more easily with a user written .M file. It should also be noted that the algorithm used by CONTROL to implement the commands for the system connections is more accurate if the system is represented in the state-space form and, moreover, the models obtained are often in a nonminimal form; therefore it is convenient to always use the command `minreal` on the model obtained.

The commands that are useful for the systems connections are

append*	augment the system state
augstate*	augment the output with the state
blkbuild*	build a system in the state space
cloop*	closed loop system
connect*	model for a block diagram
feedback*	connection in a closed loop system
parallel*	parallel connection
series*	series connection

where, as highlighted by the asterisks, all these instructions can also be used for discrete systems.

19.7.1   Series connection

Let us consider two LTI systems with transfer functions

$$G_1(s) = \frac{1}{s}, \qquad G_2(s) = \frac{4}{s+2}$$

As is well known, the series transfer function, which can be obtained by connecting the output of the first system with the input of the second one, is $G(s) = G_2(s)G_1(s)$. To obtain a MATLAB representation, a first possibility is to define

```
numw=4; denw=conv([1 0],[1 2]);
```

but the same result can be obtained directly by the command

```
[numw,denw]=series(1,[1 0],4,[1 2]);
```

which gives the numerator and denominator polynomials of $G(s)$.

In general, with reference to Figure 19.8 the system series can be defined by the instructions

```
[As,Bs,Cs,Ds]=series(A1,B1,C1,D1,A2,B2,C2,D2);
[nums,dens]=series(num1,den1,num2,den2);
```

whereas for a partial series connection (see Figure 19.9), the whole transfer function can be obtained by

```
[As,Bs,Cs,Ds]=series(A1,B1,C1,D1,A2,B2,C2,D2,us1,in2);
```

where the vectors us1 and in2 contain, respectively, the indices locating the outputs of the first system which must be connected to the inputs of the second one (obviously us1 and in2 have the same length).

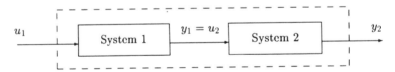

Figure 19.8 Series connection of two systems.

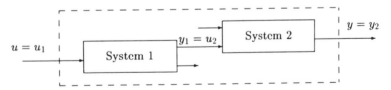

Figure 19.9 Partial series connection of two systems.

Note that, when dealing with a partial series connection, the only inputs of the whole system obtained with the command **series** are the inputs of the first system and the only outputs of the whole system are the outputs of the second system.

19.7.2 Parallel connection

Consider the systems

$$G_1(s) = 2, \qquad G_2(s) = \frac{100}{s}$$

The transfer function of their parallel connection, i.e. $G(s) = G_1(s) + G_2(s)$, can be obtained by the instruction

$$[\text{nump},\text{denp}]=\text{parallel}(3,1,2,[1\ 0]);$$

which gives the numerator and denominator polynomials of the transfer function

$$G(s) = \frac{2s + 100}{s}$$

In general, a representation of the system shown in Figure 19.10 can be obtained by the instructions

```
[Ap,Bp,Cp,Dp]=parallel(A1,B1,C1,D1,A2,B2,C2,D2);
[nump,denp]=parallel(num1,den1,num2,den2);
```

whereas a representation of the system shown in Figure 19.11 (a partial parallel connection of two systems) can be obtained by the instruction

```
[A,B,C,D]=parallel(A1,B1,C1,D1,A2,B2,C2,D2,i1,i2,u1,u2);
```

where only the inputs whose indices are detailed in i1 and i2 are assumed coincident and only the outputs whose indices are in u1 and u2 will be summed to obtain the output of the whole system (obviously, also in this case i1 and i2, as well as u1 and u2, have the same dimensions). The other inputs and outputs, differently from the series command, will be considered as inputs and outputs, respectively, of the parallel system.

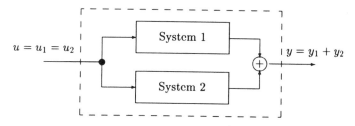

Figure 19.10 Parallel connection of two systems.

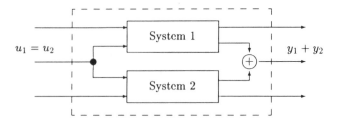

Figure 19.11 Partial parallel connection of two systems.

19.7.3   Feedback connection

Let us consider the transfer function

$$G(s) = \frac{s-1}{s^2 + 3s + 2}$$

A representation of the closed loop system with unitary negative feedback can be obtained by the instruction

```
[numc,denc]=cloop([1 -1],[1 3 2]);
```

which gives the numerator and denominator polynomials of the transfer function

$$W(s) = \frac{G(s)}{1 + G(s)} = \frac{s-1}{s^2 + 4s + 1}$$

In general, with reference to Figure 19.12, the representation of the same closed loop system can be obtained by the instructions

```
[numc,denc]=cloop(numo,deno,-1);
[Ac,Bc,Cc,Dc]=cloop(Ao,Bo,Co,Do,-1);
```

where the last letter o (c) is for the open loop (closed loop) variables. If no sign of the feedback is specified, a negative feedback will be assumed and if a number is used as the last argument of the command, only its sign, which will be the sign of the unitary feedback, will be considered.

With reference to Figure 19.13, if only some outputs, whose indices are specified in the vector us, must be fed back to the control inputs, whose indices are specified in the vector in, one can use the instruction

```
[Ac,Bc,Cc,Dc]=cloop(Ao,Bo,Co,Do,us,-in);
```

where us and in must have the same dimension.

If we wish to obtain a model for the feedback connection of two systems, by using as the last letter t for the variables of the system which is in the feedback path and can then be considered as the transducer, the following instructions can be used

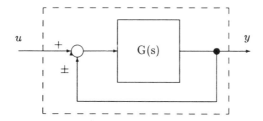

Figure 19.12 Feedback connection scheme.

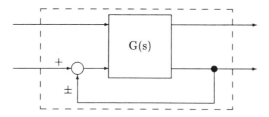

Figure 19.13 Partial feedback connection scheme.

```
[numc,denc]=feedback(numo,deno,numt,dent,signum);
[Ac,Bc,Cc,Dc]=feedback(Ao,Bo,Co,Do,At,Bt,Ct,Dt,signum);
[Ac,Bc,Cc,Dc]=feedback(Ao,Bo,Co,Do,At,Bt,Ct,Dt,in,us);
```

where the variable signum can be omitted if a negative feedback is considered, the vector in contains the indices of the 'open loop system' inputs where the output of the 'transducer' must be connected and the vector us contains the indices of the 'open loop system' outputs which must be the inputs for the 'transducer' (see Figure 19.14).

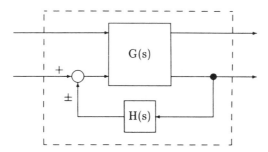

Figure 19.14 Feedback connection scheme with transducer.

19.7.4 Augmented systems

By means of the command **append** it is possible to augment the state of a given system. In particular, the instruction

$$[\texttt{A,B,C,D]=append(A1,B1,C1,D1,A2,B2,C2,D2)};$$

provides the augmented system from two given systems, i.e. builds the system

$$\begin{pmatrix} \dot{x}_1 \\ \dot{x}_2 \end{pmatrix} = \begin{pmatrix} A_1 & 0 \\ 0 & A_2 \end{pmatrix} \begin{pmatrix} x_1 \\ x_2 \end{pmatrix} + \begin{pmatrix} B_1 & 0 \\ 0 & B_2 \end{pmatrix} \begin{pmatrix} u_1 \\ u_2 \end{pmatrix}$$

$$\begin{pmatrix} y_1 \\ y_2 \end{pmatrix} = \begin{pmatrix} C_1 & 0 \\ 0 & C_2 \end{pmatrix} \begin{pmatrix} x_1 \\ x_2 \end{pmatrix} + \begin{pmatrix} D_1 & 0 \\ 0 & D_2 \end{pmatrix} \begin{pmatrix} u_1 \\ u_2 \end{pmatrix}$$

The command **augstate** is a particularization of **append**. In fact, the instruction

$$[\texttt{At,Bt,Ct,Dt]=augstate(A,B,C,D)};$$

assumes a new output vector by also considering the state as output, i.e. it builds the system

$$\dot{x} = Ax + Bu$$

$$\begin{pmatrix} y_1 \\ y_2 \end{pmatrix} = \begin{pmatrix} C \\ I \end{pmatrix} x + \begin{pmatrix} D \\ 0 \end{pmatrix} u$$

19.8   Exercises

Exercise 19.1 Assign the following transfer function in MATLAB:

$$G(s) = \frac{5 s^2 + 2 s - 1}{(s+4)(s^2 + 2 s + 1)}$$

and obtain the corresponding representations in state-space, zero and pole and residue and pole forms.

Exercise 19.2 Assign the following transfer function in MATLAB:

$$G(s) = \frac{\left(\dfrac{4}{(s+5)(s+3)} \atop 2 s^2 + 12 s + 18 \right)}{(s + 20 \pm 10j)(s^2 + 8 s + 12)}$$

and obtain the corresponding representations in state-space, zero and pole and residue and pole forms.

Exercise 19.3 Obtain a state-space representation for the electrical circuit shown in Figure 19.15. Assuming the following parameters: $R_1 = 1.5\ \Omega$, $L = 0.25$ H, $C = 0.5$ F, $R_c = 5\ \Omega$, assign the model in MATLAB and obtain the representations in transfer function, zero and pole and residue and pole forms.

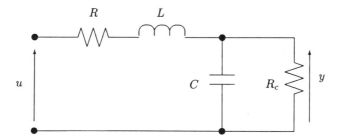

Figure 19.15 Electrical circuit.

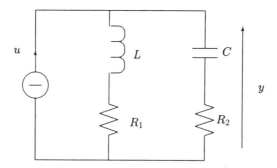

Figure 19.16 Electrical circuit.

Exercise 19.4 In the circuit of Figure 19.16 the input u is a current and the output y is a voltage. Obtain a state-space representation of the system. Assigning the circuit parameters at will, provided that the conditions $R_1 = R_2 = R$ and $R^2C/L = 1$ hold, verify that the transfer function of the circuit is a constant. Which value does this constant assume?

Exercise 19.5 Given the circuit shown in Figure 19.17 obtain a possible state-space representation. Assuming $R = 1\ \Omega$, $L = 1$ mH and $C = 5$ mF, assign the two models obtained considering first as output y and then y_1 with y_2. Represent the circuit model in transfer matrix, zero and pole and residue and pole forms.

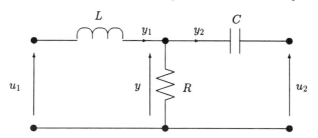

Figure 19.17 Electrical circuit with two inputs and two outputs.

Exercise 19.6 Assign the system described by the following equations in MATLAB:

$$
\begin{pmatrix} \dot{x}_1 \\ \dot{x}_2 \\ \dot{x}_3 \end{pmatrix} = \begin{pmatrix} 4.5 & 5 & -5 \\ -6.45 & -15.5 & -4.5 \\ 5 & 5 & -5 \end{pmatrix} \begin{pmatrix} x_1 \\ x_2 \\ x_3 \end{pmatrix} + \begin{pmatrix} 0 \\ 0.4 \\ 0 \end{pmatrix} u
$$

$$
y = \begin{pmatrix} -1.5 & 0 & 2 \end{pmatrix} \begin{pmatrix} x_1 \\ x_2 \\ x_3 \end{pmatrix}
$$

Compute the new state-space representations obtained with the command `ss2ss` using as transformation matrices the inverse of the controllability matrix and the observability matrix of the system.

Exercise 19.7 Write the function *cancont.m* which, from given matrices of a state-space model of an LTI system, gives as output the matrices of the system in the canonical control form (19.6).

Exercise 19.8 Given the transfer function

$$
G(s) = \frac{1}{s+1}
$$

obtain the corresponding discrete representations with the different continuous to discrete conversion methods and with sampling times 0.01 s and 2 s (with the option **prewarp**, use 1 rad s^{-1} as the critical frequency).

> **Hint.** It is interesting to compare the time response of the systems to a step input and to a sinusoidal signal with frequency 1 rad s^{-1} (see Sections 20.4, 20.2 and 24.4) and to compare the Bode plots of the different representations (see Sections 21.1 and 24.5).

Exercise 19.9 Given the transfer function

$$
G(s) = \frac{s}{s^2 + s + 25}
$$

obtain the corresponding discrete representations with the different continuous to discrete conversion methods and with sampling frequencies 3 Hz and 15 Hz (use 5 rad s^{-1} as the critical frequency in the **prewarp** command).

> **Hint.** It is interesting to compare the Bode plots of the different models obtained (see Sections 21.1 and 24.5).

Exercise 19.10 Consider the system described by the differential equations

$$
\begin{pmatrix} \dot{x}_1 \\ \dot{x}_2 \end{pmatrix} = \begin{pmatrix} -1 & 1 \\ 0 & -1 \end{pmatrix} \begin{pmatrix} x_1 \\ x_2 \end{pmatrix} + \begin{pmatrix} 1 \\ 0 \end{pmatrix} u
$$

Is the system controllable? Given the initial conditions $x_1(0) = x_2(0) = 1$, verify whether the state $(2, 1)$ is reachable. Is the origin reachable?

Exercise 19.11 Given the system (19.20)–(19.21)–(19.22), choosing the circuit parameters at will, provided that the condition $R_1C_1 = R_2C_2$ is verified, apply a state transformation such that a state variable coincides with the output. Verify that, whatever the input, the bridge reaches equilibrium, i.e. the steady state voltage on R_3 is zero.

> **Hint.** Verify that the noncontrollable part of the system is asymptotically stable.

Exercise 19.12 Assign the system described by the following equations in MATLAB:

$$\begin{pmatrix} \dot{x}_1 \\ \dot{x}_2 \\ \dot{x}_3 \end{pmatrix} = \begin{pmatrix} -0.5 & 1 & 0 \\ -1 & -0.5 & 0 \\ 0 & 1 & 0 \end{pmatrix} \begin{pmatrix} x_1 \\ x_2 \\ x_3 \end{pmatrix} + \begin{pmatrix} 1 \\ 2 \\ 0 \end{pmatrix} u$$

$$y = \begin{pmatrix} 0 & 0 & 1 \end{pmatrix} \begin{pmatrix} x_1 \\ x_2 \\ x_3 \end{pmatrix}$$

Is the system observable? If it is not completely observable, what is the non-observable part? Is the system controllable? If it is not completely controllable, what is the noncontrollable part?

Exercise 19.13 Consider the system shown in Figure 19.18: an inverted pendulum, whose mass is m, is constrained at the point A; a ring with angular moment h can only rotate around the pendulum axis; a control torque u can be applied to the ring from the pendulum. The model of the whole system can be written as

$$I\ddot{\theta} = mgl\theta - h\dot{\phi}$$
$$J\ddot{\phi} = h\dot{\theta} + u$$

where I is the moment of inertia of the pendulum and of the ring w.r.t. the point A, J is the moment of inertia of the ring w.r.t. the pendulum axis and l is the distance of the center of mass C from the point A.

Assigning the system parameters at will, determine the transfer functions between u and ϕ and between u and θ. Verify that the system is controllable by u, observable from ϕ and nonobservable from θ.

Exercise 19.14 The motion of the aerostat represented in Figure 19.19 can be described by the following linearized equations:

$$\dot{\theta} = -\frac{1}{\tau_1}\theta + u$$
$$\dot{v} = -\frac{1}{\tau_2}v + \sigma\theta + \frac{1}{\tau_2}w$$
$$\dot{h} = v$$

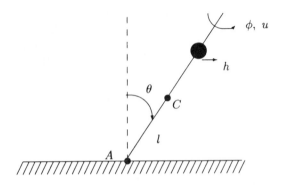

Figure 19.18 Inverted pendulum with ring.

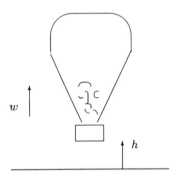

Figure 19.19 Aerostat.

where θ is the temperature change w.r.t. a reference value, u is the air heating (the control variable), h is the altitude change w.r.t. a reference value, v is the vertical speed and w is the wind speed in the same direction.

Assume $\tau_1 = 2$ s, $\tau_2 = 1$ s and $\sigma = 1$ m s^{-2}. Are the temperature variation θ and a constant wind w observable from the altitude h?

Determine the transfer functions between u and h and between w and h. Is the system controllable from u? And from w?

> **Hint.** To show if a constant wind is observable from altitude measurement, augment the state with the further equation $\dot{w} = 0$.

Exercise 19.15 The satellite motion on the equatorial plane in polar coordinates, and assuming as state variables positions and speeds, can be modeled by the linearized system

$$\dot{x}_1 = x_2$$
$$\dot{x}_2 = 3\omega_0^2 x_1 + 2\omega_0 r_0 x_4 + \frac{u_1}{m}$$

$$\dot{x}_3 = x_4$$

$$\dot{x}_4 = -\frac{2\omega_0}{r_0}x_2 + \frac{u_2}{mr_0}$$

where m is the satellite mass, r_0 and ω_0 are the radius and the angular speed of its orbit and u_1 and u_2 are the radial and tangential control forces respectively.

Assigning the parameters at will, verify whether the system is controllable in the presence of only one of the two control inputs and, if the system is not controllable, obtain a representation in the controllability Kalman form.

Exercise 19.16 The model of an inverted pendulum on a moving cart, as shown in Figure 19.20, is

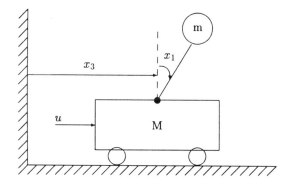

Figure 19.20 Inverted pendulum on a moving cart.

$$\begin{pmatrix} \dot{x}_1 \\ \dot{x}_2 \\ \dot{x}_3 \\ \dot{x}_4 \end{pmatrix} = \begin{pmatrix} 0 & 1 & 0 & 0 \\ \frac{M+m}{Ml}g & 0 & 0 & 0 \\ 0 & 0 & 0 & 1 \\ -\frac{mg}{M} & 0 & 0 & 0 \end{pmatrix} \begin{pmatrix} x_1 \\ x_2 \\ x_3 \\ x_4 \end{pmatrix} + \begin{pmatrix} 0 \\ -\frac{1}{Ml} \\ 0 \\ \frac{1}{M} \end{pmatrix} u$$

where x_1 and x_2 are the position and the angular speed of the pendulum, respectively, whereas x_3 and x_4 are the position and the speed of the mass M. Given the following paramenters $M = 10$ kg, $m = 1$ kg, $l = 1$ m and $g = 9.8$ m s^2, verify whether the system is controllable or not. Supposing that a measurement of the pendulum angular position is available, is the system controllable? What variables must be measured to ensure system observability?

Exercise 19.17 Given the system with transfer function

$$G(s) = \frac{1}{s-1}$$

consider a series connected controller with transfer function

$$C(s) = \frac{s-1}{s+1}$$

Compare the controllability and observability properties of the two systems obtained by connecting the controller $C(s)$ before and following $G(s)$, respectively. In particular, give a justification for the apparent discrepancy of the results obtained when using the command **series** with transfer functions and state-space models.

> **Hint.** Use the commands **series, tf2ss, ctrb** and **obsv**. It is also interesting to analyze the unforced responses of the two systems when varying the initial conditions (to this end use the **initial** command or a SIMULINK scheme). In particular, considering the case when the controller precedes the plant, comment on the choice of initial conditions such that $x_1(0) = -2x_2(0)$.

Exercise 19.18 Given the transfer functions

$$W_1(s) = \frac{s+1}{(s+10)(s+0.1)}, \qquad W_2(s) = \frac{s+0.1}{(s+10)(s+100)}$$

obtain the related control canonical state-space representations. Verify that: the series connection with $W_2(s)$ following $W_1(s)$ is controllable; the parallel connection is noncontrollable and nonobservable; the feedback configuration with $W_2(s)$ in the feedback path is controllable and observable.

Exercise 19.19 A state feedback controlled system with state observer can be rearranged as shown in Figure 19.21 where $b(s)$ and $a(s)$ are the numerator and the n_x-degree denominator polynomials of the plant transfer function, $\delta(s)$ is the $(n_x - 1)$-degree denominator polynomial of the observer, and $\beta(s)$ and $\gamma(s)$ are numerator polynomials whose degrees are equal to or less than $n_x - 1$. The problem is to design the polynomials $\delta(s)$, $\beta(s)$ and $\gamma(s)$ so that the closed loop transfer function has a desired denominator polynomial $\alpha(s)$. After some algebra it can be shown that the problem consists of finding the polynomials $\delta(s)$, $\beta(s)$ and $\gamma(s)$ which verify the Diophantine equation

$$a(s)\,(\delta(s) + \beta(s)) + b(s)\gamma(s) = \alpha(s)\delta(s)$$

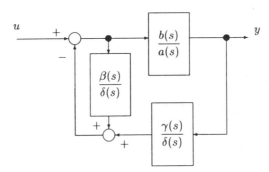

Figure 19.21 Feedback controlled system with state observer.

Assume

$$a(s) = s^2 + s, \qquad b(s) = 1, \qquad \alpha(s) = s^2 + 2s + 2, \qquad \delta(s) = s + 2$$

Compute from the Diophantine equation the coefficients of the numerator polynomials $\beta(s) = \beta_1 s + \beta_2$ and $\gamma(s) = \gamma_1 s + \gamma_2$ and, using the commands `feddback` and `series` to reduce the scheme of Figure 19.21, verify that the closed loop poles are $-1 \pm j$.

Exercise 19.20 Given the model of Exercise 19.5, from the forms defined considering both outputs, obtain those corresponding to the single outputs by means of the commands `ssselect` and `ssdelete`.

Exercise 19.21 Given the system whose model can be written as

$$\begin{pmatrix} \dot{x}_1 \\ \dot{x}_2 \end{pmatrix} = \begin{pmatrix} -10 & 0 \\ 0.1 & -1 \end{pmatrix} \begin{pmatrix} x_1 \\ x_2 \end{pmatrix} + \begin{pmatrix} 1 \\ 0 \end{pmatrix} u$$

$$y = \begin{pmatrix} 1 & 1 \end{pmatrix} \begin{pmatrix} x_1 \\ x_2 \end{pmatrix}$$

consider the opportunity to reduce the model order. Obtain the models of the 'fast' and 'slow' parts of the system by using the commands `ssdelete` or `ssselect` and `modred`. Study and compare the balanced representation of the system with that of the reduced models.

Hint. It is interesting to compare the time response of the different models to a unitary step input (see Section 20.2).

Chapter 20

Time domain response

In this chapter we present and describe the commands which allow one to analyze the time domain response of an LTI system. Recall that the output waveform of an LTI system can be analytically expressed as

$$y(t) = Ce^{At}x_0 + \int_0^t (Ce^{A(t-\tau)}B + D)u(\tau)d\tau \qquad (20.1)$$

where x_0 is the initial condition and $u(t)$ is the input. As is well known, in (20.1) the first right-hand term is the so-called unforced or free evolution, whereas the second is the so-called forced evolution. It is interesting to analyze how, by means of the MATLAB commands, it is possible to determine the analytic expression of the time response of a system starting from its transfer function and from the Laplace transform of the input signal. In other words, we are looking for the expression

$$y(t) = \mathcal{L}^{-1}(Y(s)) = \mathcal{L}^{-1}(G(s)U(s))$$

where $\mathcal{L}$ is the Laplace transform operator. To this end, let us consider the system

$$\dot{x} = \begin{pmatrix} -0.1 & 3 & 0 \\ -3 & -0.1 & 0 \\ 0 & 0 & -1 \end{pmatrix} x + \begin{pmatrix} 0 \\ 1 \\ 1 \end{pmatrix} u$$

$$y = \begin{pmatrix} 1 & 0 & 1 \end{pmatrix} x$$

Assigning the system matrices in MATLAB, the command

$$[n,d]=ss2tf(A,B,C,0);$$

returns the following transfer function of the system:

$$G(s) = \frac{s^2 + 3.2\ s + 12}{s^3 + 1.2\ s^2 + 9.2\ s + 9}$$

We now consider as input a unitary step signal whose Laplace transorm is $1/s$. By the instruction

```
[res,pol,k]=residue(n,[d 0]);
```

where the second input argument is the product between the denominator of $G(s)$ and s, we can write the Laplace transform of the output signal as a sum of partial fractions, i.e.

$$Y(s) = \frac{-0.166 \pm 0.005j}{s + 0.1 \mp 3j} + \frac{-1}{s + 1} + \frac{1.333}{s}$$

By using the inverse Laplace transform of the elementary functions we obtain

$$y(t) = (0.332 \, e^{-0.1t} \cos(3 \, t + 3.108) - e^{-t} + 1.333) \, 1(t)$$

where $1(t)$ is the unitary step function.

The previous result can easily be generalized. Let us consider a complex function $\Phi(s)$ and denote by R_i the residues and by λ_i the roots of the denominator. Supposing that we have m real and $n - m$ complex roots and, neglecting the possibility of multiple poles, the inverse transform of

$$\Phi(s) = \sum_{i=1}^{n} \frac{R_i}{s - \lambda_i}$$

can be written as

$$\Phi(t) = \sum_{i=1}^{m} R_i e^{\lambda_i t} + \sum_{j=1}^{\frac{n-m}{2}} 2e^{\alpha_j t} |R_j| \cos(\omega_j t + \underline{/R_j})$$

where each term of the second series corresponds to the complex conjugate eigenvalues $\lambda_j = \alpha_j \pm \omega_j$.

The CONTROL commands that are useful to determine the time response of an LTI system are:

impulse	impulse response
initial	unforced response
lsim	response to any input
step	response to unitary step signal

The use of these commands will be detailed in the following sections by means of different examples. Nevertheless, it is important to stress that the time simulation of a linear system can usually be more easily obtained using a SIMULINK scheme rather than the CONTROL commands.

20.1   Unforced response

The free response of an LTI system, starting from any given initial condition x0, can be obtained by the command `initial`. To this end we consider the second-order system

$$\dot{x} = \begin{pmatrix} -4 & 3 \\ -3 & -4 \end{pmatrix} x + \begin{pmatrix} 0 \\ 1 \end{pmatrix} u$$

$$y = \begin{pmatrix} 0 & 1 \end{pmatrix} x$$

Defining the following system matrices in MATLAB:

$$\texttt{A=[-4 3;-3 -4]; B=[0;1]; C=[0 1]; D=0;}$$

by the instruction

$$\texttt{initial(A,B,C,D,[2;0]);}$$

Figure 20.1 is plotted, which represents the system output (the second state variable) starting from the initial condition $x_0 = (2,0)^T$.

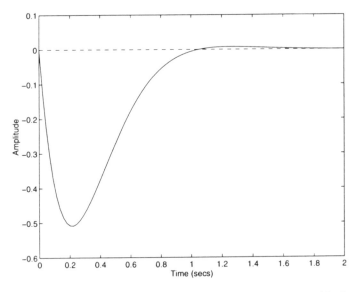

Figure 20.1 Free evolution (solid line) and steady-state value (dashed line).

Considering the same system, in order to obtain the plot of both state variables on the same diagram, we can write

$$\texttt{initial(A,B,eye(2),zeros(2,1),[2;0]);}$$

i.e. we assume both the state variables as output (the matrix C must be chosen as the identity matrix and its order equal to the state dimension, and the matrix D must be updated coherently).

In general, by the instruction

$$\texttt{[out,state,tt]=initial(A,B,C,D,x0);}$$

the values of the outputs, of the state and of the sampling time instants are computed and stored in suitable variables (out, state and tt, respectively); these can then be plotted using the command plot.

If requested, as further argument of the command initial, one can introduce a time vector t which contains the time instants at which it is desired to evaluate the state and the output of the system; this can be obtained by writing

$$\texttt{[out,state]=initial(A,B,C,D,x0,t);}$$

It is important to stress that the elements of the vector t must be equally spaced and strictly increasing. The matrices out and state will contain as many columns as the outputs and the state variables, respectively, whereas each row will correspond to a different computing time instant.

For multi-output systems, the instruction initial, without any output variable, gives, on the same window, as many curves as the number of system outputs.

20.2 Step response

The command step allows one to obtain the outputs and the state of an LTI system with zero initial condition and forced by a unitary step signal. To show how to use this command, let us consider the system whose transfer function is

$$G(s) = \frac{s+10}{s^2+8s+25} \tag{20.2}$$

To compute the step response of this system, we can use the instruction

$$\texttt{[out,state,tt]=step([1 10],[1 8 25]);}$$

which provides the output, the state and the time vectors stored in the related variables.

Suppose now that we wish to analyze how the step response of the system (20.2) changes when the zero of its transfer function is changed, without changing the system dc gain. To this end we may build a fictitious multi-output transfer matrix: we retain the same denominator and we change the coefficient of the first term in the numerator, i.e. the coefficient of s, so that the dc gain is constant but the system zero is changed. The use of the step command will then give the desired comparison. In particular, writing

```
coef=[-4 -2 -1 0 1 2 4]; den=[1 8 25];
[y,x,t]=step([coef' 10*ones(length(coef),1)],den);
mesh(coef,t,y)
```

we obtain the plot shown in Figure 20.2.

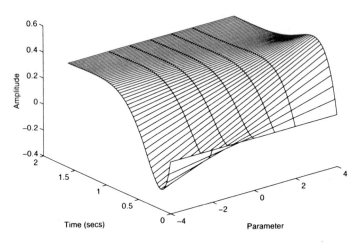

Figure 20.2 Comparison between step responses.

It is interesting to compare and justify the different time responses.

In general, the command **step** can be used in the following forms:

```
[out,state,tt]=step(A,B,C,D);
[out,state,tt]=step(A,B,C,D,ui);
[out,state]=step(A,B,C,D,ui,t);
[out,state,tt]=step(num,den);
[out,state]=step(num,den,t);
```

where, by means of the argument **ui**, it is possible to specify, for multi-input systems, which input must be considered, assuming all the others to be zero.

The following conditions hold:

- without output variables the instruction plots, on different windows, the step responses of all the possible input–output pair combinations of the system considered;
- the matrices **out** and **state** will have as many columns as the outputs and state variables, respectively, and as many rows as the time samples, i.e. the length of the vector **tt**;
- the vector **t**, if used as the input argument of the command, must be equally spaced.

20.3   Impulse response

The CONTROL instruction `impulse` allows one to obtain the unitary impulse time response of an LTI system with zero initial state conditions.

Let us consider the transfer function

$$G(s) = \frac{s + 10}{s^2 + 2\,s + 25}$$

In order to plot the impulse response of this system it is enough to write

$$\texttt{impulse([1 10],[1 2 25]);}$$

Suppose now that we wish to analyze how the impulse response changes when the zero of the transfer function is moved, without changing the dc gain of the system. Following the same procedure as presented in the previous section one can write

```
coef=[-4 -2 -1 0 1 2 4]; den=[1 2 25];
impulse([coef' 10*ones(length(coef),1)],den);
```

so that the plot shown in Figure 20.3 is obtained.

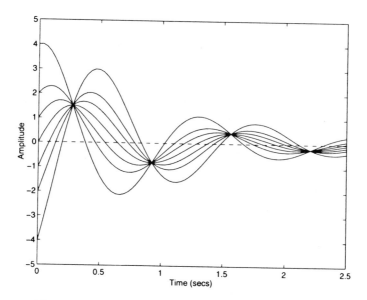

Figure 20.3 Comparison between impulse responses.

For the interested reader we propose some nice questions here: Why do the impulse responses start from different points? How can these values be determined a priori?

Why do the different waveforms cross each other at the same time instants? It could be helpful to consider the state-space representation and the evolution of all state variables.

In general, the command `impulse` can be used as follows:

```
[out,state,tt]=impulse(A,B,C,D);
[out,state,tt]=impulse(A,B,C,D,ui);
[out,state]=impulse(A,B,C,D,ui,t);
[out,state,tt]=impulse(num,den);
[out,state]=impulse(num,den,t);
```

The `impulse` command syntax is similar to that of the `step` command described in the previous section.

20.4   Response to any input

A more general problem is the computation of the output of an LTI system forced by a nonstandard input signal, which must be defined previously by means of MATLAB instructions.

Consider, for instance, the first-order system

$$\dot{x} = -x + u$$
$$y = x$$

Forcing this system with a sinusoidal input signal whose frequency is assumed to be 1 Hz, the output can be obtained by the instructions

```
freq=1; t=0:0.05:10/freq;
u=sin(2*pi*freq*t); lsim(-1,1,1,0,u,t)
```

where the last instruction gives the plot shown in Figure 20.4.

In general, each of the following instructions

```
[out,state]=lsim(A,B,C,D,u,t);
[out,state]=lsim(A,B,C,D,u,t,x0);
[out,state]=lsim(num,den,u,t);
```

provides the output and the state of the system when forced by the inputs defined in the matrix u (which has as many columns as the number of inputs and as many rows as the sampling time instants); the vector t contains the sampling time instants and x0 is the optional initial condition (if x0 is omitted, zero initial conditions are assumed).

The use of the `lsim` command is quite obvious, whereas, typically, some problems arise for the 'construction' of nonstandard inputs. To this end we present the following example. Suppose that we wish to obtain the time response of the system

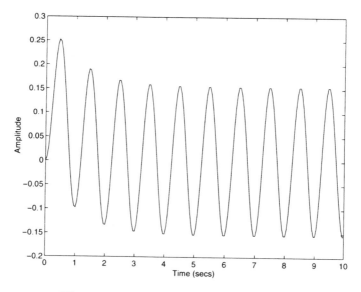

Figure 20.4 Response to sinusoidal input.

previously defined to an input signal which is zero for two seconds, then a triangular waveform of amplitude 10 and period 1 s for six seconds, and finally is constant and equals 10 for four seconds. Moreover, suppose that we look for the state evolution starting from the initial condition $x_0 = 3$. The instructions

```
amp=10; T=1; deltat=T/10; t=0:deltat:12;
h=find(t>=2); p=find(t>=T+2);
u1=zeros(1,h(1));
u2=amp/(T-deltat)*(t(h(1):p(1)-1)-t(h(1)));
u3=amp*ones(1,length(t)-length(u1)-6*length(u2));
u=[u1 u2 u2 u2 u2 u2 u2 u3];
[y,x]=lsim(-1,1,1,0,u,t,3);
plot(t,u','--',t,x);
```

will finally provide the plot shown in Figure 20.5, where both the input and the output of the system are represented.

20.5 Exercises

Exercise 20.1 Given the system with transfer function

$$G(s) = \frac{1}{(s+1)^2}$$

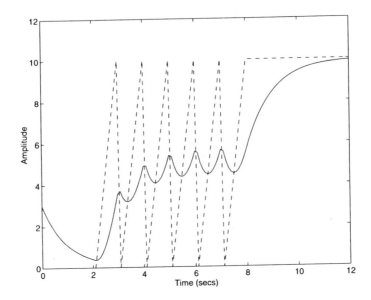

Figure 20.5 Input (dashed line) and output (solid line) of a first-order system.

compare and justify the shape of the step responses of the open and closed loop systems when a unitary negative feedback is used.

Exercise 20.2 Plot and compare the step responses of the following systems:

$$G_1(s) = \frac{1}{s+2}, \qquad G_2(s) = \frac{1}{s^2 + 0.1\ s + 1}$$

$$G_3(s) = \frac{1}{s^2 + 10\ s + 1}, \qquad G_4(s) = \frac{1}{s^3 + 4\ s^2 + 3\ s + 1}$$

and justify the different waveforms.

Exercise 20.3 Determine the open and closed loop responses of the systems

$$G_1(s) = \frac{1}{s+1}, \qquad G_2(s) = \frac{1}{s(s+2)}$$

to a unitary ramp.

> **Hint.** Use the command **step** and consider that the ramp can be obtained as the integral of a step signal and that the integrator transfer function is $1/s$; alternatively build up a ramp signal and then use **lsim**.

Exercise 20.4 Determine the time responses of the system

$$G(s) = \frac{1}{s+1}$$

to sinusoidal inputs characterized by unitary amplitude and the following frequencies: $\omega = 0.1,\ 1,\ 2,\ 10$ rad s^{-1}.

Hint. It is interesting to justify the results by means of the Bode plots (see Section 21.1).

Exercise 20.5 Determine and compare the step responses of the series and parallel connections of two systems with the same transfer function

$$G(s) = \frac{1}{s^2 + 1}$$

and justify the results.

Exercise 20.6 Given the dynamic system described by the equations

$$\begin{pmatrix} \dot{x}_1 \\ \dot{x}_2 \end{pmatrix} = \begin{pmatrix} 0 & 1 \\ -1 & 0 \end{pmatrix} \begin{pmatrix} x_1 \\ x_2 \end{pmatrix}$$

plot the system phase portrait (x_2 vs. x_1) starting from the initial condition $(1, 1)$.

Exercise 20.7 Given the first-order SISO system which is described by the equations

$$\dot{x} = -x + u, \qquad y = x$$

obtain, by means of the commands step and initial, the responses of the system to rectangular impulses whose amplitude A and time duration T are chosen so that the condition $AT = 1$ holds.

Exercise 20.8 Given the dc motor dynamic model described in Section 19.4, assume the state variables as outputs and the following parameters: $R = 1\ \Omega$, $L = 10$ mH, $J = 0.02$ kg m^{-2}, $B = 0.2$ Ns, $k_t = k_v = 0.03$. Determine the responses of the system starting from zero initial conditions and with zero load torque, for a step input voltage of amplitudes 1 V and 6 V. Starting from the steady-state conditions of the second case, determine the response of the system to a 0.1 N load torque disturbance.

Exercise 20.9 Plot the impulse response of the circuit shown in Figure 19.17.

Exercise 20.10 Plot the time evolution of the state variables of the electrical circuit shown in Figure 19.15, assuming the following parameters: $R = 100$ mΩ, $L = 400$ mH, $C = 0.1$ F, $R_c = 10\ \Omega$, and considering as inputs: an unitary step and unitary amplitude sinusoidal signals of frequencies $\omega = 1,\ 3,\ 4,\ 10$ rad s^{-1}. Compare the results.

Exercise 20.11 Given the system with transfer function

$$G(s) = \frac{s + 24}{(s + 1)(s + 2)(s + 10 \pm 10j)}$$

determine the parameters of the different standard controllers (Proportional, Integral, Proportional plus Integral, Proportional plus Integral plus Derivative) by using the Ziegler and Nichols method. Compare the step responses of the closed loop systems.

Hint. Recall that the selection of the standard controller parameters by means of the Ziegler and Nichols method is based on the following relations:

- $C(s) = K_p$ where $K_p = 0.5K_0$;
- $C(s) = K_p \left(1 + \frac{1}{T_i s}\right)$ where $K_p = 0.45K_0$ and $T_i = 0.85T_0$;
- $C(s) = K_p(1 + T_d s)$ where $K_p = 0.5K_0$ and $T_d = 0.2T_0$;
- $C(s) = K_p \left(1 + T_d s + \frac{1}{T_i s}\right)$ where $K_p = 0.6K_0$, $T_d = 0.12T_0$ and $T_i = 0.5T_0$;

$C(s)$ being the controller transfer function, $K_0 = KG(0)$, where K is the proportional action which places the closed loop poles on the imaginary axis and T_0 is the related period of the permanent oscillation. Use the command step or the commands rlocus and rlocfind (see Section 22.1) to determine the constants K_0 and T_0.

Exercise 20.12 Compute the responses of the system

$$\dot{x} = \begin{pmatrix} 0 & 1 \\ -6 & -5 \end{pmatrix} x + \begin{pmatrix} 0 \\ 1 \end{pmatrix} u$$

$$y = \begin{pmatrix} 12 & 0 \end{pmatrix} x$$

to the square wave plotted in Figure 20.6 with $T = 10,\ 1,\ 0.5$ s.

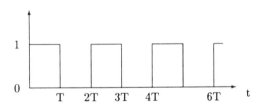

Figure 20.6 Square wave of period $2T$.

Chapter 21

Frequency domain response

In this chapter we describe the CONTROL commands which allow one to analyze the frequency domain response of an LTI system. Dealing with this topic we recall that the frequency response matrix $G(\omega)$ of a system is defined by

$$Y(\omega) = G(\omega)U(\omega) \tag{21.1}$$

where $U(\omega)$ and $Y(\omega)$ are the Fourier transforms of the input and output signals respectively.

As is well known, in the case where all the system poles have negative real parts, i.e. the system is asymptotically stable, the frequency response matrix coincides with the system transfer matrix evaluated on the imaginary axis ($s = j\omega$).

Then, for SISO systems, the frequency response is a complex number whose physical meaning is straightforward: assuming a unitary amplitude sinewave of frequency ω_0 as the input of an LTI system, at steady-state the output of the system will be a sinewave whose frequency is ω_0, whose amplitude is given by the modulus of the frequency response evaluated in ω_0, and whose phase is obtained by adding the phase of the input and the phase of the frequency response evaluated in ω_0.

The commands which are useful for the computation of the frequency response of an LTI system are:

bode	Bode plots
margin	gain and phase margins
nichols	Nichols plots
ngrid	gridding for Nichols plots
nyquist	Nyquist plots
sigma	singular values

21.1   Bode plots

The Bode plots (modulus and phase of the frequency response as a function of the frequency in logarithmic scale) can be used for the analysis of some system properties such as phase and gain margins, dc gain, bandwidth, disturbance rejection and stability of the closed loop system.

Let us consider the third-order SISO system with transfer function

$$G(s) = \frac{4}{(s+1)^3}$$

By means of the instructions

```
num=4; den=poly([-1 -1 -1]);
bode(num,den)
```

we obtain the Bode plots of the frequency response of the system considered, as shown in Figure 21.1.

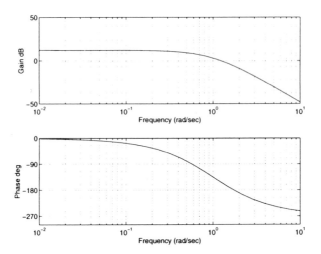

Figure 21.1 Bode plots.

In general, each of the following instructions:

```
[mod,phase,puls]=bode(A,B,C,D);
[mod,phase,puls]=bode(A,B,C,D,ui);
[mod,phase]=bode(A,B,C,D,ui,w);
[mod,phase,puls]=bode(num,den);
[mod,phase]=bode(num,den,w);
```

gives the modulus in natural values (the conversion, in decibels, can be achieved by the instruction `moddb=20*log10(mod)`), the phase, in degrees, and the vector of the frequency values (in radians per second) where the frequency response has been computed.

From the instruction syntax it is evident that one can choose a particular input `ui`, when dealing with multi-input systems, and also the frequency points (the vector `w`) in radians per second where the frequency response must be computed. Since the vector `w` is preferably defined in logarithmic scale, the command `logspace` should be used (see Section 3.2).

As in the case of the time response commands, the instructions used without output variables provide the Bode plots directly, but in this case the modulus of the frequency response is in decibels.

It is important to stress that when the dc gain of the transfer function is negative, the command `bode` gives a phase plot which starts from $+180°$ instead of, as usual, from $-180°$.

Moreover when the transfer function has poles on the imaginary axis, if the imaginary coefficients of these poles are in the vector `w` (or in the vector `puls`), MATLAB gives an error message. Nevertheless, the Bode magnitude plot will be represented, even though it will be truncated at the resonance frequency since, as is well known, the theoretical value assumed at that point by the freqeuncy response magnitude is infinite (see Exercise 21.5).

Before concluding this section we briefly present the command `sigma`. This command allows one to compute the principal gains of the transfer matrix by computing the singular values of the complex matrix

$$G(j\omega) = C(j\omega I - A)^{-1}B + D \tag{21.2}$$

for different frequency values (recall that the singular values of a complex matrix A are defined as $\sqrt{\lambda(A^*A)}$, where the asterisk indicates conjugate transpose). Comments on this command are given in this section since the singular values, when the frequency changes, can be considered as extensions to the MIMO case of the magnitude Bode plot of the frequency response of SISO systems.

The different forms of `sigma` are:

```
[singval,freq]=sigma(A,B,C,D);
[singval,freq]=sigma(A,B,C,D,'inv');
[singval]=sigma(A,B,C,D,freq);
[singval]=sigma(A,B,C,D,freq,'inv');
```

where one can use as command arguments the vector `freq` to select the frequencies (in radians per second) where $G(j\omega)$ is to be calculated, and the parameter `'inv'` is used to obtain the principal gains of the inverse matrix $G^{-1}(j\omega)$.[1]

---

[1] For further details consult the Robust Control Toolbox manual.

21.2   Nyquist plots

The Nyquist plots are a different way of representing the frequency response of an LTI system. In particular, the imaginary part of the complex number $G(j\omega)$ is plotted as a function of its real part for different frequency values.

Let us consider the following transfer function:

$$G(s) = \frac{k}{(s+1)^3}$$

Assume this to be the transfer function of an open loop system; the stability of the closed loop system with unitary negative feedback depends on the open loop gain. In particular, from the Nyquist stability theorem, we can conclude that the closed loop system becomes unstable if k is 'too large'. We now analyze this feature by using the Nyquist plots and the CONTROL commands. Writing

```
nyquist([4;10],poly([-1 -1 -1]))
```

we obtain the two diagrams represented at the top of Figure 21.2 which correspond to $k = 4$ and $k = 10$.

If the Nyquist plot is close to the so-called 'critical point' $-1 + j0$, this point is represented by a star in the same graphic window that is obtained with the command **nyquist**. Therefore, using the Nyquist theorem, it is quite simple to determine the stability of the closed loop system with unitary negative feedback. Recall that the closed loop system is stable if the Nyquist plot of the open loop frequency response $G(j\omega)$ encircles the point $-1 + j0$ anticlockwise as many times as the number of the $G(s)$ poles with positive real parts.

Using the command

```
nyquist([4;10],[1 3 3 1],1.5:0.1:2)
```

where the last argument is the frequency range chosen for the plot, we obtain the plot shown at the bottom of Figure 21.2 which is a zoom of the previous plot in the neighborhood of the critical point. As is apparent, the variation of the open loop gain k from 4 to 10 leads the closed loop system to instability.

In general, each of the instructions

```
[re,im,puls]=nyquist(A,B,C,D);
[re,im,puls]=nyquist(A,B,C,D,ui);
[re,im]=nyquist(A,B,C,D,ui,w);
[re,im,puls]=nyquist(num,den);
[re,im]=nyquist(num,den,w);
```

gives the real and imaginary parts of the frequency response (for more detailed comments on these instructions see those for the **bode** command in the previous section).

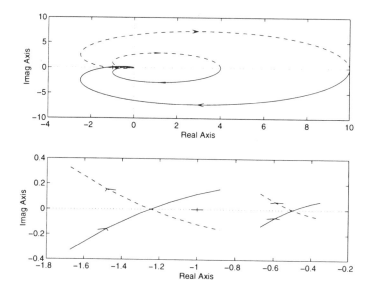

Figure 21.2 Nyquist plots.

As for the bode command, the command nyquist used without output arguments provides the Nyquist plot directly. Note that in this case a continuous line is
used for the plot related to the positive frequency range (from 0 to $+\infty$), whereas
the plot part represented with a dashed line refers to the negative frequency range
(from $-\infty$ to 0); the arrows indicate the increasing frequencies.

It is important to note that if the system has one or more poles at the origin,
the Nyquist plot obtained by the command nyquist is not correct, in the sense
that the vertical axis bounds are not appropriate. In order to obtain a 'good'
Nyquist plot the frequency vector w should be used as the input argument or, after
the instruction nyquist, the axis command should be used with the desired axis
bounds (in particular for the vertical axis).

Depending on the MATLAB version, the command nyquist gives or does not give the
error message of division per zero when used with transfer functions with poles in the
origin.

21.3   Nichols plots

The so-called Nichols plot of a frequency response represents the magnitude in
decibels as a function of the phase. As with the Nyquist plots, the Nichols plots
are characterized by a waveform that is 'graduated' for different frequency values.

Consider, for example, the transfer function

$$G(s) = 30\frac{s^2 + 7\,s + 1}{s(s + 1)^3}$$

Then, the sequence of instructions

```
num=30*[1 7 1]; den=[poly([-1 -1 -1]) 0];
nichols(num,den);
hold on, plot(-180,0,'*w'), hold off;
```

returns the Nichols plot where the so-called critical point $(-180°, 0)$ is also represented, as shown in Figure 21.3.

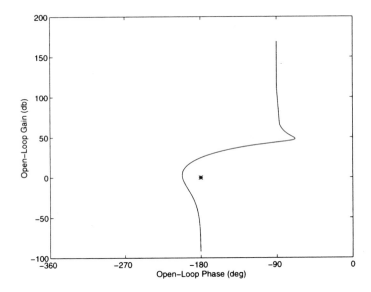

Figure 21.3 Nichols plot and 'critical point'.

In general, the instructions

```
[mod,phase,puls]=nichols(A,B,C,D);
[mod,phase,puls]=nichols(A,B,C,D,ui);
[mod,phase]=nichols(A,B,C,D,ui,w);
[mod,phase,puls]=nichols(num,den);
[mod,phase]=nichols(num,den,w);
```

give the magnitude in natural values, the phase in degrees and the vector of the
frequency points considered (in radians per second).

From the instruction structure it is apparent that it is also possible to specify the frequency points where the frequency response is to be evaluated (the vector **w**) and, for multi-input systems, which input must be considered (the input index **ui**).

As for the **nyquist** command, also in this case, by using the instruction without output variables, the Nichols plot is plotted directly.

It is important to stress that the command **nichols** always gives a phase in the interval $[-360°, 0°]$ (see Exercise 21.16).

The Nichols plots are useful for lead–lag controller design. This design can be assisted by the command **ngrid** which, if used after the **nichols** command, represents the curves corresponding to constant gain and phase of the closed loop transfer function with negative unitary feedback. The **ngrid('new')** command clears the graphic window, plots the grid and sets the **hold** command to **on** so that different Nichols plots can be represented simultaneously.

We now present a possible procedure for the design of a lag controller with the help of the **nichols** command. Let us consider the transfer function

$$G(s) = \frac{40}{(1 + 0.01s)(1 + 0.001s)^2} \qquad (21.3)$$

which can be defined in MATLAB by the instructions

```
num=40;
den=conv([0.01 1],conv([0.001 1],[0.001 1]));
```

Due to the large open loop gain the closed loop system with unitary negative feedback is unstable. This can also be deduced from the fact that the Nichols plot, represented by the instruction

```
nichols(num,den)
```

goes to the left of the point $(-180°, 0)$.

We want to design a lag compensator, i.e. a zero–pole pair controller, which stabilizes the closed loop system and which ensures a phase margin of 45° (see Section 21.4) with $\omega_t = 300$ rad s^{-1} as the crossover frequency. (As is well known, the crossover frequency is the frequency at which the frequency response magnitude is one, zero in decibels.) The transfer function of a lag controller can be written as

$$C(s) = \frac{1 + \alpha\tau s}{1 + \tau s} \qquad (21.4)$$

where $\tau > 0$ and $0 < \alpha < 1$. (The Bode plots for this kind of system justify the name of 'lag' compensator.)

The controller design can be achieved by passing through the following steps:

- The instructions

```
w=logspace(-1,4);
[m,f]=nichols(num,den,w);
mdB=20*log10(m);
```

assign the frequency vector and the related gain (in natural values) and phase
(in degrees) for the open loop transfer function previously defined. The gain
and phase values at the desired crossover frequency w_t can be achieved by
the instructions

```
[mwt,fwt]=bode(num,den,300);
mwtdB=20*log10(mwt);
```

Note that instead of the bode command we could have used the nichols
command with no difference.

We are then looking for a controller which reduces the magnitude by a quantity $1/m_{w_t} = 1/11.6$ and which, since the required phase margin is 45°,
introduces a $-180° + 45° - f_{w_t} = -30°$ degree displacement at the crossover
frequency $w_t = 300$ rad s^{-1}. These constraints can be met by using a lag
controller.

- We now have to select the lag controller parameters, i.e. α and τ in (21.4),
so that at $\omega = \omega_t$ the gain of $C(j\omega)$ is mwtdB and the phase is $-30°$. To
this end, the following nonlinear algebraic system, with α and τ as variables,
must be solved:

$$|C(j\omega_t)| = \sqrt{\frac{1 + \alpha^2\tau^2\omega_t^2}{1 + \tau^2\omega_t^2}}$$

$$\underline{/C(j\omega_t)} = \tan^{-1}(\alpha\tau\omega_t) - \tan^{-1}(\tau\omega_t)$$

The solution of this system can be achieved by using the MATLAB command
fsolve (see Section 9.2). In particular, the following *net_lag.m* file can be
written (we use the previously introduced numerical values):

```
function y=net_lag(x)
att=1/11.6; lag=-30*pi/180; puls=300;
c_puls=(1+j*x(1)*x(2)*puls)/(1+j*x(2)*puls);
y(1)=att-abs(c_puls);
y(2)=lag-angle(c_puls);
```

Given the initial guess values $\alpha = 0.01$ and $\tau = 0.3$, and using the MATLAB
instruction

```
sol=fsolve('net_lag',[0.01 0.3]);
```

after some iteration we achieve $\alpha = 0.0726$ and $\tau = 0.0716$, which are the
desired parameters of the lag controller.

- A representation of the $G(s)$ Nichols plot, the desired compensation and the Nichols plot of the open loop controlled system can be obtained by the instructions

```
plot(f,mdB,[fwt -135],[mwtdB 0]); hold on;
plot(fwt,mwtdB,'*',-135,0,'*');
nc=[sol(1)*sol(2) 1]; dc=[sol(2) 1];
[nn,dd]=series(nc,dc,num,den);
[ff,mm]=nichols(nn,dd,w); mmdB=20*log10(mm);
plot(ff,mmdB,'--'); hold off;
```

which provide the plot represented in Figure 21.4.

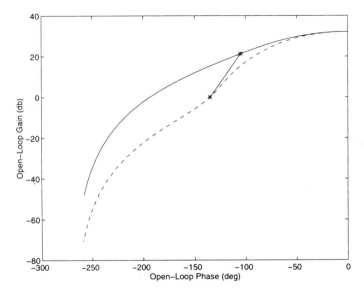

Figure 21.4   Nichols plots with lag controller: the dashed line is the Nichols plot of the controlled system and the stars limit the corrective action introduced.

21.4   Gain and phase margins

As is well known, the gain and phase margins allow analysis of the robustness property of a closed loop LTI system. In particular, we recall that the gain margin is defined as the inverse of the frequency response magnitude at the frequency where the phase equals $-\pi$. In other words, denoting by ω_π the frequency where the phase of the frequency response $G(j\omega)$ is $-\pi$, the gain margin is defined as:

$$m_g = \frac{1}{|G(j\omega_\pi)|}, \qquad m_{g_{dB}} = -|G(j\omega_\pi)|_{dB}$$

in natural values or in decibels respectively. For stable systems the gain margin can also be interpreted as the quantity by which the dc gain must be scaled so that the closed loop system with unitary negative feedback is stable but not asymptotically, i.e. it has poles on the imaginary axis of the complex plane.

The phase margin is defined as the angle which must be subtracted from the frequency response phase calculated at the crossover frequency, i.e. the frequency where the modulus is one (zero in decibels), in order to obtain $-\pi$. Denoting the crossover frequency by ω_t, the phase margin can be written as

$$m_\phi = \pi + \underline{/G(j\omega_t)}$$

For stable systems the phase margin is then the maximum delay in the open loop system at the frequency ω_t, which preserves the closed loop system stability (see Exercises 21.9 and 21.25).

So, as for the gain margin, when the phase margin is negative the closed loop system will be unstable.

Let us consider the third-order SISO system with transfer function

$$G(s) = \frac{4}{(s+1)^3}$$

By the instruction

$$\texttt{margin(4,[1 3 3 1]);}$$

we obtain the Bode plots and the gain and phase margins, as is shown in Figure 21.5.

In general, the margins can be obtained by using one of the following instructions:

```
[gm,pm,wgm,wpm]=margin(A,B,C,D);
[gm,pm,wgm,wpm]=margin(mod,phase,w);
[gm,pm,wgm,wpm]=margin(num,den);
```

where wgm is the ω_π previously defined and wpm is ω_t. As we have shown by the example, the command margin used without output arguments gives a representation of the margins on the Bode plots directly.

The gain and phase margins can also be represented on the Nyquist plots. In what follows we show how this can be achieved. From the margin definitions the gain margin will be the inverse of the modulus of the frequency response in the correpondence of the point where the curve crosses the negative real axis, whereas the phase margin will be the angle limited by the negative real axis and the line which joins the origin and the intersection between the Nyquist plot and the unitary circle. The representation of the margins on the Nyquist plot, shown in Figure 21.2, can then be obtained by writing the following instructions:

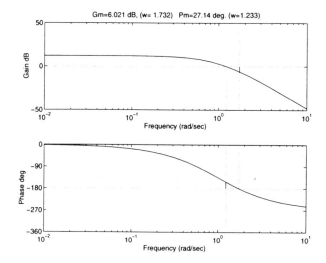

Figure 21.5 Gain and phase margins.

```
w=logspace(-0.1,2.3,300);
[re,im]=nyquist(4,poly([-1 -1 -1]),w);
p1=find(im>0)-1; p2=find((re.^2+im.^2)<1)-1;
p11=p1(1); p21=p2(1);
% Nyquist plot
plot(re,im); axis([-1.5,1.5,-1.5,1.5]); hold on
% Plots the inverse of the gain margin
plot(0,0,'*',re(p11),im(p11),'*');
plot([re(p11) 0],[im(p11) 0]);
% Plots the unitary circle
x=-1:0.08:1; y=sqrt(1-x.^2);
plot(x,[y' -y'],'.');
% Plots the phase margin
xmf=-1:0.01:1; ymf=-sqrt(1-xmf.^2);
p3=find(xmf>re(p2)); p31=p3(1);
plot(xmf(1:p31),ymf(1:p31));
plot(-1,0,'*',xmf(p31),ymf(p31),'*'); hold off
```

which provide Figure 21.6.

The representation of the margins on the Nichols plots is quite simple. In particular, one must consider, for the gain margin, the intersection of the curve with the vertical line corresponding to the angle $-180°$ and, for the phase margin, with the horizontal line corresponding to the zero value (see Exercise 21.17).

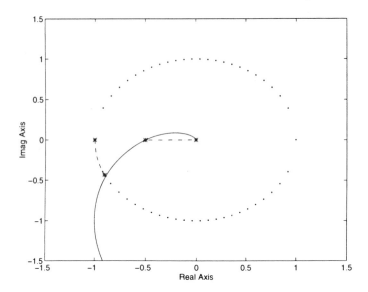

Figure 21.6 Margins on the Nyquist plot: the solid line is the Nyquist plot, the dotted line is the unitary radius circle, the dashed line is the phase margin and the dashed–dotted line is the gain margin.

21.5   Exercises

Exercise 21.1 Represent the Bode plots of the following transfer functions:

$$G_1(s) = \frac{1}{s}, \quad G_2(s) = \frac{10}{s}, \quad G_3(s) = s, \quad G_4(s) = -10\ s$$

and justify the figures obtained.

Exercise 21.2 Represent the Bode plots of the following transfer functions:

$$G_1(s) = \frac{10}{s+1}, \quad G_2(s) = \frac{10}{s-1}$$

and those of the transfer functions obtained with the opposite sign of the numerators. Justify and compare the figures obtained. Then represent the asymptotic Bode plots on the corresponding windows.

Exercise 21.3 Write the MATLAB function asint.m which traces the asymptotic Bode plots of a given second-order transfer function with two real poles.

Exercise 21.4 Represent the Bode plots of the following transfer function:

$$G(s) = 10\frac{\tau s + 1}{s + 1}$$

with $\tau = 10^4$, 10, 1, 0.1, 10^{-4} and compare the different curves.

Exercise 21.5 Represent the Bode plots of the transfer function:

$$G(s) = \frac{10}{s^2 + \alpha \, s + 1}$$

with $\alpha = 10$, 1, 0.1, 0 and compare the results. Then compare the step time responses of the different systems.

> **Hint.** To obtain the step responses use the command `step` (see Section 20.2) or build up a suitable SIMULINK scheme.

Exercise 21.6 Write a MATLAB function which represents the Bode plot of a given transfer function w.r.t. the frequency measured in Hertz instead of radians per second.

Exercise 21.7 Represent the Bode plot of the system

$$\begin{pmatrix} \dot{x}_1 \\ \dot{x}_2 \end{pmatrix} = \begin{pmatrix} 0 & 1 \\ -20 & -3 \end{pmatrix} \begin{pmatrix} x_1 \\ x_2 \end{pmatrix} + \begin{pmatrix} 1 & 1 \\ 0 & 1 \end{pmatrix} \begin{pmatrix} u_1 \\ u_2 \end{pmatrix}$$

$$\begin{pmatrix} y_1 \\ y_2 \end{pmatrix} = \begin{pmatrix} 1 & 0 \\ 0 & 1 \end{pmatrix} \begin{pmatrix} x_1 \\ x_2 \end{pmatrix}$$

Exercise 21.8 Represent the Bode plot of the transfer function

$$G(s) = \frac{100}{(s+1)(s+7)}$$

and compare it with the Bode plot of the closed loop transfer function obtained by a unitary negative feedback. Then compare the time responses of the open and closed loop systems when forced by unitary amplitude sinewaves with frequencies $\omega = 1$, 10, 100 rad s^{-1}.

> **Hint.** For the time responses use the command `lsim` or a SIMULINK scheme.

Exercise 21.9 Given the transfer function

$$G(s) = \frac{1000 \, k}{(s + 0.5 \pm 0.4j)(s + 2 \pm 10j)}$$

compute the gain and phase margins for $k = 1$. Using the commands `margin` and `bode` determine which values of k provide a stable, but not asymptotically stable, closed loop system and a phase margin of 30° respectively. Build up the related SIMULINK schemes and analyze the step responses. Using the `Transport Delay` SIMULINK block (see Section 12.4), verify that the maximum time delay on the open loop plant input, which preserves the stability of the closed loop system, can be obtained as $\Delta t = m_\phi / \omega_t$, where m_ϕ is the phase margin and ω_t is the crossover frequency.

Exercise 21:10 Compute the gain and phase margins of the system with transfer function

$$G(s) = k \frac{(s^2 + 200 \, s + 20\,000)(s + 100)}{(s + 50)^2(s^2 + 20 \, s + 500)}$$

for $k = 1, \ 10, \ 100$. Compare and discuss the results obtained.

Hint. Some help can be obtained from the root locus plot (see next chapter).

Exercise 21.11 Compare the Bode plots of the transfer function

$$G(s) = -\frac{10}{(s + 1)^2(s + 2)}$$

obtained by the commands bode and margin.

Exercise 21.12 Represent the Nyquist plots of the following transfer functions:

$$G_1(s) = \frac{10}{s + 1}, \quad G_2(s) = \frac{10}{s^2 + s + 1}, \quad G_3(s) = \frac{10}{s(s^2 + s + 1)}$$

Hint. For the last system use the command axis([-15 1 -5 5]).

Exercise 21.13 Trace the Nyquist plot of the transfer function

$$G(s) = \frac{k}{(s - 1)(s + 5)}$$

with $k = 1$ and $k = 10$. Which values of k correspond to an asymptotically stable closed loop system?

Exercise 21.14 Given the transfer function

$$G(s) = k \frac{s^2 + 2 \, s + 101}{s^3 + 7 \, s^2 + 7 \, s - 15}$$

determine, using the Nyquist stability theorem and the Nyquist plot, which values of k ensure an asymptotically stable closed loop system.

Exercise 21.15 Given the transfer function

$$G(s) = \frac{k}{s(s + 1)^2}$$

using the command nyquist, compute for which value of k the closed loop system is stable but not asymptotically stable.

Hint. For each value of k, after the command nyquist and using the command find, obtain the first point characterized by a positive imaginary part and then evaluate the real part at that point.

Exercise 21.16 Given the following transfer functions:

$$G_1(s) = \frac{s+1}{s^2 + 0.1\ s + 1}, \qquad G_2(s) = -\frac{10}{s^2 + 0.1\ s + 1}$$

compare the Bode plots and the Nichols plots of the two systems. In one case the Nichols plot is not correct: determine which of the transfer functions causes the error and why.

Exercise 21.17 Write a MATLAB function which, given a SISO system in the transfer function or state-space form, provides a Nichols plot where the gain and phase margins are highlighted.

Exercise 21.18 Given the transfer function

$$G(s) = \frac{150}{s(s+10)(s+20)}$$

using the command `nichols` design a lead controller which provides a 45° phase margin with 8 rad s^{-1} as the crossover frequency.

Exercise 21.19 Given the transfer function

$$G(s) = 10\frac{s^2 + 7s + 1}{s(s+1)^3}$$

using the command `nichols` design a lead controller which provides a 30° phase margin with 10 rad s^{-1} as crossover frequency. Then compute the gain of the Proportional controller which provides the same phase margin.

Using the commands `cloop` and `step`, compare the step responses of the closed loop systems with and without the controllers. Build up a SIMULINK scheme of the closed loop systems and put a `Saturation` block (see page 224) before the plant to simulate the actuator saturation. Using ±1 as saturation bounds, what is the influence of the saturation on the closed loop step responses previously represented?

> **Hint.** From the SIMULINK schemes, store and then also represent the plant input.

Exercise 21.20 From the considerations presented in Section 21.3, write a MATLAB function which, given a SISO system represented by means of a transfer function or in the state-space form, and given the desired phase margin and crossover frequency, returns the lag controller parameters, if this kind of controller can satisfy the desired constraints or otherwise displays an error message.

> **Hint.** A step-by-step refinement procedure is suggested. First suppose that the system is given in the transfer function form and write the function which provides the attenuation and the delay required; a test on these values should verify if a lag controller can be used. In the positive case a function similar to *net_lag* (see Section 21.3) must be run, where the command `global` (see Section 8.2) is used. If the system is given in the state-space form the problem can be solved using the commands `nargin` (see Section 8.2) and `ss2tf` (see Section 19.2); it could be convenient to check the compatibility of the input matrices sizes using the command `abcdchk` (see Section 25.1).

Exercise 21.21 From the previous exercise and from the considerations presented in Section 21.3, write a MATLAB function such that, given a SISO system represented by a transfer function or in the state-space form and given the desired phase margin and crossover frequency, provides the type and the parameters of the required controller (lag, lead or lead–lag) or an error message if a single controller is not enough.

> **Hint.** Solve the previous exercise and from a test on the required compensation at the crossover frequency determine the type of suitable controller and the corresponding function to be run.

Exercise 21.22 Given the transfer function

$$G(s) = \frac{10}{s(s+1)(s+3\pm j)}$$

design a controller such that the 3 dB bandwidth of the closed loop system is 0.52 rad s^{-1}.

> **Hint.** Before the command `nichols` use `ngrid('new')`.

Exercise 21.23 Given the system with transfer function

$$G(s) = \frac{1+0.2s}{s(1+0.01s)^2}$$

design two Proportional controllers such that the closed loop system has the 3 dB bandwidth equal to 30 rad s^{-1} and 570 rad s^{-1} respectively. Analyze the systems using the root locus (see next chapter), the Bode plots and the Nyquist plots. Then, building a suitable SIMULINK scheme, compare the step time responses of the two closed loop systems until 1 s. What happens if the controller gain is increased?

> **Hint.** It could be useful, and sometimes necessary, to use the commands `ngrid('new')`, `nichols`, `find`, `zoom`, `ginput`, `rlocus`, `bode` and `nyquist` with suitable `axis`.

Exercise 21.24 Given the system with transfer function

$$G(s) = \frac{1}{s(1+0.01s)^2}$$

design a Proportional controller such that the rise time of the step response of the closed loop system is $t_r = 0.07$ s. Then build up a SIMULINK scheme and check by means of a time domain simulation whether the constraints have been verified.

> **Hint.** Recall that the rise time t_r is approximately related to the 6 dB bandwidth (in Hertz) B_6 of the closed loop system by the relation $t_r B_6 = 0.45$. Use the command `find`. Moreover, it is convenient to use the command `ngrid('new')` before representing the Nichols plot.

Exercise 21.25 Given a system with the following transfer function:

$$G(s) = \frac{100}{(s+1)^2}$$

design a controller such that the steady-state output error to a unitary ramp is 10%, the phase margin is 40° and the crossover frequency is 10 rad s^{-1}.

Build up a SIMULINK scheme for the closed loop system and determine whether the proposed controller verifies the constraints. Then introduce a finite time delay $\tau = 0.05$, 0.1, 0.2 s before the controller and analyze how the system performance changes.

> **Hint.** The controller will have a pole at the origin, a suitable gain and will consist of more series connected lead–lag compensators; for their design it could be useful to consult the abacus. The finite time delay can be realized by the SIMULINK block `Transport Delay` (see Section 12.4.)

Chapter 22

Root locus

Given a transfer function $G(s)$, the root locus is a representation on the complex plane of the zeros of the function $1+kG(s)$ when the parameter k changes. The root locus can be used to study the effect of the open loop gain variation on the closed loop pole position, and then, indirectly, how k influences the time and frequency responses.

The CONTROL commands described in this chapter are:

pzmap*	zeros and poles positions
rlocfind*	gain selection
rlocus*	root locus plot
sgrid	grid for ζ and ω_n

where the asterisk indicates that the command can also be used for discrete systems.

22.1 Root locus plot

Consider the following transfer function:

$$G(s) = \frac{s+1}{s^2 + 5s + 6}$$

and suppose that we wish to represent its zero and poles on the complex plane. By means of the instruction

$$\text{pzmap([1 1],[1 5 6]);}$$

we obtain a plot on the complex plane of the zero -1 (represented by a circle) and of the poles $(-3, -2)$ (by stars).

In general, the command pzmap can be used in the following forms:

```
[pol,zer]=pzmap(A,B,C,D);
[pol,zer]=pzmap(num,den);
```

where, if no ouput variables are indicated, the zeros and the poles of the system will be represented on the complex plane by circles and stars respectively. Conversely, the poles and the zeros of the system will be assigned to the vectors pol and zer. If the system considered is multi-input, the transmission zeros will be represented on the complex plane, along with the poles (see Section 19.5.5).

As stated above, to show how the closed loop poles 'move' depending on the open loop gain, we can use the root locus plot. Let us consider, for example, the following open loop transfer function:

$$G(s) = \frac{s+4}{(s+1)(s+2)}$$

Writing the instructions

```
num=[1 4];
den=conv([1 1],[1 2]);
rlocus(num,den);
```

we obtain the root locus plot shown in Figure 22.1.

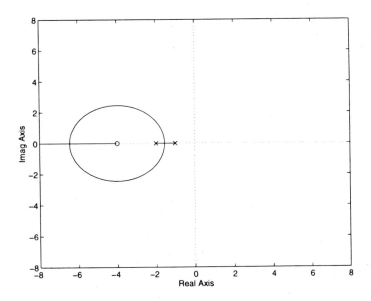

Figure 22.1 Root locus.

In general, the command rlocus can be used in the following forms

```
[rt,k]=rlocus(A,B,C,D);
rt=rlocus(A,B,C,D,k);
[rt,k]=rlocus(num,den);
rt=rlocus(num,den,k);
```

which, given the transfer function $G(s)$, or one of its state-space representations, provide the zeros of $1 + kG(s)$ for different values of k, which can also be specified as the input argument. The output variable k can be omitted and if no output variables are used the root locus plot for positive k values is obtained. Note that in the root locus plot a different color is used for each branch.

As is well known, if an LTI system has two dominant poles, it is possible to relate the step time response parameters to the pole positions in the complex plane. To this end we need to represent on the complex plane the loci at constant damping ratios ζ (which are line segments lying in the real negative half-plane, starting from the origin and with different slopes which depend on ζ), and the ones at constant natural frequencies ω_n (which are circles centered at the origin and with radii ω_n).

Let us consider the transfer function

$$G(s) = \frac{s+4}{s(s+1)(s+2)}$$

which is obtained from the previous $G(s)$ by adding a pole at the origin. Writing the instructions

```
n=[1 4]; d=poly([0 -1 -2]);
rlocus(n,d); sgrid([.4 .8],1);
```

the root locus represented in Figure 22.2 is obtained, where, due to the use of the sgrid command, the curves corresponding to the damping ratios $\zeta = 0.4$, $\zeta = 0.8$ and natural frequency $\omega_n = 1$ are also plotted.

The instruction sgrid(zita,wn) then allows us to represent on the complex plane the loci for the desired constant values of the damping ratios and natural frequencies, detailed in the vectors zita and wn. Other forms of the use of this command are:

```
sgrid;
sgrid('new');
sgrid(zita,wn,'new');
```

where in the first case the loci corresponding to constant values of ζ and ω_n are plotted, in the second case the screen is cleared and the command hold is set on before plotting the curves (it could be useful before plotting the root locus by the command rlocus) and in the third case the last effects are combined with the plot of the locus for some desired values of ζ and ω_n.

There exists a discrete command equivalent to sgrid, i.e. zgrid, which will be described in Section 24.5.

When using the root locus for the controller design, it could be useful to know which value of the open loop gain provides a specific position of the closed loop poles. This value can be obtained graphically by one of the following instructions

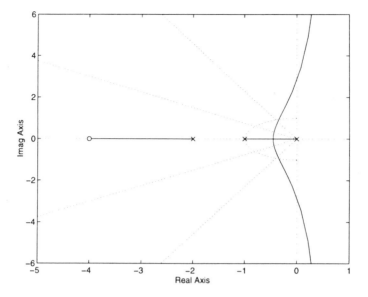

Figure 22.2 Root locus with curves for constant values of ζ and ω_n.

```
[k,pol]=rlocfind(A,B,C,D);
[k,pol]=rlocfind(num,den);
```

These instructions must be written after the root locus has been plotted. The cursor will appear on the screen and the vectors **k** and **pol** are assigned by pushing the mouse button with the cursor on the root locus point where we desire to assign the poles. If the cursor is not positioned exactly on a point of the root locus, the output value will be referred to the nearest point of the locus.

The same result can be obtained analytically by the instructions

```
[k,pol]=rlocfind(A,B,C,D,poldes);
[k,pol]=rlocfind(num,den,poldes);
```

where the vector **poldes** collects the specified pole locations. Also in this case, if the vector **poldes** does not belong to the root locus, the nearest point of the root locus will be considered instead.

22.2 Exercises

Exercise 22.1 Plot the root locus of the following open loop transfer functions:

$$G_1(s) = \frac{1}{s+1}, \qquad G_2(s) = \frac{1}{s(s+2)}, \qquad G_3(s) = \frac{1}{s^2+s+1}$$

$$G_4(s) = \frac{1}{s(s^2+s+1)} \qquad G_5(s) = \frac{1}{s(s^2+2\,s+1.25)}, \qquad G_6(s) = \frac{1}{(s+2)(s^2+s+1)}$$

$$G_7(s) = \frac{1}{(s+1)(s^2+1)}, \qquad G_8(s) = \frac{s+0.5}{(s+1)(s^2+1)}, \qquad G_9(s) = \frac{s+1.5}{(s+1)(s^2+1)}$$

Compare and justify the results obtained. With reference to Figure 22.3, compare

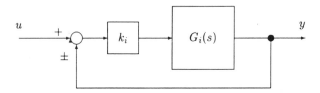

Figure 22.3 Closed loop system with proportional controller.

and justify, by means of the root locus plot, the open and closed loop step responses assuming, respectively

$$\begin{array}{lll} k_1 = 9, & k_2 = 1,\ 10^4 & k_3 = 100 \\ k_4 = 1, & k_5 = 1, & k_6 = 2 \\ k_7 = 1, & k_8 = 1, & k_9 = 1 \end{array}$$

Exercise 22.2 Given the system with transfer function

$$G(s) = k\,\frac{s^2+2\,s+101}{s(s+1)(s+5)}$$

determine the ranges of k for which the closed loop system is asymptotically stable.

Exercise 22.3 For each of the following systems:

$$G_1(s) = \frac{1}{s(s+1)(s+2)}, \qquad G_2(s) = \frac{s+0.5}{s^2(s^2+8\,s+20)}$$

plot on the same graphic window the root locus and the corresponding asymptotes.

Hint. Recall that, if n and m are the number of poles p_i and zeros z_i, respectively, the asymptotes of the root locus constitute a star of $n-m$ lines centered at the real axis point

$$\sigma = \frac{\sum_{i=1}^{n}\mathcal{R}(p_i) - \sum_{i=1}^{m}\mathcal{R}(z_i)}{n-m}$$

where $\mathcal{R}(x)$ is the real part of x, and which form the angles

$$\theta = \frac{(2\alpha + 1)\pi}{n - m}, \qquad \alpha = 0, 1, ..., n - m - 1$$

with the real axis. (Positive open loop gain and unitary negative feedback have been supposed.)

Exercise 22.4 Plot the locus of the closed loop poles of the open loop transfer functions

$$G_1(s) = \frac{s + 1.2}{s^3 + 0.95\, s^2 + s + 0.95}, \qquad G_2(s) = \frac{s + 2}{s^3 + 2\, s^2 + 2\, s}$$

considering both negative and positive feedback. Describe the effects of the positive feedback.

Exercise 22.5 Given the system with the following transfer function:

$$G(s) = \frac{1}{(s + 1)(s + 10)}$$

using the command `rlocus` design a Proportional Integral (PI) controller such that the closed loop system has two dominant poles, a step response characterized by 5% overshoot (which corresponds to $\zeta = 0.7$) and settling time 9.2 s (which corresponds to -0.5 as the real part of the poles). Compare the step responses of the closed loop systems obtained with and without the designed controller. Is it possible to verify the same constraints using a simple Proportional controller? And using an Integral controller?

> **Hint.** Use the commands `series`, `sgrid`, `cloop` and `step`. Try first to verify the constraints with Proportional and with Integral controllers.

Exercise 22.6 Given the discrete systems with transfer functions

$$D_1(z) = k\frac{z + 0.2}{(z - 1)(z - 0.5)}, \qquad D_2(z) = k\frac{z - 0.2}{(z - 1)(z - 0.5)}$$

determine for which values of k the closed loop systems are asymptotically stable.

> **Hint.** Recall that a discrete LTI system is stable if and only if all its poles are inside the unitary circle. Use the commands `rlocus`, `abs` and `find`.

Exercise 22.7 Given the system represented in Figure 22.4 determine the values of k which ensure the asymptotic stability of the system. Fixing a value for k determine, using a SIMULINK scheme, the step response of the system.

> **Hint.** Use the command `c2d`, choosing a suitable sampling period.

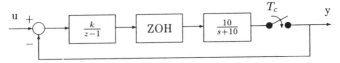

Figure 22.4 Closed loop system.

Exercise 22.8 (Edge theorem) Consider the following polynomial:

$$p(s, \pi) = s^2 + \pi_1 \, s + 1 + \pi_2$$

where $\pi_1 \in [1, 2]$ and $\pi_2 \in [0, 1]$. Defining the 'vertex' polynomials

$$
\begin{aligned}
p_{11} &= s^2 + s + 1, & p_{12} &= s^2 + s + 2 \\
p_{22} &= s^2 + 2\,s + 2, & p_{21} &= s^2 + 2\,s + 1
\end{aligned}
$$

verify that the zeros of $p(s, \pi)$, when π_1 and π_2 change in the previously defined sets, always belong to the region of the complex plane bounded by the root locus plots of $p_{11} + kp_{12}$, $p_{12} + kp_{22}$, $p_{22} + kp_{21}$ and $p_{21} + kp_{11}$.

Hint. Note that if $p(s, \pi)$ is the characteristic polynomial of a class of LTI systems depending on the unknown parameters π, then the stability of all the class is ensured by the stability of the 'vertex' polynomials.

Chapter 23

State feedback

In this chapter we describe the CONTROL commands that are useful for the state feedback controller design of LTI systems. In the following we assume that the reader is familiar with the main concepts of classical control theory, and some modern control theory must be known. More in-depth study can be carried out by considering the Robust Control Toolbox and the μ-Analysis and Synthesis Toolbox.

The instructions which will be described are:

acker*	poles placement for SISO systems
estim	estimator matrices
lqe	Kalman filter (KF) gain
lqed	discrete KF for continuous system model
lqe2	KF with Schur method
lqew	KF design with output noise matrix
lqr	linear quadratic (LQ) feedback design
lqrd	discrete LQ design with continuous objective function
lqry	LQ with output weighting
lqr2	LQ with Schur method
place*	pole placement
reg	estimator and feedback matrices

where the asterisk indicates that the command can also be used for discrete systems.

We now detail the use of these commands, dealing with the following topics: state feedback, observer design, LQ optimal control, the Kalman filter and the separation principle.

23.1   Feedback gain design

As is well known from classical control theory, given a system which is completely controllable, it is possible by means of a linear state feedback to place the closed loop poles anywhere in the complex plane. In particular, considering the controllable system

$$\dot{x} = A\,x + B\,u$$

and a linear state feedback such as

$$u = -K x$$

the closed loop system can be written in the following form:

$$\dot{x} = (A - BK)\,x$$

and the matrix K allows one to place the closed loop poles where required.

As an example, consider the system described by the equations

$$
\begin{aligned}
\dot{x}_1 &= x_2 \\
\dot{x}_2 &= -2\,x_2 + u \\
y &= x_1.
\end{aligned}
\tag{23.1}
$$

The system root locus is obtained by the instruction

rlocus([0 1;0 -2],[0 1]',[1 0],0)

In particular, this plot represents the poles of the closed loop system obtained by considering the feedback of the system output only, i.e. in this case the feedback of the second state variable x_1. From this root locus one can deduce that it is not possible to place the closed loop poles anywhere by changing only the output feedback gain.

The pole placement can, conversely, be obtained by means of a full state feedback. If, for instance, we are looking for the k_1 and k_2 values such that the poles of the closed loop system shown in Figure 23.1 are $-2 \pm j$, we can use the command

K=place([0 1;0 -2],[0 1]',[-2+j -2-j]);

which gives the vector $K = (k_1, k_2) = (5, 2)$ as the result.

In general, by means of the instructions

```
K=place(A,B,poldes);
K=acker(A,B,poldes);
```

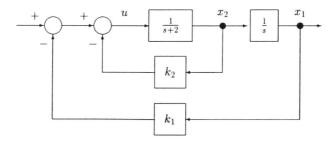

Figure 23.1 Linear state feedback.

we obtain the feedback gains matrix K such that the closed loop poles are those detailed in the vector `poldes`. In particular, the command `place` can also be used for multi-input systems (note that all the inputs are considered as control inputs), whereas `acker` can only be used for single input systems.

The command `place` computes the feedback gains matrix by means of the formula

$$K = (\alpha - a) \left(\Phi^{-1} \right)^T M_{cont}^{-1}$$

where α and a are the vectors containing the coefficients of the characteristic polynomials desired and assigned respectively, Φ is the lower triangular Toeplitz matrix whose first column is $(1 \quad a_1 \quad a_2 \quad \ldots \quad a_{n-1})^T$ and M_{cont} is the controllability matrix of the system considered.

The instruction `acker` uses, instead, the following Ackermann formula:

$$K = q^T \alpha(A)$$

where q is the last row of the inverse of the controllability matrix and $\alpha(A)$ indicates the matrix obtained by substituting the matrix A for the variable s in the characteristic polynomial.

23.2 The observer

The state feedback control design assumes that the whole state of the system is available. In practical situations, this hypothesis is far from being verified. In these cases, in order to estimate the state from a knowledge of the outputs and inputs of the system, an observer must be introduced into the control loop.

A widely used class of observers consists of a copy of the system model 'corrected' by a term which depends on the error between the measured and estimated outputs. This kind of observer can then be represented as

$$\dot{\hat{x}} = A\hat{x} + Bu + L(y - C\hat{x} - Du)$$

which can be rewritten as

$$\dot{\hat{x}} = (A - LC)\hat{x} + (B - LD)u + Ly$$

It is obvious that the observer dynamic equations are quite similar to those of a closed loop system with linear state feedback. Therefore, the design of the matrix L, i.e. the observer design, using the duality principle, can be obtained by the instruction

```
L=place(A',C',poldes)';
```

where one must put the poles which should characterize the observer dynamics in the vector `poldes`.

Let us consider, for example, the system of (23.1). By the instructions

```
A=[0 1;0 -2]; B=[0 1]';
C=[1 0]; D=0;
rank(obsv(A,C))
```

we can conclude that the system is observable since the observability matrix is full rank. The eigenvalues of the dynamic matrix are 0 and -2. Suppose that we are looking for an observer whose poles are $-10 \pm j$. By the instruction

```
L=place(A',C',[-10+j -10-j])';
```

we obtain the vector $L = (18, 25)$ and writing

```
Ao=A-L*C; Bo=[B-L*D L];
Co=eye(2); Do=zeros(2,2);
```

we assign the observer matrices so that the output of the observer consists of the estimated state variables.

In order to verify the observer performance, let us compare the actual and estimated state evolutions in the presence of different initial conditions for the system and the observer. Assuming that the system is forced by a step input the instruction

```
[y,x,t]=step(A,B,C,D,1,0:0.01:2);
```

provides the actual state **x** assuming zero system initial conditions. The observer state evolution with initial condition $(0.2, -0.4)$ and a graphical comparison between the actual and the estimated state can be obtained by the instructions

```
xo0=[0.2 -0.4];
[yo,xo]=lsim(Ao,Bo,Co,Do,[ones(size(t))' y],t,xo0);
plot(t,x,t,xo,'--');
```

which give the result shown in Figure 23.2. A dotted line has been used for the estimated variables.

It is important to stress that the simulation proposed here could also be achieved quite simply by using a SIMULINK scheme.

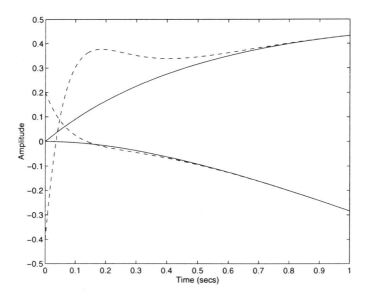

Figure 23.2 Actual (solid lines) and estimated (dashed lines) state variables.

23.3 Linear quadratic optimal control

The linear quadratic (LQ) optimal control technique can be used to design a linear state feedback which minimizes the performance index

$$J = \int_0^\infty (x^T Q x + u^T R u) dt$$

where the symmetric so-called 'weighting matrices' Q and R are non-negative definite and positive definite respectively.

As is well known this problem has a solution if (A, B) is controllable or at least if the noncontrollable part of the system is asympotically stable. Under this hypothesis the optimal feedback gain can be written as

$$K = R^{-1} B^T P$$

where P is the definite positive solution of the algebraic Riccati equation

$$A^T P + P A - P B R^{-1} B^T P + Q = 0$$

Consider the mechanical system represented in Figure 23.3, where $m_1 = m_2 = m$. By choosing as state variables the mass positions and their velocities, a possible state-space model of this system is

$$
\begin{pmatrix} \dot{x}_1 \\ \dot{x}_2 \\ \dot{x}_3 \\ \dot{x}_4 \end{pmatrix} = \begin{pmatrix} 0 & 0 & 1 & 0 \\ 0 & 0 & 0 & 1 \\ -\frac{2k}{m} & \frac{k}{m} & -\frac{b}{m} & 0 \\ \frac{k}{m} & -\frac{2k}{m} & 0 & -\frac{b}{m} \end{pmatrix} \begin{pmatrix} x_1 \\ x_2 \\ x_3 \\ x_4 \end{pmatrix} + \begin{pmatrix} 0 \\ 0 \\ 0 \\ \frac{1}{m} \end{pmatrix}(u+d)
$$

where u is the control input, i.e. the force applied to the second mass, and d is a disturbance.

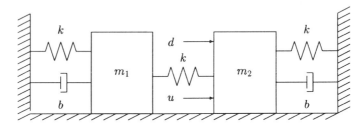

Figure 23.3 Mechanical mass-spring system.

Let us choose the following system parameters: $m = 1$ kg, $k = 1000$ N m^{-1} and $b = 5$ Ns m^{-1}. We want to design an LQ feedback control which minimizes the relative position between the masses; in other words we desire that the two masses remain as close as possibie by applying a force to the second mass only. Therefore, a suitable choice for the performance index is

$$
J = \int_0^\infty (q(x_1 - x_2)^2 + ru^2)dt
$$

We can assign the system matrices by means of the instructions

```
A=[0 0 1 0;0 0 0 1;-2000 1000 -10 0;1000 -2000 0 -10];
B=[0 0 0 1]'; C=[eye(2) zeros(2,2)]; D=zeros(2,1);
```

In our case the matrices Q and R can be assigned by the commands

```
q=1e6; r=1e-2;
Q=[q -q 0 0;-q q 0 0;zeros(2,4)];
```

and, consequently, the LQ design is achieved by the instruction

```
K=lqr(A,B,Q,r);
```

We now compare the performance of the closed and open loop systems in the presence of a disturbance. First we define the vectors that are representative of the time instants and of the disturbance d (for this signal we assume, for a 2 s simulation time duration, a constant signal equal to 1000 for the first half of the simulation and zero after):

```
t=0:0.001:2; d=[1e3*ones(1,1000) zeros(1,1001)];
```

The desired comparison can be achieved by the commands

```
Ac=A-B*K;
[y,x]=lsim(A,B,C,D,d,t);
[yc,xc]=lsim(Ac,B,C,D,d,t);
plot(t,x(:,1),t,x(:,2),'--');
figure(2);plot(t,xc(:,1),t,xc(:,2),'--');
```

which give the waveforms shown in Figures 23.4–23.5.

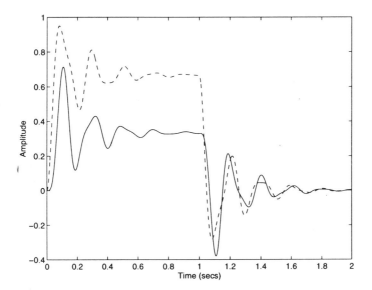

Figure 23.4   Mass positions for the noncontrolled system: x_1 solid line and x_2 dashed line.

It can be seen that the mass positions are changed dramatically due to the feedback since the closed loop control tries to reduce the variation of the relative position of the masses. In particular, the control is quite effective during the transient when the distubance disappears. Better performance could be achieved by reducing r or by increasing q but the cost is an increased control signal.

In general, the matrices K and P, and the vector E, containing the closed loop dynamic matrix eigenvalues, can be obtained by the command

```
[K,P,E]=lqr(A,B,Q,R);
```

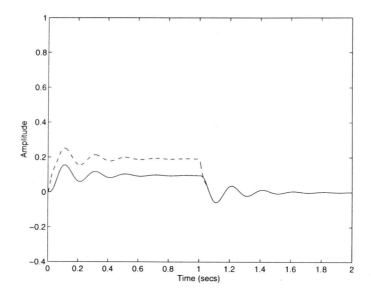

Figure 23.5   Mass positions for the controlled system: x_1 solid line and x_2 dashed line.

Often, in practical situations, the minimization of a weighted state norm added to a weighted input norm is not enough to satisfy the constraints required. In particular, it could be possible that direct interaction between the input and the state is to be considered or that the output rather than the state must be controlled. In other words we could be interested to the minimization of performance indices of the form

$$J = \int_0^\infty (x^T Q x + 2u^T N x + u^T R u)dt$$

or

$$J = \int_0^\infty (y^T Q y + u^T R u)dt$$

which can be achieved with the instructions

```
[K,P,E]=lqr(A,B,Q,R,N);
[K,P,E]=lqry(A,B,C,D,Q,R);
```

respectively.

Often, even though the design has assumed a continuous-time plant, a discrete-time implementation of the controller is required. In this case, after that the weighting matrices have been chosen assuming continuous-time performance, the discrete-time feedback gains matrix K can be obtained by the command

$$[K,S,E]=\texttt{lqrd(A,B,Q,R,N,Tc)};$$

where `Tc` is the sampling rate, `S` is the solution of the discrete time Riccati equation and `E=eig(Ad-Bd*K)`, with `Ad` and `Bd` being the matrices of the discretized system obtained by the instruction `c2d` with `Tc` as the sample time. The matrix `N` can be omitted if there is no input–state coupling.

The same results achieved by the command `lqr` can be obtained by `lqr2`, which uses the so-called Schur method which is slower but numerically more robust for nondiagonalizable systems.

23.4 Kalman filter

Suppose that the dynamic model of an LTI system is affected by noise which can be modeled as Gaussian white noise, i.e. the whole system can be modeled as

$$\begin{aligned} \dot{x} &= Ax + Bu + Gw \\ y &= Cx + Du + v \end{aligned}$$

where w and v are the process and measurement noises respectively. As is well known, the optimal estimator of the state variable in the stochastic environment is the Kalman filter. The structure of this filter is

$$\dot{\hat{x}} = A\hat{x} + Bu + L(y - C\hat{x} - Du)$$

where the gain matrix L, which is in general time varying, becomes constant if a steady-state estimator is considered. (This hypothesis is always assumed by the CONTROL commands for the Kalman filter design.)

This structure is similar to that of a classical observer, but the design method is quite different (for details see any book on control theory). To obtain an optimal state estimation, the gain L can be achieved by writing

$$[L,P,E]=\texttt{lqe(A,G,C,Q,R)};$$

where Q and R are the covariance matrices of the process and measurement noises, which are supposed to have zero mean and to be uncorrelated.

As is well known, to apply the Kalman filter to an LTI system the pair (A, C) must be detectable, i.e. the nonobservable part of the system must be asymptotically stable. Moreover, the matrix Q must be nonnegative definite and the matrix R must be positive definite.

To show an example of the application of the CONTROL commands dealing with the Kalman filter design, let us consider the inverted pendulum on a moving cart whose scheme is shown in Figure 19.20. The linearized model of this system is

$$\begin{pmatrix} \dot{x}_1 \\ \dot{x}_2 \\ \dot{x}_3 \\ \dot{x}_4 \end{pmatrix} = \begin{pmatrix} 0 & 1 & 0 & 0 \\ \frac{M+m}{Ml}g & 0 & 0 & 0 \\ 0 & 0 & 0 & 1 \\ -\frac{mg}{M} & 0 & 0 & 0 \end{pmatrix} \begin{pmatrix} x_1 \\ x_2 \\ x_3 \\ x_4 \end{pmatrix} + \begin{pmatrix} 0 \\ -\frac{1}{Ml} \\ 0 \\ \frac{1}{M} \end{pmatrix} u \qquad (23.2)$$

where x_1 and x_2 are the position and angular speed of the pendulum, whereas x_3 and x_4 are the position and linear speed of the mass M respectively.

We now show a possible way of obtaining the model of (23.2). First, one must consider that the coordinates of the center of mass of the pendulum and the mass m are $x_G = x_3 + l \sin x_1$ and $y_G = l \cos x_1$ and that the potential and kinetic energies of the whole system can be written as

$$ T = \frac{M}{2}\dot{x}_3^2 + \frac{m}{2}\left(\dot{x}_3^2 + 2\,\dot{x}_1\dot{x}_3 l\cos x_1 + l^2\dot{x}_1^2\right) $$
$$ V = mgl\cos x_1 $$

respectively. Defining x_1 and x_3 as generalized coordinates, and the Lagrangian function $L = T - V$, the application of the Lagrange equations

$$ \frac{d}{dt}\left(\frac{\partial L}{\partial \dot{x}_1}\right) - \frac{\partial L}{\partial x_1} = 0, \qquad \frac{d}{dt}\left(\frac{\partial L}{\partial \dot{x}_3}\right) - \frac{\partial L}{\partial x_3} = u $$

yields the following equations:

$$ ml\,\cos x_1\,\ddot{x}_3 + ml^2\,\ddot{x}_1 - mgl\,\sin x_1 = 0 $$
$$ (M + m)\ddot{x}_3 + ml\,\cos x_1\,\ddot{x}_1 - ml\,\dot{x}_1^2\,\sin x_1 = u $$

These equations, linearized in the neighborhood of the origin, provide

$$ m\,\ddot{x}_3 + ml\,\ddot{x}_1 - mg\,x_1 = 0 $$
$$ (M + m)\,\ddot{x}_3 + ml\,\ddot{x}_1 = u $$

which, assuming $x_2 = \dot{x}_1$ and $x_4 = \dot{x}_3$ and after some algebra, result in the model (23.2).

For the sake of simplicity, starting from (23.2), assume that the position and speed of the mass M are available and then let us consider the first two differential equations which do not depend on the third and fourth state variables. In other words, we consider the system

$$ \dot{x}_1 = x_2 $$
$$ \dot{x}_2 = \alpha x_1 - \beta u + w $$
$$ y = x_1 + v $$

where $\alpha = (M + m)g/Ml$ and $\beta = 1/Ml$ are constant, w is a noise which could represent, for instance, the air currents, and v is the measurement noise on the pendulum position. Assuming $M = 2$ kg, $m = 0.1$ kg and $l = 0.5$ m, it follows that $\alpha = 20.6$ and $\beta = 1$. We are interested in estimating, by means of a Kalman filter, the pendulum position and speed. To this end, defining the matrices

```
A=[0 1;20.6 0]; B=[0;-1];
C=[1 0]; D=0;
G=[0 1]';
```

by the instruction

$$L=lqe(A,G,C,10,1);$$

one computes the gains matrix of the filter, assuming that the process and measurement noise covariances are 10 and 1 respectively. We can now define the filter matrices as

$$Af=A-L*C; \ Bf=[B-L*D \ L];$$
$$Cf=C; \ Df=zeros(2,2);$$

We leave it to the careful reader to check the Kalman filter time domain performance, which can be obtained either in MATLAB, similarly to that presented in Section 23.2, or by building a suitable SIMULINK scheme.

The filter matrices can also be calculated using the command estim. In fact, the command lqe allows the design of the Kalman filter gains but does not build the whole dynamic model of the filter.

In particular, the instruction

$$[Af,Bf,Cf,Df]=estim(A,B,C,D,L);$$

provides the matrices of the following filter:

$$\dot{\hat{x}} = (A - LC)\hat{x} + Ly = A_f\hat{x} + B_fy$$
$$\left(\begin{array}{c} \hat{y} \\ \hat{x} \end{array}\right) = \left(\begin{array}{c} C \\ I \end{array}\right)\hat{x} = C_f\hat{x} + D_fy$$

whereas, using estim as follows:

$$[Af,Bf,Cf,Df]=estim(A,B,C,D,L,meas,in);$$

it is possible to choose as input arguments the vectors meas and in where the indices of the outputs (measurements) and the noncontrol inputs of the system, which both constitute the inputs for the Kalman filter, can be defined respectively.

If the process and measurement noises are correlated, N being the correlation matrix, one should use the command lqe as follows:

$$[L,P,E]=lqe(A,G,C,Q,R,N);$$

The same results obtained by the command lqe can also be achieved by using the command lqe2, which adopts the so-called Schur method, is slower than lqe and is more robust if a nondiagonalizable system is considered.

The instruction

$$[L,P,E]=lqew(A,G,C,J,Q,R);$$

computes the Kalman filter gains for the system

$$\dot{x} = Ax + Bu + Gw$$
$$y = Cx + Du + Jw + v$$

assuming that the process and measurement noises are uncorrelated.

The command lqed allows the design of a discrete Kalman filter, which is suitable for practical applications, starting from a system model represented in the continuous-time domain (see also the command lqrd in the previous section).

23.5   Separation principle

As is well known the separation principle allows one to design the linear state feedback and the state estimator independently. In other words, one can design a linear state feedback (by means of pole placement or LQ techniques, for instance) assuming that the state variables are available. Then the same feedback gains can still be used when a state estimator (a classical observer or a Kalman filter) is inserted into the control loop.

The matrices of the whole system, i.e. the LTI system with the estimator and a linear feedback of the estimated state, can be obtained by the instructions

        [Ac,Bc,Cc,Dc]=reg(A,B,C,D,K,L);
        [Ac,Bc,Cc,Dc]=reg(A,B,C,D,K,L,meas,in,cont);

where the second instruction must be used if we wish to specify which outputs, noncontrol inputs and control inputs are involved in the design of the whole system.

23.6   Exercises

Exercise 23.1 Given a double integrator system, design a linear state feedback which places the closed loop poles in $-1 \pm j$.

Exercise 23.2 Consider the system described by the equations

$$\begin{pmatrix} \dot{x}_1 \\ \dot{x}_2 \end{pmatrix} = \begin{pmatrix} -1 & 1 \\ 0 & 2 \end{pmatrix} \begin{pmatrix} x_1 \\ x_2 \end{pmatrix} + \begin{pmatrix} 1 \\ 0 \end{pmatrix} u$$

Using the place command, determine the state feedback gains such that the closed loop poles are placed in $(-3, 2)$ and successively in $(-1, -2)$. Why does the command give an error message in the second case only?

Exercise 23.3 Consider the system with transfer function

$$G(s) = \frac{1}{s(s+1)(s+2)}$$

Design a state feedback such that the closed loop poles are $(-2 \pm 3j, -10)$. Then compare the step time response of the closed and open loop systems.

Exercise 23.4 Consider the state-space representation

$$\begin{pmatrix} \dot{x}_1 \\ \dot{x}_2 \end{pmatrix} = \begin{pmatrix} -2 & 1 \\ 0 & -4 \end{pmatrix} \begin{pmatrix} x_1 \\ x_2 \end{pmatrix} + \begin{pmatrix} 1 \\ 1 \end{pmatrix} u$$

$$y = \begin{pmatrix} 1 & -1 \end{pmatrix} \begin{pmatrix} x_1 \\ x_2 \end{pmatrix}$$

Design a controller such that the closed loop system has a zero steady state output error w.r.t. a step reference input and its poles are $(-1 \pm j, -10)$.

Hint. Design the linear state feedback augmenting the state, i.e. considering as a further state variable the input of the integrator which must be inserted in order to satisfy the steady-state constraint.

Exercise 23.5 Given the system whose transfer function is

$$G(s) = \frac{s + 7}{(s + 2)(s + 5 \pm j)}$$

design a state feedback such that the closed loop poles are $(-1 \pm j, -16)$. Then, supposing that the state is not available, design an observer whose dynamics are 'sufficiently' faster than those of the system. Compare the closed loop system step responses with and without the observer, assuming that the initial state of the observer is different from that of the system.

Exercise 23.6 Write a MATLAB function which, given as input the matrices of a SISO system whose output coincides with a state variable, provides the matrices of a reduced order observer, i.e. an observer which estimates only the state variables that are different from the system output.

Hint. This exercise is not trivial and its solution needs some analytical results. In particular, one must consider that, partitioning the system in the following form:

$$\begin{pmatrix} \dot{x}_1 \\ \dot{x}_2 \end{pmatrix} = \begin{pmatrix} A_{11} & A_{12} \\ A_{21} & A_{22} \end{pmatrix} \begin{pmatrix} x_1 \\ x_2 \end{pmatrix} + \begin{pmatrix} B_1 \\ B_2 \end{pmatrix} u,$$

$$y = (I \quad 0) \begin{pmatrix} x_1 \\ x_2 \end{pmatrix}$$

the reduced order observer dynamics can be written as

$$\dot{\hat{x}}_2 = A_{22}\hat{x}_2 + A_{21}x_1 + B_2 u + L(\dot{x}_1 - A_{11}x_1 - A_{12}\hat{x}_2 - B_1 u)$$

Exercise 23.7 Given the system

$$\dot{x} = 2\,x + u, \qquad y = x$$

design a state feedback which minimizes the performance index

$$J = \int_0^\infty (y^2 + ru^2)\,dt$$

choosing different values for the constant r. Compare the step responses of the different closed loop systems.

Assuming the same values for r, what happens if 2 is replaced by -2 in the differential equation?

Exercise 23.8 Given the system

$$\dot{x} = \alpha \, x + \beta u$$

with α and β to be assigned at will, design a state feedback which minimizes the performance index

$$J = \int_0^\infty (q \, x^2 + r \, u^2) dt$$

with different values of the positive constants q and r, but such that the same ratio q/r is preserved. How does the feedback gain change when q and r vary in such a way? How does the closed loop pole move when $q \to 0$ (or $r \to \infty$)? Which values approach the closed loop pole when $q \to \infty$ (or $r \to 0$) in the case where α is negative or positive respectively?

Exercise 23.9 Consider the dynamic system described by the differential equations

$$\begin{pmatrix} \dot{x}_1 \\ \dot{x}_2 \end{pmatrix} = \begin{pmatrix} 0 & 1 \\ 0 & -1 \end{pmatrix} \begin{pmatrix} x_1 \\ x_2 \end{pmatrix} + \begin{pmatrix} 0 \\ 1 \end{pmatrix} u$$

where x_1 can be considered as a position, x_2 as a speed and the control variable u as a force. Design the two state feedback gains vectors which minimize the performance index

$$J = \int_0^\infty (x^T Q_i x + u^T u) dt$$

with $Q_1 = diag(10, 0)$ and $Q_2 = diag(0, 10)$ respectively. Compare the step responses of the two closed loop systems.

Exercise 23.10 With reference to the dynamic system described in the previous exercise, design three optimal controls such that the state variable x_1 tracks the following reference position:

$$\tilde{x}(t) = 1 - e^{-2t}$$

considering the following as the performance index:

$$J = \int_0^\infty (q \, (x_1 - \tilde{x}_1)^2 + r \, u^2) dt$$

and assuming that $r = 0.01$ and $q = 10^2$, 10^4, 10^6.

> **Hint.** To solve this exercise theoretical knowledge of the solution of the tracking problem with optimal control is needed (a suitable feedforward compensator must be added to the feedback action). The speed reference signal is obviously the time derivative of the position reference signal.

Exercise 23.11 Consider the dynamic system described by the equations

$$\begin{pmatrix} \dot{x}_1 \\ \dot{x}_2 \end{pmatrix} = \begin{pmatrix} -1 & 0 \\ 0 & -2 \end{pmatrix} \begin{pmatrix} x_1 \\ x_2 \end{pmatrix} + \begin{pmatrix} 1 + \varepsilon \\ 1 \end{pmatrix} u.$$

Assuming $\varepsilon = 0$, design a state feedback which minimizes the performance index

$$J = \int_0^\infty \left((x_1 - x_2)^2 + ru^2 \right) dt$$

with $r = 1$, 10^{-2}, 10^{-4}. Verify that, using the feedback gains designed in nominal conditions, the closed loop systems become unstable for $\varepsilon = -8$, -0.9, -0.2 respectively.

Chapter 24

Discrete functions

In this chapter we briefly describe the CONTROL commands that are useful for the analysis and design of discrete-time systems. It is important to stress that, as has been already highlighted, some of the previously considered commands can be used for both continuous- and discrete-time systems. In particular, following an alphabetic order, these instructions are:

```
acker      append    augstate  cloop  feedback
margin     parallel  place     pzmap  rlocfind
rlocus     series    ss2ss     ss2tf  ss2zp
ssdelete   ssselect  tf2ss     tf2zp  zp2ss
zp2tf
```

Besides these commands, some specific instructions for discrete-time systems exist; and these will be considered in this chapter. Their descriptions will be very short when their use is similar to that of the related continous-time command.

24.1 Discrete to continuous conversion

In Section 19.3 we presented the command which allows one to obtain the conversion of a system representation from the continuous-time domain to the discrete-time one. The inverse procedure can be obtained by the commands d2c and d2cm:

```
d2c     discrete to continuous conversion
d2cm    discrete to continuous conversion with method
```

In particular, the instruction

$$[\texttt{Ac,Bc}]=\texttt{d2c(Ad,Bd,Tc)};$$

is the inverse of the instruction c2d.

Dealing with the d2cm command, it is important to stress that this command, as with the related continuous one c2dm, can be used in one of the following forms:

$$[\text{Ac},\text{Bc},\text{Cc},\text{Dc}]=\text{d2cm}(\text{Ad},\text{Bd},\text{Cd},\text{Dd},\text{T},'method');$$
$$[\text{numc},\text{denc}]=\text{d2cm}(\text{numd},\text{dend},\text{T},'method');$$

depending on the kind of system representation available. The conversion method can be one of those already described in Section 19.3 (zoh, tustin, prewarp, matched), except for the option foh, which can only be used for continuous-time systems.

24.2   Properties

The following instructions:

dcovar	covariance matrices of the response to white noise
ddamp	damping ratios and natural frequencies
ddcgain	dc discrete gain
dgram	controllability and observability gramians
dsort	eigenvalues sort by magnitude

will be described in this section. The corresponding continuous commands are detailed in Section 19.5.

- dcovar Discrete equivalent of the command covar, allows one to compute the co-variance matrices of the output and state of a discrete-time system in the presence of white Gaussian noise.
- ddamp This command can be used in one of the following forms:

$$\text{mod}=\text{ddamp}(\text{A});$$
$$[\text{mod},\text{wn},\text{zita}]=\text{ddamp}(\text{A},\text{Tc});$$

where the vector mod contains the modulus of the discrete eigenvalues λ_d, the vector wn contains the continuous equivalent natural frequencies, i.e.

$$\omega_n = \frac{|\log \lambda_d|}{T_c} \tag{24.1}$$

and the vector zita contains the continuous equivalent damping ratios, i.e.

$$\zeta = -\cos\left(\angle \log \lambda_d\right) \tag{24.2}$$

It is important to stress that, for the input argument A, the same considerations as are given in Section 19.5.1 hold, i.e. if A is a square matrix it is considered to be the dynamic matrix of a state-space representation, if A is a row vector it is considered to be a vector containing the coefficients of the characteristic polynomial and if A is a column vector it is considered to be a vector containing the roots of the characteristic polynomial.

In order to justify (24.1) and (24.2) we give some theoretical considerations here. These relations can be achieved with two different approaches. The first one starts from the following well-known continuous–discrete complex variables equivalence

$$z = e^{sT_c} \tag{24.3}$$

Given a complex conjugate poles pair $\alpha \pm j\beta$, the natural frequency and the damping ratio are defined as

$$\omega_n = \sqrt{\alpha^2 + \beta^2}, \quad \zeta = -\frac{\alpha}{\sqrt{\alpha^2 + \beta^2}}$$

i.e. as the modulus and the opposite of the cosine of the phase of the complex number $\alpha + j\beta$, respectively. From these considerations and from (24.3) it follows that

$$\omega_n = |s| = \frac{|\log \lambda_d|}{T}$$

which gives (24.1), and

$$\zeta = -\cos\left(\underline{/s}\right) = -\cos\left(\underline{/\log \lambda_d}\right)$$

which gives (24.2).

The same results can also be obtained using the analytical expressions for the impulse responses of continuous- and discrete-time LTI systems. In particular, the pseudo-periodic response related to each complex conjugate poles pair, in the continuous-time domain, can be written as

$$e^{\alpha t} \cos(\beta t) = e^{-\zeta \omega_n t} \cos\left(\omega_n \sqrt{1 - \zeta^2}\, t\right)$$

and in the discrete-time domain:

$$|\lambda_d|^n \cos\left(\underline{/\lambda_d}\, n\right)$$

where we have considered the modulus and the phase of the complex discrete eigenvalue λ_d. From the last two equations, imposing that the continuous-time response in T_c equals the first sample ($n = 1$) of the discrete-time response and after some algebra, we obtain

$$-\zeta \omega_n = \frac{\log |\lambda_d|}{T_c} \tag{24.4}$$

$$\omega_n \sqrt{1 - \zeta^2} = \frac{\underline{/\lambda_d}}{T_c} \tag{24.5}$$

Substituting $\zeta \omega_n$ from (24.4) in (24.5), computing ω_n and substituting this value again in (24.4) to obtain ζ, obtains the relations (24.1) and (24.2).

- ddcgain By the instructions

$$gn=ddcgain(A,B,C,D);$$
$$gn=ddcgain(num,den);$$

we obtain the dc gain corresponding to the discrete transfer matrix, i.e. its value for $z = 1$.
- **dgram** The equivalent discrete version of the command **gram**, allows one to compute the controllability and observability gramians for discrete-time systems.
- dsort By the instruction

$$[pord,indp]=dsort(p);$$

we obtain the vector pord, which contains the elements of the vector p sorted according to decreasing values of their modulus and the vector indp, which contains the indices of the original positions of the elements of p.

24.3   Model reduction

The following instructions:

dbalreal	balanced representation
dmodred	model reduction

comprise the discrete version of the continuous commands described in Section 19.6. In particular dbalreal, the discrete equivalent of balreal, allows one to obtain a balanced representation, i.e. a representation such that the discrete controllability and observability gramians coincide whereas dmodred, the discrete equivalent of the command modred, allows one to reduce the number of state variables of a given discrete system.

24.4   Time domain response

The instructions

dinitial	unforced response given the initial state
dimpulse	impulse response
dlsim	time response to given input
dstep	step response

which refer to the equivalent continuous commands described in Chapter 20, allow one to obtain the discrete-time response of an LTI system. Their use is similar to the related continuous commands, but for the discrete commands it is not possible to specify time as the output variable. For instance, the different ways of using the `dstep` command are:

```
[out,state]=dstep(A,B,C,D);
[out,state]=dstep(A,B,C,D,ui);
[out,state]=dstep(A,B,C,D,ui,N);
[out,state]=dstep(num,den);
[out,state]=dstep(num,den,N);
```

where N is the number of samples to be considered for the time response. The different forms of the command `dstep` can also be applied to the commands `dinitial`, `dimpulse` and `dlsim`.

In general, the time domain response of an LTI system such as

$$\begin{aligned} x(n+1) &= Ax(n) + Bu(n) \\ y(n) &= Cx(n) + Du(n) \end{aligned}$$

can be written as

$$y(n) = CA^n x_0 + \sum_{i=0}^{n-1}(CA^{n-i-1}B + D)u(i) \qquad (24.6)$$

where the first term is the unforced response and the second term is the forced response.

As for continuous-time systems, also for LTI discrete systems, it is possible to use the CONTROL commands to obtain the analytic expression of the time response from knowledge of the discrete transfer function of the system and from the $\mathcal{Z}$-transform of the input signal. To this end consider the continuous-time system with transfer function

$$G(s) = \frac{s^2 + 3.2\,s + 12}{s^3 + 1.2\,s^2 + 9.2\,s + 9}$$

A sampled data version of this system, with sampling time $T_c = 1$ s, can be obtained by the instructions

```
n=[1 3.2 12]; d=[1 1.2 9.2 9];
[A,B,C,D]=tf2ss(n,d);
Tc=1; [Ad,Bd]=c2d(A,B,Tc);
[nd,dd]=ss2tf(Ad,Bd,C,D);
```

which give the discrete transfer function

$$G(z) = \frac{1.26\,z^2 + 1.47\,z + 0.31}{z^3 + 1.42\,z^2 + 0.16\,z - 0.30}$$

Given the $\mathcal{Z}$-transform of the input signal $u(n)$ which is defined as

$$U(z) = \mathcal{Z}(u(n)) = \sum_{n=0}^{\infty} u(n)z^{-n}$$

we want to obtain the following analytic expression for the output:

$$y(n) = \mathcal{Z}^{-1}(Y(z)) = \mathcal{Z}^{-1}(G(z)U(z))$$

Let us consider as input a step signal whose $\mathcal{Z}$-transform is $z/(z-1)$. Since all the $\mathcal{Z}$-transforms of the elementary discrete functions have the variable z as a numerator factor, we use the following instruction to obtain the partial fraction decomposition:

```
[res,poles,k]=residue(nd,conv(dd,[1 -1]));
```

which allows one to write the partial fraction decomposition of $Y(z)/z$ so that we can then write $Y(z) = z \times fractions$. In other words, after obtaining the residues by the previous instruction, we can write the $\mathcal{Z}$-transform of the system response as

$$Y(z) = \frac{(-0.17 \pm 0.01j)\, z}{z + 0.90 \mp 0.13j} - \frac{z}{z + 0.37} + \frac{1.33\, z}{z - 1}$$

Using the inverse $\mathcal{Z}$-transforms of the elementary discrete functions we obtain

$$y(n) = (0.33\,(0.91)^n\,\cos(0.14\,n + 3.12) - (-0.37)^n + 1.33)\,1(n)$$

where $1(n)$ indicates the unitary discrete step function.

Given a discrete complex function $\Phi(z)$, we recall that, denoting the residues by R_i and the denominator roots by λ_i, assuming that m denominator roots are real and $n-m$ are complex conjugate and that no multiple root exists, the inverse transform of

$$\Phi(z) = z \sum_{i=1}^{n} \frac{R_i}{z - \lambda_i}$$

can be written as

$$\Phi(n) = \sum_{i=1}^{m} R_i \lambda_i^n + \sum_{j=1}^{\frac{n-m}{2}} 2|R_j||\lambda_j|^n \cos(\angle\lambda_j\, n + \angle R_j)$$

where we incorporate the complex conjugate eigenvalues $\lambda_j = \alpha_j \pm j\omega_j$.

The chosen sampling time $(T_c = 1s)$ is not sufficient to 'exactly' reconstruct the response of the continuous original system. In fact, the continuous- and the related discrete-time step responses can be compared by writing

```
[yc,xc,tc]=step(n,d,0:.1:20);
[yd,xd]=dlsim(Ad,Bd,C,D,ones(21,1));
plot(tc,yc,0:20,yd,'*');
```

which provide the plot shown in Figure 24.1, where the stars indicate the discrete samples. A better reconstruction can be achieved by reducing the sampling time.

Note that in the last set of instructions one could also use the command `dstep` instead of `dlsim` without specifying the input signal as the instruction argument.

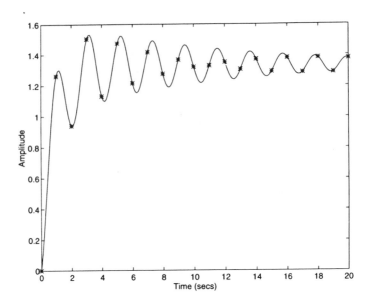

Figure 24.1 Step responses of the continuous (solid line) and the sampled data (star points) systems.

24.5 Frequency domain response

The following CONTROL instructions:

dbode	discrete Bode plots
dnichols	discrete Nichols plots
dnyquist	discrete Nyquist plots
dsigma	discrete singular values
zgrid	grid for ζ and ω_n

are useful for the frequency response analysis of discrete LTI systems, and refer to the continuous domain commands described in Chapter 21.

Recall that the sampling operation in the time domain results, in the frequency domain, in a repetition of the frequency response with period given by the inverse of the sampling time period. In particular, the following correspondence can be written:

$$f_c(t) = \sum_{n=1}^{\infty} f(t)\delta(t - nT_c) \longrightarrow F_c(j\omega) = \frac{1}{T_c} \sum_{n=-\infty}^{+\infty} F\left(j\omega + jn\frac{2\pi}{T_c}\right)$$

The aliasing, i.e. the overlapping of the frequency response replicae, is the well-known problem of this time–frequency domain relation. Usually, the spectrum of a continuous function can be assumed to be band limited, i.e. there exists a frequency f_{lim} (the so-called Nyquist frequency) such that the spectrum can be assumed to be zero outside the frequency range $(-f_{lim}, +f_{lim})$. In this case, in order to avoid the aliasing and considering the spectrum symmetry with respect to the zero frequency, a sampling frequency larger than $f_c = 2\ f_{lim}$ must be used.

Let us now detail the syntax of the commands listed at the beginning of this section:

- dbode Discrete equivalent of the command bode (see Section 21.1), allows one to obtain, by the following instructions:

```
[mod,phase,puls]=dbode(A,B,C,D,Tc);
[mod,phase,puls]=dbode(A,B,C,D,Tc,ui);
[mod,phase]=dbode(A,B,C,D,Tc,ui,w);
[mod,phase,puls]=dbode(num,den,Tc);
[mod,phase]=dbode(num,den,Tc,w);
```

  the frequency response of the discrete transfer function

$$G(z) = C(zI - A)^{-1}B + D$$

  The considerations presented for the command bode are still appropriate for the command dbode, but, when using the latter, the sampling period is required as the input argument. In particular, the vectors mod and phase will contain, respectively, the following variables:

$$mod(\omega) = \left| G(e^{j\omega T_c}) \right|$$
$$phase(\omega) = \angle G(e^{j\omega T_c})$$

  whereas the vector puls will contain the frequencies in radians per second where the frequency response is computed. The values of the vector puls are in the range $(0, \pi/T_c)$. Assuming that π/T_c is the Nyquist frequency, i.e. a sufficently small sampling period is used, this interval is enough to completely represent the frequency response of the discrete system.
  Let us consider the transfer function

$$G(s) = \frac{s + 0.2}{s^3 + 0.2\ s^2 + 4\ s + 0.4}$$

  We want compare the magnitude Bode plot of this system with that of the corresponding sampled data system. Assume 1 s as the sampling period and a frequency range from 0.1 rad s^{-1} to 10 rad s^{-1}. By the following instructions:

```
n=[1 0.2]; d=[1 0.2 4 0.4];
w=logspace(-1,1); [m,f]=bode(n,d,w);
[A,B,C,D]=tf2ss(n,d);
Tc=1; [Ad,Bd]=c2d(A,B,Tc);
[md,fd]=dbode(Ad,Bd,C,D,Tc,1,w);
plot(w,m,w,md,'.');
```

we obtain Figure 24.2, where the frequency responses of the two systems are plotted. Note that differences between the plots are apparent only beyond π rad, i.e. beyond the Nyquist frequency; the replicae of the discrete frequency response are also apparent.

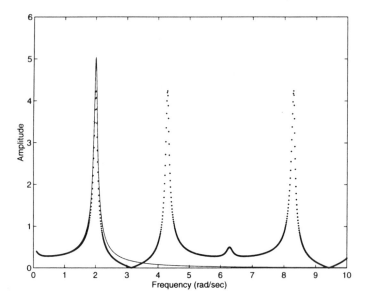

Figure 24.2   Bode magnitude plots of a continuous system (solid line) and of the associated sampled data system (dotted line).

- dnichols Discrete equivalent of the command nichols (see Section 21.3), allows one to obtain the Nichols plots of the discrete transfer function $G(z)$. This command can be used in the following forms:

```
[mod,phase,puls]=dnichols(A,B,C,D,Tc);
[mod,phase,puls]=dnichols(A,B,C,D,Tc,iu);
[mod,phase]=dnichols(A,B,C,D,Tc,iu,w);
[mod,phase,puls]=dnichols(num,den,Tc);
[mod,phase]=dnichols(num,den,Tc,w);
```

where the vector puls will contain frequency values in radians per second between 0 and π/T_c.

- dnyquist Discrete equivalent of the command nyquist (see Section 21.2), allows one to obtain the Nyquist plot of the discrete transfer function $G(z)$. This command can be used in the following forms:

```
[re,im,puls]=dnyquist(A,B,C,D,Tc);
[re,im,puls]=dnyquist(A,B,C,D,Tc,ui);
[re,im]=dnyquist(A,B,C,D,Tc,ui,w);
[re,im,puls]=dnyquist(num,den,Tc);
[re,im]=dnyquist(num,den,Tc,w);
```

where the vector `puls` will contain frequency values in radians per second between 0 and π/T_c. The Nyquist plots so obtained can be directly compared with those of the continuous systems.

- `dsigma` Discrete equivalent of the command `sigma` (see Section 21.1), allows one to compute the singular values of the complex matrix

$$G(\omega) = C(e^{j\omega T_c}I - A)^{-1}B + D \qquad (24.7)$$

when ω changes. The forms of this command are:

```
[singval,freq]=dsigma(A,B,C,D,Tc);
[singval,freq]=dsigma(A,B,C,D,Tc,'inv');
[singval,freq]=dsigma(A,B,C,D,Tc,freq);
[singval,freq]=dsigma(A,B,C,D,Tc,freq,'inv');
```

where, besides the arguments which must also be used for the continuous command `sigma`, the sampling period `Tc` must also be specified. The vector `freq` contains values which range from 0 to the Nyquist frequency. Moreover, when used, the parameter `'inv'` allows one to obtain the singular values of the inverse of $G(\omega)$.

- `zgrid` Discrete equivalent of the command `sgrid` (see Chapter 22.1), allows one to plot curves with constant natural frequencies and damping ratios on the complex plane. The analytic expressions for these curves can be obtained by manipulating the following relations:

$$-\zeta\omega_n = \frac{\log|\lambda_d|}{T_c}$$

$$\omega_n\sqrt{1-\zeta^2} = \frac{\angle\lambda_d}{T_c}$$

which, after some algebra, can be rewritten as

$$-\frac{\zeta}{\sqrt{1-\zeta^2}}\angle\lambda_d = \log|\lambda_d| \qquad (24.8)$$

$$\omega_n = \frac{1}{T_c}\sqrt{(\angle\lambda_d)^2 + \log^2|\lambda_d|} \qquad (24.9)$$

From (24.8) we can deduce that all curves characterized by a constant value of ζ cross the point $(1+j0)$ of the complex plane; these curves are called *cardioids*[1] (see Exercise 7.10), due to their particular shape. Moreover, the locus at constant ω_n is obtained by the points of the complex plane whose modulus and phase verify (24.9).

---

[1]'Cordum' in Latin means 'heart'.

Plotting on the complex plane the discrete root locus by the command `rlocus` or the poles and zeros by the command `pzmap` and using the command `zgrid`, the curves at constant values of ζ and ω_n are represented. For instance, the root locus of the following transfer function:

$$G(z) = \frac{z + 0.1}{z^3 - 0.1\,z^2 - 0.2\,z + 0.05}$$

is shown in Figure 24.3. The loci with constant damping ratios are obtained by varying ζ from 0 to 1 with constant step 0.1, whereas the ones with constant natural frequency values are obtained by varying the product $\omega_n T_c$, obtained from (24.9), from 0 to π with constant step $\pi/10$.

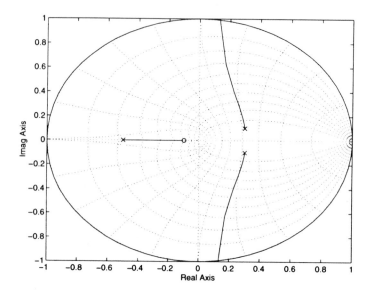

Figure 24.3 Loci for constant ζ and ω_n and root locus for $G(z)$.

Analogous to the command `sgrid`, the following instructions can be used:

```
zgrid(zita,wn);
zrid('new');
zgrid(zita,wn,'new');
```

where in the first case the loci at constants ζ and ω_n are plotted, in the second case the screen is cleaned and the command `hold` is set to on before plotting the curves (this option could be useful before plotting the root locus by the command `rlocus`) and in the third case the last effects are combined with the plot of the locus corresponding to some desired values of ζ and ω_n.

It could be useful, when using the `zgrid` command for the controller design, to recall that the 'continuous' complex plane areas corresponding to a maximum value of the rise time, i.e. half-planes at the left-hand side of the vertical line whose points have $-\zeta\omega_n$ as the real value, become circles in the 'discrete' complex plane, as emerges from (24.4).

24.6 State feedback design

The CONTROL instructions that are useful for the state feedback and estimator design in the discrete-time domain are:

destim	discrete estimator matrices
dlqe	discrete Kalman filter
dlqew	general discrete Kalman filter
dlqr	discrete LQ feedback
dlqry	discrete LQ feedback with output weighting
dreg	matrices with LQ feedback and estimator

The related continuous versions of these commands are described in Chapter 23.

- `destim` Discrete equivalent of the command `estim`, allows one to obtain the matrices of a state-space form of a discrete Kalman filter.
- `dlqe` Discrete equivalent of the command `lqe`, from a state-space representation of a given system, allows one to obtain, by means of the instructions

$$[L,M,P,E]=dlqe(A,G,C,Q,R);$$
$$[L,M,P,E]=dlqe(A,G,C,Q,R,N);$$

the Kalman filter gain matrix L, the covariance matrix M of the estimation error computed before the measurement updating, i.e. the solution of the discrete Riccati equation

$$0 = S - A^T S A + A^T S B (R + B^T S B)^{-1} B S^T A - Q$$

the covariance matrix P of the estimation error after the measurement updating and the vector E, which contains the estimator eigenvalues. The last three input arguments can also be omitted (see Section 23.4).

- `dlqew` Discrete equivalent of the command `lqew`, allows one to obtain, by means of the instruction

$$[L,M,P,E]=dlqew(A,G,C,J,Q,R,N);$$

the same variables as for the previous command (for further details see Section 23.4).

- `dlqr` Discrete equivalent of the command `lqr`, by means of the instructions

$$[\texttt{K,S,E]=dlqr(A,B,Q,R)};$$
$$[\texttt{K,S,E]=dlqr(A,B,Q,R,N)};$$

provides: the matrix K, which contains the state feedback gains which minimize, respectively, the following performance indices:

$$J = \sum_{n=0}^{\infty} \left(x(n)^T Q x(n) + u(n)^T R u(n) \right)$$

$$J = \sum_{n=0}^{\infty} \left(x(n)^T Q x(n) + 2u(n)^T N x(n) + u(n)^T R u(n) \right)$$

the positive definite solution S of the discrete Riccati equation and, in the vector E, the eigenvalues of the closed loop system. The output variables S and E, if not required, can be omitted.

- `dlqry` Discrete equivalent of the command `lqry`, by means of the instruction

$$[\texttt{K,S,E]=dlqry(A,B,C,D,Q,R)};$$

provides: the state feedback gains matrix K, which minimizes the performance index

$$J = \sum_{n=0}^{\infty} \left(y(n)^T Q y(n) + u(n)^T R u(n) \right)$$

the positive definite solution S of the discrete Riccati equation and, in the vector E, the eigenvalues of the closed loop system. The output variables S and E can be omitted if not required.

- `dreg` Discrete equivalent of the command `reg`, allows one to obtain the matrices of a state-space representation of a controller consisting of a Kalman filter and the feedback of the estimated state.

24.7   Exercises

Exercise 24.1 Given the system with transfer function

$$G(z) = \frac{z + 0.4}{(z - 0.2)(z - 0.3 \pm 0.1j)}$$

compute, using the command `residue`, the analytic expression of the time response of the system to a step and to a ramp input.

Exercise 24.2 Given the system with transfer function

$$G(z) = \frac{z - 0.1}{(z + 0.8)(z - 0.1 \pm 0.2j)}$$

obtain the state-space form by the command `tf2ss` and compute the unforced response of the system assuming as initial conditions $(0.1, -0.1, 0)$, and the first twenty samples of the state evolution for an impulse, a step and a ramp input assuming zero initial conditions. Justify the state time waveforms for the ramp input.

Plot the root locus of the same transfer function and determine the open loop gain value such that the closed loop system has $\zeta = 0.3$ and $\omega_n = 2.3$.

> **Hint.** For the time response to a ramp input see the hint in Exercise 20.3 and adapt it to discrete-time systems. For the justification of the state evolution for a ramp input consider the state-space representation of the system. For the second part of the exercise use the commands `rlocus`, `zgrid` and `rlocfind`.

Exercise 24.3 Given the system with transfer function

$$G(s) = \frac{1}{(s + 0.1)(s^2 + 1)}$$

compare its step responses from zero to 70 s and its Bode plots with those of the discrete systems obtained with the different continuous to discrete conversion methods (see Section 19.3) and with sampling periods $T_c = 0.1,\ 1,\ 3,\ 6$ s.

Exercise 24.4 Given a system with the following continuous transfer function:

$$G(s) = \frac{1}{1 + 10\,s}$$

assuming $T_c = 5$ s as the sampling period, design a discrete controller

$$C(z) = k\frac{z - a}{z - b}$$

such that the closed loop system has a zero steady-state output error in the presence of a step input, and the closed loop poles are inside the region bounded by the curves corresponding to $\zeta = 0.8$ and the circle with radius 0.5.

> **Hint.** Converting $G(s)$ in the corresponding discrete transfer function and defining the controller pole on the basis of the steady-state constraint and the controller zero as desired, determine k using the root locus plot.

Exercise 24.5 Given the system described by the following dynamic equations:

$$\begin{pmatrix} x_1(n+1) \\ x_2(n+1) \end{pmatrix} = \begin{pmatrix} 0.3 & 0.04 \\ 1 & 0 \end{pmatrix} \begin{pmatrix} x_1(n) \\ x_2(n) \end{pmatrix} + \begin{pmatrix} 1 \\ 0 \end{pmatrix} u(n)$$

$$y(n) = \begin{pmatrix} 0 & 2 \end{pmatrix} \begin{pmatrix} x_1(n) \\ x_2(n) \end{pmatrix} .$$

design a state feedback such that the system output in the presence of a step input reaches the steady state in a finite time (*deadbeat* response). Plot the state variables and the output of the system.

> **Hint.** Recall that, when dealing with a system with no poles in $z = 1$, in order to allow the system response to reach the steady state in a finite time when forced by a step input, the system must have all poles at $z = 0$. To obtain this result, approximate these zeros with very 'small' numbers.

Exercise 24.6 Write a MATLAB function which, given a system represented in the state-space form and without any pole at $z = 1$, provides the feedback gain vector such that the closed loop system output in the presence of a step input reaches the steady state in a finite time.

Then write the MATLAB function for a system with a pole at $z = 1$.

> **Hint.** Recall that, in order to be characterized by a deadbeat response in the presence of a step input, a system with a pole at $z = 1$, must have a zero at $z = 1$ and all poles at $z = 0$. A dynamic controller and a state feedback are then required.

Exercise 24.7 Given the discrete system described by the equations

$$x_1(n + 1) = 0.2\, x_1(n) + 3u(n)$$
$$y(n) = x_1(n)$$

design a controller such that the closed loop system has a zero steady-state error to a step input and such that the following performance index:

$$J = \sum_{n=0}^{\infty} \left(100x_1(n)^2 + u(n)^2 \right)$$

is minimized.

> **Hint.** Augment the state, considering also the following equation, which describes an integrator before the process (see Exercise 23.4):
>
> $$x_2(n + 1) = x_2(n) + r(n) - y(n)$$
>
> where $r(n)$ is the reference input. Note that for the whole closed loop system the output coincides with the first state variable and the input $u(n)$ with the second state variable. On the other hand, the open loop augmented system now has two inputs (u and r), but the performance index considers only u: what does this mean?

Chapter 25

Utility functions

In this chapter we describe some commands which can be useful for programming operations in MATLAB. These instructions will not be treated in depth, and for further details on their use the reader can refer to the on-line help or to the manuals.

25.1 Modeling

- **abcdchk** Checks the dimensional consistency of the matrices of a state-space representation. It could be useful to test the dimensions of the matrices used as the input arguments of a self-written MATLAB functions.

- **drmodel** Builds a random discrete-time model in the state-space or transfer function form, given the desired system order.

- **ord2** Provides a second-order system model in the state-space or transfer function form, given the damping ratio and the natural frequency.

- **pade** Provides the state-space or transfer function representation which approximates a time delay, given the desired model order.

- **rmodel** Builds a random continuous-time model in the state-space or transfer function form, given the desired system order.

- **tfchk** Checks whether a given transfer function is proper or not.

- **tzreduce** Provides the model reduction by computing the transmission zeros.

25.2 Time response

- **dexresp** Provides an example of the use of the CONTROL commands devoted to the discrete-time and frequency responses. It can be used, for instance, in the following way:

343

dexresp('dnichols');

which gives an example of the use of the command **dnichols**.

- **dmulresp** Works as the previous command but on multi-variable systems.

- **dtimvec** Provides a discrete-time vector that is suitable for the simulation of a discrete system represented in the state-space form. The syntax is

tdis=dtimvec(A,B,C,x0,err,numpt);

where **A**, **B** and **C** are the system matrices, **x0** is the initial condition, **err** is the maximum percentage error of the output w.r.t. its initial or maximum value and **numpt** is the desired size of the vector **t**.

- **exresp** Provides an example of the use of the CONTROL commands devoted to the continuous-time and frequency responses. It can be used, for instance, in the following way:

exresp('bode');

which demonstrates one possible use of the command **bode**.

- **ltitr** Computes the state at the successive time step, given the dynamic and input matrices, the input value and the initial condition.

- **mulresp** Works as the command **exresp** but on multi-variable systems.

- **timvec** Provides a discrete-time vector that is suitable for the simulation of a continuous system represented in the state-space form. The syntax is

tdis=timvec(A,B,C,x0,err,numpt);

where **A**, **B** and **C** are the system matrices, **x0** is the initial condition, **err** is the maximum percentage error of the output w.r.t. its initial or maximum value and **numpt** is the desired size of the vector **t**.

25.3   Frequency response

- **dfrqint** Provides a frequency vector suitable for the Bode plot of a discrete-time system represented in the state-space or transfer function form. In particular, the command can be used as follows:

freqdis=dfrqint(A,B,C,D,Tc,numpt);
freqdis=dfrqint(num,den,Tc,numpt);

where Tc is the sampling period and **numpt** is the desired size of the vector **freqdis**.

- `dfrqint2` Provides a frequency vector that is suitable for the Nichols or Nyquist plot of a discrete-time system represented in the state-space or transfer function form.

- `fbode` Is a faster but less accurate version of the command `bode`.

- `freqint` Provides a frequency vector that can be used for the Bode plot of a continuous-time system represented in the state-space or transfer function form. In particular, the command can be used as follows:

$$\texttt{freq=freqint(A,B,C,D,numpt);}$$
$$\texttt{freq=freqint(num,den,numpt);}$$

where `numpt` is the desired size of the vector `freq`.

- `freqint2` Provides a frequency vector that is suitable for the Nichols or Nyquist plot of a continuous-time system represented in the state-space or transfer function form.

- `freqresp` Given a state-space or transfer function representation of a system, `freqresp` computes its frequency response at the desired frequency values.

- `ltifr` Computes the frequency response $G(\omega) = (j\omega I - A)^{-1}b$ of a single input system at the desired frequency values.

25.4 Lyapunov and Riccati equations

- `lyap` Given a stable matrix A and a symmetric definite positive matrix Q, the command

$$\texttt{P=lyap(A,Q);}$$

provides the solution of the Lyapunov equation

$$AP + PA^T = -Q$$

The solution of the generalized Lyapunov equation

$$AP + PB = -Q$$

can be obtained by the instruction

$$\texttt{P=lyap(A,B,Q);}$$

Note that the solution of this equation is only unique if the sum of each eigenvalue of A and each eigenvalue of B is nonzero. In the other case an error message will appear.

The same results, achievable with the command `lyap`, can be obtained by using the command `lyap2`, which is faster but can only be applied to diagonalizable systems.

- `dlyap` Provides the solution of the discrete Lyapunov equation

$$APA^T - P = -Q$$

- `are` Provides the solution of the algebraic Riccati equation

$$A^TP + PA - PBP + Q = 0$$

Part IV

Appendices

Appendix A

Advanced graphic functions

A.1 Graphic objects in MATLAB

In Chapter 7 we discussed high-level functions to plot graphics in MATLAB in a simple way. We now devote our attention to analyzing MATLAB graphic capabilities in depth. As always happens, the use of high-level functions greatly simplifies the handling of rather common situations, but it does not allow us to treat special situations, such as changing the size or the typeface of characters on the plot.

To study low-level graphic functions we need a preliminary concept: the *graphic object*. A graphic object is a structure that is characterized by some attributes.[1] An example can help to make the concept clear. When we enter the command `plot`, MATLAB performs the following operations:

1. creates a graphic window, named **figure**;
2. inside the window, creates a reference frame, named **axes** (note that this is quite different from the *command* `axis`);
3. inside the reference frame, traces a set of lines, obviously named **line**, by which the curve to plot is drawn.

Now consider all these operations ordered not only chronologically, but also hierarchically; then we may say that the 'son' of **figure** is **axes**, whose 'son' is in turn **line**, or, dually, the 'father' of **line** is **axes**, whose 'father' is in turn **figure**. Note that the correspondence parents–children is one-to-many: each object has a single parent, but may have many children. Now it should be clear how the command `subplot` (see Section 7.6) works: it assigns to the same **figure** more children of type **axes**, each of them defining a graphic subwindow.

The entities described are 'objects', in the sense mentioned above, thus they are characterized by a set of attributes: for instance, some of the attributes of a **figure** are:

---

[1]MATLAB manuals use the word 'property' rather than 'attribute', but we have preferred the latter term, which is standard in object-oriented programming terminology.

- Position of the figure on the page;
- Format of the page on which to print the figure;
- Figure orientation (horizontal or vertical).

The attributes of the object may be both listed and changed. To this end it is necessary to identify a particular graphic object. Graphic objects are identified in MATLAB by numeric indices, named *pointers* or *handles*: for instance, the number of the **figure** is itself the pointer: 1 for the first figure, 2 for the second and so on. Hence, manipulating the first figure means working on the object whose handle is 1. The command **gcf** (Get Current Figure) returns the pointer of the current figure. An analogous command exists for **axes** objects: **gca**. For other objects at lower hierarchy levels there is no command of the same kind, thus it would seem impossible to obtain a pointer, for instance to a given set of solid lines to turn them into dotted lines. To avoid this drawback, any graphic object has two peculiar attributes: **Parent**, which contains the pointer to the parent of the object and, dually, **Children**, which contains the list of (pointers to) children of the object. Since line objects are children of the current axes, as mentioned above, we can surely find the pointer to the line to modify among the **Children** of the current axes. To pick the pointers we use the command **get**, which allows the user to read and store the value of an attribute in a variable. Its syntax is

p=get(*pointer*,'*AttributeName*');

which stores in the variable p the value of the attribute *AttributeName* related to the object indicated by the pointer *pointer*. Thus, the list of the children of the current axes is available with

p=get(gca,'Children');

An object may be canceled, if its pointer is available, by the command

delete(*pointer*)

An example is useful to acquire familiarity with these commands. Suppose we start a MATLAB session by typing

```
t=0:.1:10;
t=t';
y=sin(t);
plot(t,y)
```

Now,

pa=gca

returns the pointer to the current axes (obviously, the same result could be obtained with pa=get(gcf,'Children')); with

$$\text{pl=get(pa,'Children')}$$

we obtain the pointer to the line drawn on the screen (the graph of the sine). To alter this object we use the command `set`, which is the dual of `get`:

$$\text{set(pl,'LineStyle','-.')}$$

traces a dash-dotted line, giving the same result as the command `plot(t,y,'-.')`. The general syntax of the command `set` is

$$\mathbf{set}\,(\textit{pointer},\,'\textit{AttributeName}',\,\textit{NewValue})$$

Note that the command `get`(*pointer*) returns a list of the attributes of the object indexed by the pointer and the corresponding values, while `set`(*pointer*) returns a list of the values that can be given to the attributes. Note that the values of the attributes may be written both in capital and lower-case letters, although the preferred form, which improves readability, is the one used in this appendix.

Continuing our example, let us type

$$\text{y1=cos(t);}$$
$$\text{plot(t,[y y1])}$$

Executing the previous commands again, we now note that *pl* is a vector with two elements, i.e. the pointers to the two curves. Thus it is clear that using the command `get` to inspect the `Children` property can produce such a large a list of pointers as to render the interpretation of the result awful. To what does each pointer point? The answer to this question is given by the property `Type`, which is a property of any object and specifies the type of the object itself. In our example, `get(pl(1),'Type')` returns `line`, while `get(pa,'Type')` results in `axes`. By the way, many high-level graphic commands may return the corresponding pointers: for instance

$$\text{pl=plot(t,[y y1])}$$

plots two curves and returns the corresponding pointers in the vector *pl*.

Now we can start a detailed analysis of graphic objects. MATLAB defines ten graphic objects, listed below along with the high-level functions to create them:

1. **Root**: is the root of the hierarchy and corresponds to the computer screen.
2. **Figure**: as stated before, is a single window. There is no limit to the number of windows that can be opened; all of them are children of **root**. They are created by the function `figure`.
3. **Axes**: this object, too, has been mentioned before; it is a portion of the window used to plot graphics.[2] Its parent is **figure** and it is created with the command `axes`.

---

[2]The reader may ask why not use the whole window for the plot. We will see later that a **figure** not only contains plots, but also control buttons and menus.

4. **Line**: this is the graphic primitive to draw 2D and 3D plots. They are children of **axes**. They are created by the commands `plot`, `plot3`, `contour` and `contour3`.

5. **Patch**: are filled polygons. Their parent is **axes**. They are created by the high-level commands `fill` e `fill3`, not described in this text.

6. **Surface**: as **line** defines line segments joining two points and is the basic element to trace curves, so **surface** graphic objects define elementary quadrilaterals whose vertices are four given points. They are children of **axes** and are created by the commands `surf`, `mesh` and `pcolor`.

7. **Image**: results from associations of matrix elements and entries of the current color map. They are children of **axes** and are created with the function `image`.

8. **Text**: these are character strings to be inserted in the graphics. They are children of **axes**.

9. **Uicontrol**: are menu controls that when selected execute a given action. Their use is discussed in Appendix B. They are children of **figure**.

10. **Uimenu**: these are objects to create pull-up menus in graphic environments. See Appendix B for their use. They are children of **figure**.

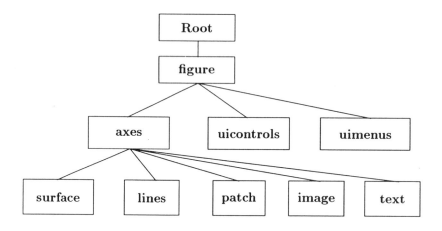

Figure A.1 Graphic object hierarchy.

Object hierarchy is summarized in Figure A.1. Note that the hierarchy organization is useful both to delete objects (on deleting an object its children are deleted, too) and to allow the children to inherit properties: for instance the position of all the children of **axes** is determined by the coordinate system established by their parent.

Some attributes are common to all graphic objects:

- `Interruptible`

- UserData
- Visible
- Type
- Clipping
- Parent
- Children

The attributes `Type`, `Parent` and `Children` have already been presented before. The attribute `Visible` controls graphic object visualization: when set to the value `'off'` this attribute forbids the visualization of the corresponding graphic object. `Clipping` will be described shortly when discussing the properties of **line** objects. Finally, `Interruptible` and `UserData` will be presented in Appendix B.

We are now ready to analyze each graphic object and to discuss its properties.

A.2   Root

The commands given in this section refer to the object **root**, i.e. is the computer screen, and are seldom used. Most screen attributes cannot be changed, while the remainder are more easily handled by high-level functions. Nevertheless, we will give a partial list of screen attributes. Since the pointer to the root is 0, the reader can obtain the complete list with the command **get(0)**.

- `CurrentFigure` returns the pointer to the current **figure**.

- `Diary` and `DiaryFile` specify whether to start to record the current MATLAB session on a file, and the name of the file itself. See the command **diary** on page 93.

- `Echo` activates the command **echo** (see page 99) to display the commands during the execution of an M-file.

- `Format` and `FormatSpacing` define the screen representation format of the data according to the table on page 10 and the spacing according to the formats `compact/loose`.

- `PointerWindow` returns the pointer to the window containing the mouse cursor.

- `Units` defines measure units in the current coordinate system. The origin is in the lower left corner. Possible choices are

  - `pixels`
  - `normalized`, in which the coordinates of the upper right corner are $(1,1)$
  - `inches`
  - `centimeters`
  - `points`

- **PointerLocation** returns the current coordinates of the mouse pointer using measure units defined by the previous command.

- **ScreenSize** returns the size of the screen in the same units.

A.3   Figure

As stated before, the command **figure** creates or selects a graphic window. When used in the form **p=figure** it defines a new window and stores its pointer in the variable p. We have already discussed the command **set** to change object attributes, but it is possible to set attributes directly when creating the window by using the command

$$p\texttt{=figure}('\textit{AttributeName}', \textit{Attribute Value}, \dots)$$

The most interesting attributes are those used to define the colors of the window, the current color map and the position of the figure on the paper. Here is a partial list of **figure** object attributes:

- **Color** defines the background color of the window. The definition of the colors obeys the rules given on page 79.
- **Colormap** defines the current color map. It is equivalent to the command **colormap** presented on page 79.
- **Units** defines the measure units of the coordinate system for the window. Possible measure units are those listed in the previous section.
- **Pointer** defines the shape of the mouse cursor. Possible values are

  - **'arrow'** is the standard value, the classic arrow;
  - **'crosshair'** a cross;
  - **'watch'** a clepsydra;
  - **'cross'** a double cross;
  - **'topl'**, **'topr'**, **'botl'**, **'botr'** various arrows;
  - **'circle'** a circle;
  - **'fleur'** a snowflake.

- **CurrentPoint** returns the coordinates of the last point where the mouse button has been pressed or released. The measure unit is set by the **Units** attribute. The pointer to the object whose coordinates are those of the current point is given by the attribute **CurrentObject**.
- **MenuBar**, whose value is either **'figure'** or **'none'**, activates or deactivates the window pull-up menu. **Name** defines the window name, which is the character string appearing on the right of the heading **Figure No.** n; **NumberTitle** activates or deactivates the heading **Figure No.** n; the values allowed are **on** and **off**.

- InvertHardcopy has the value 'on' or 'off'; graphics appear by default on a black background on the screen and on a white background when printed on paper; selecting the value 'off' the background is black on paper, too.
- Position defines the position of the window on the screen. Its value is a vector of four elements: the two coordinates of the lower left corner, the width and the height of the window.
- The following attributes define how to print the figure on paper: PaperType defines the paper format according to the following possible values:
  - 'a4letter' A4 format;
  - 'usletter' letter format;
  - 'uslegal' legal format.

  PaperUnits defines the sheet measure units; PaperSize returns the size of the sheet defined by the PaperType attribute, according to the units defined by PaperUnits; its value cannot be changed. PaperPosition is like Position, but is referred to the sheet; finally PaperOrientation may assume the values 'landscape' and 'portrait', which print the figure horizontally and vertically respectively.
- Resize is a switch to enable window-size alterations.
- CurrentAxes returns the pointer to the current axes; this can be done with the command gca, as described above.

A.4 Axes

Dealing with **axes** objects we go further into the development of MATLAB graphics. We first stated that the **root** controls the computer screen, while **figure** controls the structure and the size of the graphic window both on the screen and on the paper. **axes** graphic objects are more closely related to the plot: in fact they control the position of the plot within the window and define the axis, both for 2D and 3D plots. Moreover, they also define the viewpoint for 3D graphics.

The creation of an **axes** object is done by the command **axes**, whose syntax is the same as that of **figure**; it can be used in three forms:

```
p=axes
axes(p)
p=axes('AttributeName',AttributeValue,...)
```

The most interesting attributes concern the following properties:

- The position of a plot inside a window is defined by the Position attribute, whose value is a four-element vector, as seen in the previous section, with measure units defined by the attribute Units; the default value for the units attribute is 'normalized', so that default measurements are relative to the size of the window. It is, moreover, possible, by setting the attribute `AspectRatio`, to impose aspect ratios both for the axes and for their values. The value of this attribute is a two-element vector whose first entry defines the ratio between the length of the vertical axis and that of the horizontal one, while the second entry defines the length of the unity on the vertical axis with respect to that on the horizontal axis, i.e. controls the scale factor ratio. For instance, the command

```
set(gca,'AspectRatio',[4/3 1])
```

defines the axes aspect ratio 4:3 (the common video aspect ratio) and the same scale factors on both axes (like the command `axis('equal')`); try this command on the sine function drawn before; moreover, also try the command `set(gca,'AspectRatio',[1 1])`, which returns a square frame (like that produced by `axis('square')`).

- `NextPlot` may assume the values `'new'`, `'add'` or `'replace'` and define how a next `plot` will be interpreted. Indeed, with the default value `'replace'` the next plot deletes the previous one, while `'add'` adds a new plot to the current **axes**, as when we use `hold on`. The value `'new'` is the least frequently used: it adds a new **axes** object to the current **figure** without changing the position of the previous plot: in this way the two graphics are superimposed but belong to two different **axes**.

- Color definition inside an **axes** object is performed with four different attributes. The background color of the plot is defined by the attribute `Color` (note that the same attribute when referred to the **figure** object defines the background color of the whole window). For 2D plots `ColorOrder` defines the sequence of colors to use for multiple plots. The default order is the one indicated in Table 7.1. A possible order is a whole color map, as in the following example:

```
t=(0:.1:10)';
y=sin(t);
Y=[];
for i=1:64, Y=[Y sin(i/64*t)];end
plot(t,y);
set(gca,'ColorOrder',cool,'NextPlot','add');
plot(t,Y)
```

note that the first graphic has been traced to set scale factors, and note the use of the value `'add'` for the attribute `'NextPlot'` to override the default value `'replace'` which would result in the creation of a new **axes** object, thus losing the color definition.

For a 3D object two attributes are used for the definition of color axes, as given on page 79; `'CLim'` defines the limits of the color axis, like the high-level command `caxis`, while `CLimMode` selects whether or not to use automatic axis scaling.

- 2D graphics traced by the command `plot` are surrounded by a rectangle; to cancel the rectangle, leaving the Cartesian axes alone, one sets the attribute `Box` to `'off'`. The same attribute in 3D context is used to insert the surface in a parallelogram, when set to `'on'`.
- The grid is traced by the command `grid`, but can be altered by means of the attribute `GridLineStyle`: try to plot a graphic, draw a grid and then type

$$\text{set(gca,'GridLineStyle','-','LineWidth',5)}$$

The attribute `LineWidth` defines the width of the lines in the current axes, including the `Box`. See the last point of this section to remove grid lines along particular directions.

- The length of the ticks on the axes is set by the attribute `TickLength` and their direction by `TickDir`: the default value for this attribute is `'in'`, which draws ticks from the box towards the inside of the plot, while `'out'` defines the opposite orientation. To control the ticks independently on the vertical and horizontal axes, see the last point of this section.
- The attribute `View` defines the viewpoint, as does the command `view` discussed in Section 7.7.
- The character strings in a graphic may be
  1. title;
  2. axis labels;
  3. text and comments;
  4. numeric (or alphabetic, see next point) values to be used as labels for tick marks along the axes.

Only the elements at point 4 are attributes of the reference frame (the **axes**), while the others are objects themselves, in particular **text** objects and, as such, are children of the current **axes** (see Figure A.1). The values on the axes are drawn with the font defined by the attribute `'FontName'`, whose value is one of the possible fonts defined in the current Windows configuration, with dimension set by the attribute `FontSize`, and typographic face defined by the attribute `FontWeight`, which can assume the values `'light'`, `'normal'`, `'demi'` or `'bold'`; it is, moreover, possible to trace underlined or 'canceled' (i.e. crossed by a horizontal line) characters by setting the attributes `FontUnderline` and `FontStrikeThrough`, respectively, to the value `'on'`. Character slanting is controlled by the attribute `FontAngle`, which admits one of the three values `'normal'`, `'italic'` or `'oblique'`.

Pointers to text and comments are stored in the `Children`, attribute of the **axes** object, the pointer to the title in the attribute `Title`; axis labels will be discussed in the next point, which deals with axis properties definition.

- The three axes are controlled independently by a set of attributes with the same syntax and meaning for each axis. Thus we will confine ourselves to the description of the attributes controlling the x-axis alone, extension to the y- and z-axes being trivial. XColor defines the color of the x-axis; XDir defines the direction of the axis: 'reverse' gives increasing values from right to left, while the default value is 'normal'. The grid lines originating from the x-axis (i.e. along the y direction in 2D graphics) can be eliminated with set(gca,'XGrid','off') (of course, we can restore gridding by setting the same attribute to the value 'on'). Setting the attribute XScale to the value 'log' defines a logarithmic x-axis scale, like that obtained by using semilogx; the scaling is now also extendable to 3D plots. Axis limits are set by XLim and automatic scaling by XLimMode; these attributes work like the attributes CLim and CLimMode discussed earlier. The pointer to the axis label is stored in XLabel. The values on the axes are handled by two attributes: XTickLabels is a vector of strings (a character matrix) and defines 'what to write', while XTick is a numeric vector and defines 'where to write'; an example can assist understanding this point. The mean orbital velocities of the planets of the solar system are given in the following table (in km h^{-1}):

Mercury	172.41
Venus	126.11
Earth	107.25
Mars	86.87
Jupiter	47.01
Saturn	34.70
Uranus	24.51
Neptune	19.55
Pluto	17.06

A MATLAB representation of these data is obtained with the following lines

```
y=[172.41 126.11 107.25 86.87 47.01 34.70 ...
24.51 19.55 17.06];
bar(y);
planets=str2mat('Mercury','Venus','Earth',...
'Mars','Jupiter','Saturn','Uranus',...
'Neptune','Pluto');
set(gca,'Xlim',[0.5 9.5],'XTick',1:9,...
'XTickLabels',planets);
```

It is possible to define auto scaling with XTickLabelMode and XTickMode.

A.5   Line

Line objects can be created by the command line, although they are usually produced by the functions plot or plot3, which, moreover, set the parameters of the current axes.

The number of properties of line objects is considerably smaller than that of axes objects. Consider the following example:

```
t=(0:.1:10)';
y1=sin(t);
y2=cos(t);
p=plot(t,[y1 y2]);
set(p(1),'Color','r','LineStyle','-',...
'LineWidth',10);
set(p(2),'Color','g','LineStyle','o',...
'MarkerSize',20,'Clipping','off');
```

This example uses almost all the attributes of **line**. The definitions of color and line type can easily be understood. Note the attribute LineWidth, which defines the width of solid lines and, dually, MarkerSize, which specifies the size of dots. Moreover, by assigning the value 'off' to the attribute Clipping we allow the curve to exit the axes rectangle.

Other **line** attributes are XData, YData and ZData, whose values are the vectors of coordinates of the lines drawn, and EraseMode, which defines how to plot and cancel the lines; it may assume one of four values: with 'normal' when a line is deleted the whole corresponding screen region is analyzed for refreshing, thus avoiding the deletion of possible overlapping lines; using 'background' the line is deleted by painting it with the color of the background, thus painting any possible overlapping object; 'none' does not erase the line even when the line object is moved or deleted; finally, 'xor' allows a cancellation without altering overlapping objects; this could seem the best solution, but a drawback arises: if the figure background is not black the color of the curve is changed. Continuing the previous example, try the following lines:

```
set(p(1),'EraseMode','xor')
set(gca,'Color','w')   note that the line is no more red
set(p(1),'EraseMode','none')   now is red again
delete(p(1))   the curve is not deleted
```

As a more complex example, consider the function file BOOM.M:

```
function boom(map)
fig_p=figure;
ax_p=axes;
g=9.81;
plot(0,0,'.','erasemode','none','markersize',2);
pnt_p=get(ax_p,'children');
set(ax_p,'xlim',[-5000 5000],'ylim',...
[0 30000],'visible','off');drawnow;
while(1),
  v0=randn(1)*1000;
```

```
theta0=rand(1)*pi/2+pi/4;
i0=fix(rand(1)*64)+1;
c0=map(i0,:);
set(pnt_p,'color',c0);
v0x=v0*cos(theta0);
v0y=v0*sin(theta0);
t=0;y=0;x=0;
while (y>=0)&(x>-5000)&(x<5000)&(y<30000),
  t=t+.1;
  x=v0x*t;
  y=v0y*t-.5*g*t^2;
  set(pnt_p,'xdata',x,'ydata',y);
  drawnow
end
end
```

A possible call to this function is boom(hot).

A.6   Surface

Surface objects are the 3D analogue of **line** and may be generated by the command surface, although they are more frequently the result of the high-level commands mesh, surf and surfl.

The attributes EraseMode, LineStyle, LineWidth and MarkerSize have the same syntax and meaning as the corresponding line attributes described in the previous section. EdgeColor and FaceColor allow one to define edge and face colors for each element defining the surface; the value of these attributes may be either a specific color or one of the strings 'none', 'flat' or 'interp', which do not paint, use a single color and interpolate the color for each face, respectively, as stated on page 80. Finally, MeshStyle is used to trace the edges by using the rows only, the columns only or all the elements of the matrix containing the data to be represented.

As in the case of line objects, the data to be represented are stored in the attributes XData, YData and ZData, while color data are stored in the attribute CData.

As an example, try the following commands:

```
p=surf(peaks); note that surf returns the pointer
set(p,'LineStyle','-','FaceColor','w');
set(p,'MeshStyle','row')
```

A.7 Image

Image objects are characterized by a direct association between numeric values and color map entries: we can think of them as a visualization of color maps. An object of type **image** is created by the command `image(C)`, where the entries of the matrix C are usually integers between 1 and 64 (remember the predefined maps are made up of sixty-four colors). To visualize the `hot` we type

```
image(1:64), colormap(hot)
```

Obviously, each 2D image can be considered to be a collection of colors, and this explains the name of the objects discussed in this section. Load the demo file EARTH.MAT, which contains the two variables X and *map*

```
load earth
image(X)
colormap(map)
```

The same result could have been obtained by using the command `pcolor`, as seen on page 81, but this requires more time to complete the picture. Actually, `pcolor` creates a **surface** object and defines a viewpoint 'from above', i.e. an elevation angle of 90°, while `image` creates a different object, of type **image**. The number of attributes of the latter object is dramatically smaller than that of the former: only `CData`, `XData` and `YData` are present; the last two matrices specify the position of rows and columns in the image.

A.8 Text

Text objects are all the character strings in the graphics. We have already mentioned three objects of this type when discussing **axes** attributes: `Title`, `Xlabel` and `Ylabel` define the pointer to title and axes labels. We have discussed on page 68 how to insert comments in plots by using the command `text`. Actually, this command creates a **text** object, and also returns its pointer when used in the form `p=text(x,y,'`*string*`')`. The attributes `FontAngle`, `FontName`, `FontSize`, `FontWeight`, `FontStrikeThrough` and `FontUnderline` have been discussed before; the color of the text is defined by the attribute `Color`. There is, moreover, a set of attributes to define the position of the string; these are: `Units` (note that to define a position, units must always be defined first); `Position`, defining the two (or three, for 3D graphics) text coordinates; `HorizontalAlignment` and `VerticalAlignment` with values `'left'`, `'center'` or `'right'` for the first, `'top'`, `'middle'` or `'bottom'` for the second, defining the text alignment with respect to the point defined in the attribute `Position` and finally `Rotation` to specify one of the seven predefined angles (in degrees) by which the text should be rotated (0, ±90, ±180, ±270). It is possible, moreover, to set the attribute `EraseMode`, whose syntax and meaning have already been discussed when we were dealing with **line** objects.

Appendix B

Graphic user interface design

In this appendix we will deal with the design of a Graphic user interface (GUI) in MATLAB. We will present a set of commands which allow a MATLAB program to be completely mouse-driven through buttons, menus, windows, etc. A complete example is the MATLAB demo, available in any MATLAB version.

Three main topics are covered in this appendix: how to create control buttons, how to add menus to windows and how to handle mouse action (cursor movements, left- and right-hand buttons). The first two points require the discussion of two new graphic objects, while attributes not described in the previous appendix will be presented in the final section.

B.1 Controls

MATLAB has eight different objects devoted to user interface control; they are created by the function `uicontrol` and share a set of common attributes, the most important being `Style`, which is used to define the type of control object to create. Here is the list of control objects, along with the values of the `Style` attribute needed to create them:

1. **Push Button**: are the classic buttons available in any Windows application and are activated by pressing the left-hand mouse button; to create them, set the `Style` attribute to the value `'push'`.

2. **Check Box**: these controls also are familiar to Windows users, which consist of a list of items whose active status is indicated by a cross; the value of `Style` to create them is `'checkbox'`.

3. **Radio Button**: are analogous to the previous controls, but are mutually exclusive, i.e. on selecting an item from the list the others are deselected; this operation is not automatic and the instruction to obtain it is `Style` value: `'radio'`.

4. **Static Text**: are simple comments and character strings, and they cannot be altered; are they defined by the value `text`.

5. **Editable Text**: unlike the previous commands, the text can now be edited and altered by the user; they are created by giving the value `'edit'` to the attribute `Style`.

6. **Slider**: are the classic Windows 'sliding bars' and are used to let a variable assume continuous values in a given interval; they are used together with **text** to define extreme values of the interval and the actual value of the variable; to create them `Style` assumes the value `'slider'`.

7. **Frame**: defines a frame that groups related controls in logic units; it is defined by the value `'frame'`.

8. **Pop-up Menu**: are window pop-up menus, a simplified version of the window menus that we will discuss in the next section; they are defined by the value `'popup'` of `Style`.

A preliminary discussion of the attributes shared by these objects[1] is useful in order to simplify the following detailed presentation.

We have already stated that the controls are created by the command `uicontrol`, by setting the attribute `Style`; since the objects are related to a single graphic window, the first parameter to supply is the window pointer. For instance, to create a **Push Button** in the current window, we use

$$pb=uicontrol(gcf,'Style','push');$$

note that `uicontrol` also returns the pointer to the button. Then we must provide the position within the window and the size of the control object: this is done by the attribute `Position`, which again has a `Units` attribute associated with it, to define current measure units. With reference to the previous example, type

$$set(pb,'Units','normalized','Position',[0\ 0\ 0.2\ 0.2]);$$

thus the button extends for 20% of the whole **figure** in both directions (see page 353). The labels on the commands are defined by the attribute `String` and (for **Static Text** objects only) are centered by `HorizontalAlignment`, whose possible values are `'center'` (default), `'left'` and `right'`. The character type may be selected from the **Options** of the MATLAB main window. Background and label color is defined by the attributes `BackgroundColor` and `ForegroundColor` respectively:

```
set(pb,'String','Close','ForegroundColor','w',...
'BackgroundColor','b')
```

---

[1]Note that the term 'object' is used, strictly speaking, in an improper sense; actually, **uicontrol** is *the only object* described in this section, while the controls introduced above are different *instances* of this object, but, to simplify exposition, we have preferred to drop the concept of 'instance', referring to each control as a single object.

Obviously, the main feature of the control objects is their ability to execute some actions when activated; this is the role of the attribute `Callback`, whose values are the MATLAB commands to execute, stored as a vector of strings, for instance

<code>set(pb,'Callback','close(gcf)')</code>

closes the window when the button is selected. It is now clear that this attribute is the most significant, and we will see that it is also the most complex to define, both from a syntax point of view (dealing with strings, each apostrophe must be a double apostrophe, which causes loss of readability in the code) and from a semantic point of view, due to problems of variable visibility which will be discussed later.

Finally, the attributes `Value`, `Min` and `Max` define the value assumed by the object and the extreme admissible values; this is not intended for **Push Buttons**, whose value is 1 when selected and 0 otherwise, but is crucial for objects like **Slider**, which is capable of assuming a set of values.

Let us now begin a detailed analysis of the different controls; the exposition is performed via examples which describe how any control works and the problems that could possibly be encountered.

Push Buttons are mainly used for actions that happen only once, such as closing a window, plotting a graph, starting to run a program. An example has already been given above and a more complex case is given in the sequel. Let us consider the function BOOM.M, presented on page 359 and create a button to run the function from a graphic window, using the color map `hot`; of course we must set the attribute `Callback` to the value `'boom(hot)'`. A problem immediately arises: the function BOOM itself creates a new **figure** and new **axes**, and the **push button** cannot be associated with a figure that is still to be created. The solution is to create **figure** and **axes** first, then the button, and finally to force the function to use current **figure** and **axes**; to do this, let us change the first lines of the file BOOM.M and store the result in BOOM1.M as follows:

```
function boom1(map)
fig_p=gcf;
ax_p=gca;
```

we leave the rest of the file unchanged.

At this point the list of commands from the MATLAB prompt:

```
pf=figure;
axes;
pb=uicontrol(pf,'Style','push','Units',...
'normalized','Position',[0.01 0.01 ...
0.2 0.1],'String','Fire!','Callback',...
'boom1(hot)');
```

creates a window with the button 'Fire!', by pressing which the fireworks start. This is still a crude version of a mouse-driven program: to stop the execution we must use the CTRL-C key combination, which is a *debacle* in a mouse-driven program. If we want to create one more button to halt the program, the following problems should be considered: 1) the pressure of this button must change the state of a variable halting the execution, then the while loop in boom1 must be broken by this variable; 2) the same variable cannot be passed as an input parameter to the function boom1 because when the first button starts the execution its callback sets the value of all the input parameters, which can no longer be altered: a possible solution is to use a global variable; 3) the program execution must allow a second button to alter its stop variable, in other words the program must accept interruptions.

Here is the solution: let us define the global variable f_end and impose that the first button must accept interruptions: the first lines of boom1 become

```
function boom1(map)
global f_end
f_end=0;
fig_p=gcf;
```

The next command while(1) becomes while(~f_end). In the MATLAB window we must type

```
pf=figure;
axes;
f_end=0;
global f_end
pb=uicontrol(pf,'Style','push','Units',...
    'normalized','Position',[0.01 0.01 0.2 0.1],...
    'String','Fire!','Callback','boom1(hot)',...
    'Interruptible','yes');
pb1=uicontrol(pf,'Style','push','Units',...
    'normalized','Position',[0.41 0.01 0.2 0.1],...
    'String','End','Callback','f_end=1;');
```

Note that we have first introduced the attribute Interruptible, which enables interruptions in a sequence of commands defined in the attribute Callback to process actions in a different callback. In particular, the interruption takes place when one of the commands

- drawnow
- figure, gcf, gca
- getframe
- pause

is encountered in the callback to interrupt.

Each time one of these commands is encountered, MATLAB suspends the execution in order to analyze the events that have been stored in a temporary queue; these events are processed if the Interruptible attribute has the value 'yes' and otherwise are ignored.

We have described how to use global variables to halt a program. Later we will see that many pointers to control objects can be defined as global variables so as to allow the controls to influence each other; this can result in three inconvenients: first, if the user types clear global all the pointers are lost; second, the workspace is poorly organized, as it is full of variables that, from the user point of view, are dummy variables; finally, problems arise when more than one window must be opened, unless we define a set of global variables for each window.

Later we will give a more efficient (although more complex) way of managing these situations. For the moment we continue to use global variables in order to gain more familiarity with the control objects.

In the previous example we used two push buttons to start and halt the program boom1; actually these two operations are mutually exclusive, thus it is more natural to use controls that stress this characteristic: the **radio buttons**. As mentioned before, although used almost always in a mutually exclusive fashion, they have to be programmed in order to guarantee such behavior. Let us start by creating two **radio buttons**, then we will detail the operations to make them mutually exclusive. Moreover, it is better to modify our function to obtain a 3D visualization of the fireworks. Let us start from this last point: we create the file BOOM3D.M introducing a second random angle ϕ_0:

```
function boom3d(map)
global prd_end
fig_p=gcf;
ax_p=gca;
g=9.81;
pnt_p=plot3(0,0,0,'.','erasemode','none',...
'markersize',2);
set(ax_p,'xlim',[-5000 5000],'ylim',[-5000 5000],...
'zlim',[0 30000],'visible','off');drawnow;
while(get(prd_end,'Value')==0),
  v0=randn(1)*1000;
  theta0=rand(1)*pi/2+pi/4;
  phi0=rand(1)*pi/2+pi/4;
  i0=fix(rand(1)*64)+1;
  c0=map(i0,:);
  set(pnt_p,'color',c0);
  v0x=v0*cos(theta0)*cos(phi0);
  v0y=v0*cos(theta0)*sin(phi0);
  v0z=v0*sin(theta0);
```

```
t=0;y=0;x=0;z=0;
while (z>=0)&(x>-5000)&(x<5000)&...
(y>-5000)&(y<5000)&(z<30000),
  t=t+.1;
  x=v0x*t;
  y=v0y*t;
  z=v0z*t-.5*g*t^2;
  set(pnt_p,'xdata',x,'ydata',y,'zdata',z);
  drawnow
end
end
```

This time the global variable is a pointer, in particular the pointer to the **radio button** to halt the program, as will be illustrated below. Now we create the window and the buttons:

```
pf1=figure(1);
pa1=axes;
prd_start=uicontrol(pf1,'Style','radio','Units',...
'norm','Pos',[0.01 0.06 .2 .05], 'Value',0,...
'String','Start');
prd_end=uicontrol(pf1,'Style','radio','Units',...
'norm','Pos',[0.01 0.01 .2 .05], 'Value',1,...
'String','Stop');
```

By the way, note that the attribute `Position` has been abbreviated to 'Pos': all the graphic objects allow abbreviation unless ambiguities occur.

For the **radio buttons** the value of the attribute `Value` is the value defined in the `Max` attribute when the button is selected and `Min` if deselected; default values for these two attributes are 1 and 0, respectively; hence, according to the previous commands, the 'Stop' button is initially active, while 'Start' command is deselected.

We must now define the `Callback` attribute for both buttons; the former when selected must first deselect the other, then run the program boom3d; for the 'Stop' button it is sufficient to deselect the other button, since the stop condition only checks the value of the second button, whatever its callback is. Finally, we must remember to make the first button interruptible. The resulting code is

```
set(prd_start,'Callback',['set(prd_end,''Value'',''...
0);boom3d(hot);'],'Interruptible','yes');
set(prd_end,'Callback','set(prd_start,''Value'',0)');
global prd_end
uicontrol(pf1,'Style','push','units','norm','Pos',...
[.79 .01 .2 .09],'String','Close','Callback',...
'close(pf1)')
```

The last command introduces a **push button**, this time with a more appropriate role: simply to close the window. Note that its pointer has not been assigned to a variable, since it is never manipulated.

To avoid confusion between the 'Close' button (that closes the window) and the 'Stop' button (that ends the program execution) we may give the **radio buttons** pair a textual heading with a **Static Text** control:

```
uicontrol(pf1,'Style','text','units','norm','pos',...
[.01 .11 .2 .05],'String','Run')
```

Finally, we group these controls in a single unit by using **Frame**:

```
uicontrol(pf1,'Style','frame','units','norm','pos',...
[0 0 .22 .17])
```

Note that the only thing to take care of is the computation of the size of the control. Usually, in designing the program it is better to store *all* the pointers, for debugging purposes, then to clean them in the final program. Their availability in the testing phase is always an advantage; for instance, if we are not satisfied with the colors (this point may appear inessential, but in the final program it is the aesthetics that make the difference), it is possible to type the following lines if one has at one's disposal the pointers of the **Static text** object in the variable **ptx_exec** and that of **frame** in **pfr_exec**:

```
set(pfr_exec,'back','r');
set(ptx_exec,'back','b','fore','w');
set(prd_start,'back','w','fore','k');
set(prd_end,'back','w','fore','k')
```

The program simulates a random launch of fireworks in 3D. The visualization of each projectile ceases as soon as it goes beyond the axis limits; this is apparent in the 2D case, but now the result could appear rather confusing: it would be better to visualize axes enclosed by a parallelogram (see the attribute Box, on page 357). To this end we will use two **Check Box** controls. The first one makes the axes visible, while the second draws the bounding box. Then we will add a frame and a heading:

```
pck_box=uicontrol(pf1,'Style','check','units','norm',...
'pos',[0.79 .84 .2 .05],'String','Box','Callback',...
['set(gca,''box'',''off'');if (get(',...
'pck_box,''Value'')==1),set(gca,''box'',''on''); end']);
pck_axes=uicontrol(pf1,'Style','check','units','norm',...
'pos',[0.79 .89 .2 .05],'String','Axes','Callback',...
['axis(''off'');if (get(pck_axes,''Value'')==1)',...
'axis(''on''); end']);
ptx_axes=uicontrol(pf1,'Style','text','units','norm',...
'pos',[0.79 .94 .2 .05],'String','Control');
pfr_axes=uicontrol(pf1,'Style','frame','units','norm',...
'pos',[0.78 .83 .22 .17],'back','r');
```

Note that the number of apostrophes increases dramatically (and the readability decreases in the same way) as soon as the `Callback` commands become less trivial: this is why it is generally better to use function files in the `Callback`; on the other hand, this would lead to a large number of global variables being used and thus it is necessary to explore a different solution.

Note, by the way, that in the example that we are developing, anything is related to **figure** 1; if we replace all the pointers `pf1` with the command `gcf` we may refer to any active window.

Up to now we have used only the `hot` color map, but we know that MATLAB has many other predefined maps. Let us now address the problem of dynamically changing the maps from the active window. It is clear that to create a set of **radio buttons** is a formidable task, so let us turn our attention to a new control object: the **Pop-up menu**.

For the sake of simplicity, let us start with two maps only, `hot` and `cool`; moreover let us suppose that we may select only one map for each run. Above we have used global variables as a means of making `boom3d` data available to the program at each time instant, and we have discarded input parameters since they are set once and for all in the callback. But there is a different solution, based on the consideration that graphic objects, too, are available at each time instant (if we know their pointers, of course). In particular, the easiest objects to address are current **figure** and **axes**, whose pointers are returned by the functions `gcf` and `gca`, respectively.

Thus, if we associated data (pointers, colormaps, etc.) to graphic objects, we could avoid the use of global variables to 'keep track' of data; moreover we would increase the generality, since each program is linked to the current window, along with its data: we have an autonomous entity, a stand-alone application defined by the figure–program pair.

To store data into a graphic object we use the attribute `UserData`, mentioned on page 353 which we have not discussed in detail before as it could appear too extravagant: now that we are in need of such an attribute, its presence seems more natural. It has been presented as an attribute common to all the objects; it is simply a 'free cell', where we may store any data. Its relevance can be deduced by the following lines for solving the color map problem. To select `hot` or `cool` color maps we define a **Pop-up Menu** object as follows:

```
pmn_col=uicontrol(gcf,'style','popup','units','norm',...
'pos',[.01 .9 .2 .3],'string','hot|cool');
```

Note that the menu items are separated by the vertical bar |; moreover, the dimensions defined by the `Position` attribute are referred to the menu *when it is open*. The `Callback` attribute will associate the maps with the object whose pointer is most easily obtained: the current **figure**; then the program `boom3d` will set the current color map to the matrix stored in the `UserData` attribute of the current **figure**. Thus we have avoided global variables, the variables involved are all local, or, more correctly, are related to specific instances of graphic objects; moreover, each figure may have its own map.

We define the `CallBack` as follows:

```
set(pmn_col,'CallBack',['if (get(pmn_col,',...
'''value'')==1),set(gcf,''userdata'',',','hot),',...
'elseif (get(pmn_col,''value'')==2),',...
'set(gcf,''userdata'',cool),end']);
```

Note the use of the attribute `Value`: its value is the index of the item selected from the menu, 1 for the first item, 2 for the second, and so on. The function `boom3d` will pick the current color map from the `UserData` attribute of the current figure, then its first lines are to be changed as follows:

```
function boom3d(map)
global prd_end
fig_p=gcf;
ax_p=gca;
map=get(fig_p,'userdata');
```

Finally, we initialize the `UserData` attribute to the value `hot` (otherwise we are forced to select a map before the first run):

```
set(gcf,'UserData',hot)
```

Note that as the input parameter is now useless, the function heading can be

```
function boom3d
```

and the **radio button** callback must be modified.

Up to now to change color maps we must stop execution, select a different map from the menu and restart the program, but the previous shots are lost.[2] We are now going to discuss a modification to define more maps and to switch among them during run time.

To achieve the first goal we must alter the value of the attribute `String`:

```
set(pmn_col,'String',['hsv|gray|hot|cool|',...
'bone|copper|pink|prism|jet|flag'])
```

The number of lines to store in the `Callback` is now too large, so it is better to introduce a new function, MAPFC.M:

```
function mapfc(n)
maps=str2mat('hsv','gray','hot','cool',...
'bone','copper','pink','prism','jet','flag');
eval(['set(gcf,''UserData''',',maps(n,:),');']);
```

---

[2]This behavior is inevitable, as we have set to 'none' the value of the attribute **Erasemode**, see page 359.

A less compact, although more readable, solution makes use of if,...,elseif,...,end:

```
if n==1,
    set(gcf,'UserData',hsv);
elseif n==2,
    set(gcf,'UserData',gray);
elseif n==3,
    set(gcf,'UserData',hot);
    .....
```

The Callback is now simply

```
set(pmn_col,'val',3,'call',...
['mapfc(get(pmn_col,''value''))')]);
```

The initial value 3 (hot map) is due to the initial value of the UserData attribute in the current **figure** defined above.

Changing the colors during run time is simple: it is sufficient to alter the file BOOM3D.M by moving the line map=get(fig_p,'userdata'); inside the while loop.

In our program, the elevation and azimuth angles of the shots are generated at random. Let us now define a control object to vary the azimuth angle ϕ_0 continuously. As we already know, to change variables continuously we must use a control of type **Slider**. Since we have no more room in our window, figure 1, we will create a new window which contains only the azimuth angle control:

```
pf2=figure(2);
set(pf2,'Units','centimeters','Pos',[5 5 10 2],...
'Name','Azimuth control','MenuBar','none',...
'NumberTitle','off');
```

Inside this window we define a **Slider** whose value (the azimuth angle in degrees) is given to the UserData attribute in figure 2:

```
psl_angle=uicontrol(pf2,'style','slider','u','n','p',...
[.1 .01 .8 .5],'Min',0,'Max',180,'val',90,'Call',...
'set(pf2,''UserData'',get(psl_angle,''val''))');
set(pf2,'User',90)
```

Again, the last instruction initializes the UserData value. Note the abbreviations used in the above lines: 'u' for 'Units', 'n' for 'normalized', 'p' for 'Position', etc. (take loss of readability into account!) This time the attributes Min and Max play a crucial role: they define the angle range of variation.

A last modification affects the function boom3d, whose line

```
phi0=rand(1)*pi/2+pi/4;
```

is to be replaced with

```
phi0=get(2,'userdata')*pi/180;
```

Now running the program we see that the trajectories of the shots belong to a vertical plane that can be rotated at will with the **Slider** control. Some refinements remain to be done, i.e. to add some text; let us start with the creation of labels for the extreme values:

```
uicontrol(pf2,'style','text','u','n','p',...
[.01 .1 .08 .3],'string',num2str(get(psl_angle,'min')));
uicontrol(pf2,'style','text','u','n','p',...
[.91 .1 .08 .3],'string',num2str(get(psl_angle,'max')));
```

To display the current value of the angle we use an **Editable Text** control, so as to be able to both read and input the value of the current angle from keyboard:

```
uicontrol(pf2,'style','text','u','n','p',...
[.3 .6 .4 .2],'string','Azimuth angle');
ped_angle=uicontrol(pf2,'style','edit','u','n','p',...
[.71 .6 .11 .2],'string',num2str(get...
(psl_angle,'val')),'call',...
['q=str2num(get(ped_angle,''string''));',...
'set(pf2,''user'',q);set(psl_angle,''val'',q);']);
```

To input the angle from keyboard it is sufficient to type the value in the cell and click the mouse button with the mouse cursor anywhere outside the cell, or to press the RETURN key, or the key combination CTRL-RETURN; the latter method is mandatory for multi-line **Editable Text** objects, for which a single RETURN is simply a line feed, not a data acquisition. Note that the only actions defined in the control `CallBack` are to assign the value (actually it is a string that must be converted into a numeric value) of the control to the `UserData` attribute of figure 2, and to update the **slider**; on the other hand, any variation in the **slider** bar should also update the value of the `String` attribute of the **Editable Text** object:

```
set(psl_angle,'call',...
['set(pf2,''userdata'',get(psl_angle,''val''));',...
'set(ped_angle,''string'',',...
'num2str(get(psl_angle,''val'')))']);
```

All the operations considered up to now can be collected in the single M-file `b3dinit`; further modifications may be considered in order to achieve more generality, for instance not using fixed figure numbers to avoid overlapping with previously defined figures.

We conclude this section by giving for the sake of clarity, the whole of the files B3DINIT.M and BOOM3D.M: note that some lines are slightly different in order to achieve more generality.

B3DINIT.M

```
pf1=figure;
set(pf1,'user',hot)
pa1=axes;
prd_start=uicontrol(pf1,'Style','radio','Units',...
'norm','Pos',[0.01 0.06 .2 .05], 'Value',0,...
'String','Start');
prd_end=uicontrol(pf1,'Style','radio','Units',...
'norm','Pos',[0.01 0.01 .2 .05], 'Value',1,...
'String','Stop');
set(prd_start,'Callback',...
'set(prd_end,''Value'',0);boom3d;',...
'Interruptible','yes');
set(prd_end,'Callback',...
'set(prd_start,''Value'',0)');
global prd_end
pps_ch=uicontrol(pf1,'Style','push','Units',...
'norm','Pos',[.79 .01 .2 .09],'String',...
'Close','Callback','close(pf2);close(pf1)');
ptx_exec=uicontrol(pf1,'Style','text','Units',...
'norm','Pos',[.01 .11 .2 .05],...
'String','Execution');
pfr_exec=uicontrol(pf1,'Style','frame',...
'Units','norm','Pos',[0 0 .22 .17]);
set(pfr_exec,'back','r');
set(ptx_exec,'back','b','fore','w');
set(prd_start,'back','w','fore','k')
set(prd_end,'back','w','fore','k')
pck_box=uicontrol(pf1,'Style','check','Units',...
'norm','Pos',[0.79 .84 .2 .05],'String','Box',...
'Callback',['set(gca,''box'',''off'');',...
'if (get(pck_box,''Value'')==1)',...
'set(gca,''box'',''on''); end']);
pck_axes=uicontrol(pf1,'Style','check','Units',...
'norm','Pos',[0.79 .89 .2 .05],'String','Axes',...
'Callback',['axis(''off'');',...
'if (get(pck_axes,''Value'')==1)',...
'axis(''on''); end']);
ptx_axes=uicontrol(pf1,'Style','text','Units',...
'norm','Pos',[0.79 .94 .2 .05],'String','Control');
pfr_axes=uicontrol(pf1,'Style','frame','Units',...
'norm','Pos',[0.78 .83 .22 .17],'back','r');
pmn_col=uicontrol(pf1,'Style','popup','Units',...
```

```
'norm','Pos',[.01 .9 .2 .3]);
set(pmn_col,'String',['hsv|gray|hot|cool|',...
'bone|copper|pink|prism|jet|flag'],'val',3,'call',...
['mapfc(get(pmn_col,''value''))']);
pf2=figure;
set(pf2,'units','centimeters','pos',[5 0 10 1.5],...
'Name','Azimuth control','MenuBar','none',...
'NumberTitle','off');
psl_angle=uicontrol(pf2,'style','slider','u',...
'n','p',[.1 .01 .8 .4],'min',0,'max',180,'val',...
90,'call',['set(pf2,''userdata'','',...
'get(psl_angle,''val''));set(ped_angle,',...
'''string'',num2str(get(psl_angle,''val'')))']);
set(pf2,'user',90)
uicontrol(pf2,'style','text','u','n','p',...
[.01 .1 .08 .3],'string',num2str(get(psl_angle,'min')));
uicontrol(pf2,'style','text','u','n','p',...
[.91 .1 .08 .3],'string',num2str(get(psl_angle,'max')));
uicontrol(pf2,'style','text','u','n','p',...
[.3 .6 .4 .3],'string','Azimuth angle');
ped_angle=uicontrol(pf2,'style','edit','u','n','p',...
[.71 .6 .11 .3],'string',num2str(get...
(psl_angle,'val')),'call',...
['q=str2num(get(ped_angle,''string''));',...
'set(pf2,''user'',q);set(psl_angle,''val'',q);']);
```

BOOM3D.M

```
function boom3d
global prd_end
fig_p=gcf;
ax_p=gca;
g=9.81;
pf2=fig_p+1;
pnt_p=plot3(0,0,0,'.','erasemode','none',...
'markersize',2);
set(ax_p,'xlim',[-5000 5000],'ylim',...
[-5000 5000],'zlim',[0 30000]);
while(get(prd_end,'Value')==0),
map=get(fig_p,'userdata');
  v0=randn(1)*1000;
  theta0=rand(1)*pi/2+pi/4;
  phi0=get(pf2,'userdata')*pi/180;
  i0=fix(rand(1)*64)+1;
```

```
c0=map(i0,:);
set(pnt_p,'color',c0);
v0x=v0*cos(theta0)*cos(phi0);
v0y=v0*cos(theta0)*sin(phi0);
v0z=v0*sin(theta0);
t=0;y=0;x=0;z=0;
while (z>=0)&(x>-5000)&(x<5000)&...
(y>-5000)&(y<5000)&(z<30000),
  t=t+.1;
  x=v0x*t;
  y=v0y*t;
  z=v0z*t-.5*g*t^2;
  set(pnt_p,'xdata',x,'ydata',y,'zdata',z);
  drawnow
end
end
```

B.2   Menu

In this section we deal with **uimenu** objects: these are windows menus (and possibly submenus, too) that appear as standard window menus. We will develop the material via an example, simpler than the previous one, to show the characteristics of this kind of object.

Let us start with a file using the `pcolor` command (see page 81). Consider an 8×8 matrix whose entries are the integers from 1 to 64 and let us associate with it a matrix of the same size whose ijth entry is 1 if the element of the same place in the first matrix divides 3, 0 otherwise; we now visualize this second matrix using pcolor and an interpolated shading:

```
C=1:64;
C=reshape(C,8,8);
C1=rem(C,3)==0;
pcolor(C1);
shading interp
```

To obtain a satisfactory result a high-resolution screen with 256 colors, at the least, is required.

We now want to alter the color options by adding the menu **Colors** to the standard window menus

```
p=uimenu(gcf,'Label','&Colors');
```

Thus the menu **Colors** appears in the current window, after the standard **Help** menu. Note that the letter C is underlined and, as in all Windows applications, the key combination ALT-C is a shortcut to selecting this menu quickly: this is due to the character **&** that we have used in defining the value of the Label attribute. Now we must fill this menu with some material, in particular more **uimenu** objects (to obtain submenus), whose parent is now the **Colors** menu; specifically, we are interested in changing color maps and shading. Let us start with color maps:

```
p1=uimenu(p,'label','&Maps');
```

We now create a submenu to select among various color maps:

```
uimenu(p1,'label','&Gray','callback','colormap(gray)');
uimenu(p1,'label','&Hot','callback','colormap(hot)');
uimenu(p1,'label','&Bone','callback','colormap(bone)');
uimenu(p1,'label','&Copper','callback',...
'colormap(copper)');
uimenu(p1,'label','&Jet','callback','colormap(jet)');
uimenu(p1,'label','C&ool','callback','colormap(cool)');
```

Note the use of the **&** character in the last line; moreover the attribute CallBack, by now well known to the reader, has had its *debut*; its use is exactly the same as that described in the previous section.

We now address the problem of handling color shading through submenus:

```
p2=uimenu(p,'label','&Shading');
uimenu(p2,'label','Flat',...
'callback','shading flat');
uimenu(p2,'label','Faceted','Separator','on',...
'callback','shading faceted');
uimenu(p2,'label','Interp','Separator','on',...
'callback','shading interp');
```

The new attribute introduced in these lines is Separator, whose role is to trace menu items separated by lines.

Two more attributes are still to be described: Checked and Position. The latter differs from the homonymous attribute described up to now, because to define the initial coordinates and extension of a menu object makes no sense. To understand the meaning of this attribute we introduce a new option. We have considered the integers divisible by 3; now define a menu dividing different numbers, in particular the primes between 2 and 10 (any others would result in too small a number of points), i.e. 2, 3, 5 and 7. We now define a menu **Dividing** as follows:

```
p3=uimenu(gcf,'Label','&Dividing','Position',1);
```

The role of the Position attribute is to define an order for **uimenu** objects; this can be verified by typing the same command without the Position attribute (*having previously deleted* the previous object):

```
delete(p3)
p3=uimenu(gcf,'Label','&Dividing');
```

Now we define a submenu of p3 to select a prime between 1 and 10; we know how to do this now, it is sufficient to use submenus with the CallBack property suitably defined. One more problem arises: how do we know which is the current prime? In our example we can detect this by direct inspection, but if we consider more primes, for instance the first fifty primes, how do we distinguish between 19 and 23? The solution is to use a check mark to check the current item; this is done by setting the attribute Checked to 'on'; each time an item is selected its CallBack must deactivate the other check marks and check itself:

```
p3f(1)=uimenu(p3,'Label','By 2','call',...
['set(p3f,''checked'',''off'');',...
'set(p3f(1),''checked'',''on'');',...
'C1=rem(C,2)==0;pcolor(C1)']);
p3f(2)=uimenu(p3,'Label','By 3','Checked','on','call',...
['set(p3f,''checked'',''off'');',...
'set(p3f(2),''checked'',''on'');',...
'C1=rem(C,3)==0;pcolor(C1)']);
p3f(3)=uimenu(p3,'Label','By 5','call',...
['set(p3f,''checked'',''off'');',...
'set(p3f(3),''checked'',''on'');',...
'C1=rem(C,5)==0;pcolor(C1)']);
p3f(4)=uimenu(p3,'Label','By 7','call',...
['set(p3f,''checked'',''off'');',...
'set(p3f(4),''checked'',''on'');',...
'C1=rem(C,7)==0;pcolor(C1)']);
```

Remark: the example presented in this section behaves quite differently on low 16 colors and high 256 colors resolution screens. Those who wish to use the latter can add the following command:

```
pb=uicontrol(1,'style','push','units',...
'normalized','position',[.01 .01 .2 .05],...
'string','Spin Map','callback','spinmap');
```

Note that by changing the prime number the menu color is also redefined. This is due to the command pcolor, which resets color maps and shading. The reader who wishes to overcome this problem is now in position to do so by replacing this command with a series of commands set, XData, etc.

The sequence of commands shown can be used in different contexts: with a high-resolution screen the reader may try the color maps of the example mentioned on page 361, as follows:

```
load earth
image(X)
axis('equal')
axis('off')
p1=uimenu(gcf,'label','&Maps');
uimenu(p1,'label','&Gray','callback','colormap(gray)');
uimenu(p1,'label','&Hot','callback','colormap(hot)');
uimenu(p1,'label','&Bone','callback','colormap(bone)');
uimenu(p1,'label','&Copper','callback','colormap(copper)');
uimenu(p1,'label','&Jet','callback','colormap(jet)');
uimenu(p1,'label','C&ool','callback','colormap(cool)');
uimenu(p1,'label','&Map','callback','colormap(map)');
pb=uicontrol(1,'style','push','units','normalized',...
'position',[.01 .01 .2 .05],'string','Spin Map',...
'callback','spinmap');
```

B.3   Using the mouse

Before discussing the attributes to handle mouse actions, recall of the attribute
CurrentPoint is in order. This attribute, both in the **figure** case and in the **axes**
case, stores the value of the current mouse cursor coordinates. The difference is
that in the first case the coordinates are expressed according to the current **figure**
Units, while in the second case the units are those of the current **axes**, and the
point is considered in three dimensions, even for 2D plots; in fact:

$$get(gca,'CurrentPoint')$$

returns a 2×3 matrix whose rows are (x, y, z) coordinate triplets; in the 2D case
the two rows are equal but for the third coordinate, which is fictitious and is by
default 0 for the first row and 1 for the second. In the 3D case the matrix contains
the coordinates of two points, the extremal coordinates of the segment orthogonal
to the screen intersecting the axes box; note that the coordinates are computed
even if the pointer is outside the current axes.

In this way we are able to know, at any time, the position of the pointer with
respect to two different reference frames: the first one can be called the 'absolute
coordinate system' and is expressed in **figure** coordinates, which in turn can be
expressed in different measure units by changing the value of the Units attribute
of the current figure, while the second can be called the 'user coordinate system',
expressed in current **axes** coordinates, i.e. those appearing on the axes according
to the data to be plotted. We now need some CallBacks to be associated with
mouse actions.

Figure objects have three attributes which are devoted to handle mouse actions:
WindowButtonDownFcn and WindowButtonUpFcn, which are callbacks activated by
pressing and releasing the mouse button, and WindowButtonMotionFcn, which is
related to mouse cursor motion; as a simple example, type:

```
p=plot(0,0,'.','Erasemode','none');
set(gcf,'WindowButtonMotionFcn',...
['h=get(gca,''CurrentPoint'');',...
'set(p,''xdata'',h(1,1),''ydata'',h(1,2))'])
```

This simple program leaves a visible track of cursor motion.

The other objects have the only attribute `ButtonDownFcn`, i.e. a callback activated when the mouse button is pressed near the object.

The capacity to distinguish between button pressing and releasing is characteristic of the **figure** object only, and is usually used to define actions such as 'zooming'. If, for instance, we want to define a zoom on the Mandelbrot set discussed on page 82, we can use the following commands. By the way, note that the MATLAB function `zoom` (see page 73) cannot be used in this context; since a fractal object has infinitely more detail levels than any zoom requires new computations and it is not sufficient to alter the current axis limits:

```
set(gcf,'UserData',[-2 -1.25;0.5 1.25]);
set(gcf,'WindowButtonDownFcn',['pc=get',...
'(gca,''CurrentPoint'');qc=get(gcf,''UserData'');',...
'qc(1,1:2)=pc(1,1:2);set(gcf,''UserData'',qc)']);
set(gcf,'WindowButtonUpFcn',['pc=get',...
'(gca,''CurrentPoint'');qc=get(gcf,''UserData'');',...
'qc(2,1:2)=pc(2,1:2);set(gcf,''UserData'',qc)']);
```

Thus the matrix defined in the `UserData` attribute of the current **figure** contains the coordinates of the point where the mouse button has been pressed and of that where it has been released. It is now sufficient a button to start computation of the Mandelbrot set zoom:[3]

```
uicontrol(gcf,Style,'push','u','n',...
'p',[.8 0 .2 .1],'String','Zoom',...
'CallBack','mand(get(gcf,''User''))')
```

The function `mand` is a variation of that presented on page 82, with less points in order to reduce the computational burden:

```
function mand(P)
x=linspace(P(1,1),P(2,1),50);
y=linspace(P(1,2),P(2,2),50);
[X,Y]=meshgrid(x,y);
C=X+j*Y;
W=100*ones(size(C));
```

---

[3]Of course this could simply be the last command in `WindowButtonUpFcn`, but since an accidental click would force a useless and time-consuming computation, it is better to have a sort of 'confirm' button.

```
Z=zeros(size(C));
for n=1:50,
Z=Z.*Z+C;
h=find(abs(Z)<2);
if ~isempty(h),
W(h)=n*ones(size(h));
else
break
end;
end;
pcolor(X,Y,W);shading interp
```

Finally, we can use **Static Text** controls to visualize current coordinates after mouse actions; then the main program becomes the following:

```
set(gcf,'UserData',[-2 -1.25;0.5 1.25]);
axis([-2 0.5 -1.25 1.25]);
set(gcf,'WindowButtonDownFcn',['pc=get',...
'(gca,''CurrentPoint'');qc=get(gcf,''UserData'');',...
'qc(1,1:2)=pc(1,1:2);set(gcf,''UserData'',qc);',...
'set(px_corr,''String'',[''Real part: '',',...
'num2str(qc(1,1)),'' '',num2str(qc(2,1))]);',...
'set(py_corr,''String'',',...
'[''Imaginary coefficient: '',',...
'num2str(qc(1,2)),'' '',num2str(qc(2,2))]);'])
set(gcf,'WindowButtonUpFcn',['pc=get',...
'(gca,''CurrentPoint'');qc=get(gcf,''UserData'');',...
'qc(2,1:2)=pc(2,1:2);set(gcf,''UserData'',qc);',...
'set(px_corr,''String'',[''Real part: '',',...
'num2str(qc(1,1)),'' '',num2str(qc(2,1))]);',...
'set(py_corr,''String'',',...
'[''Imaginary coefficient: '',',...
'num2str(qc(1,2)),'' '',num2str(qc(2,2))]);'])
uicontrol(gcf,'Style','push','u','n',...
'p',[.8 0 .2 .1],'String','Zoom',...
'CallBack','mand(get(gcf,''User''))')
px_corr=uicontrol(gcf,'style','text','u','n','p',...
[0 0 .6 .05],'string','Real part: -2 0.5');
py_corr=uicontrol(gcf,'style','text','u','n','p',...
[0 .95 .7 .05],'string',...
'Imaginary coefficient: -1.25 1.25');
```

The reader can try different modifications of this scheme (tracing a rectangle while the cursor moves on the screen, modifying values from the keyboard by using **Editable Text**, color map selections with a new **uimenu**).

We conclude this appendix with an example that uses all the material presented in the book.

The problem is to stabilize an inverted double pendulum using graphic commands to display the result (Figure B.1) and the robustness with respect to input disturbances.

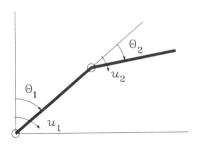

Figure B.1 Inverted double pendulum.

Let θ_1 be the angle between the first link and the local vertical and θ_2 the angle between the two links; suppose each actuator (motor) can give torque u_i, $i = 1, 2$. The double pendulum equations, linearized around the vertical (unstable) equilibrium position are, assuming length and mass unitarity for both pendula,

$$
\dot{x} = \begin{pmatrix} 0 & 1 & 0 & 0 \\ 9.8 & 0 & -9.8 & 0 \\ 0 & 0 & 0 & 1 \\ -9.8 & 0 & 29.4 & 0 \end{pmatrix} x + \begin{pmatrix} 0 & 0 \\ 1 & -2 \\ 0 & 0 \\ -2 & 5 \end{pmatrix} u
$$

$$
y = \begin{pmatrix} 1 & 0 & 0 & 0 \\ 0 & 0 & 1 & 0 \end{pmatrix} x
$$

where the state vector is $x = (\theta_1, \dot{\theta}_1, \theta_2, \dot{\theta}_2)$, the inputs are the control joint torques and the output the joint angles.

As a control law we consider a state feedback; let us build the SIMULINK scheme shown in Figure B.2. Note the presence of a sum block on the input to add disturbances.

Let us design an LQ compensator with unitary weighting matrices, $Q = I_4$, $R = I_2$, by using the Control Toolbox command K=lqr(A,B,Q,R);

We must now visualize the results. To display the motion of the double pendulum, we initially draw it in any initial configuration, then change the pendulas' position at each time instant by changing their attributes 'XData' and 'YData' according to the results of the SIMULINK integrator. Then the output y must input an M-function to update these attributes; the initialization phase is done by

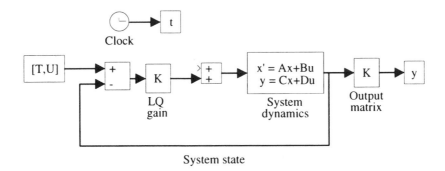

Figure B.2 SIMULINK scheme of the inverted double pendulum with state feedback.

```
p=plot(0,0,'-','linewidth',4);
axis([-2 2 0 2]);
set(p,'erasemode','xor');
```

The MATLAB function MOVEPEND.M, computes the Cartesian coordinates of the link end points and changes the corresponding 'XData' and 'YData' attributes:

```
function y=movepend(x)
p=get(gca,'children');
set(p,'ydata',[0 cos(x(1)) cos(x(1))+cos(x(2))],...
'xdata',[0 sin(x(1)) sin(x(1))+sin(x(2))]);
y=x;
```

The SIMULINK scheme is shown in Figure B.3.

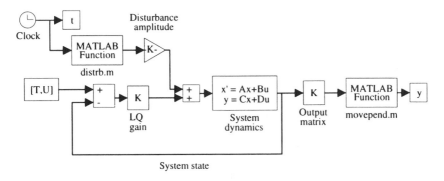

Figure B.3 SIMULINK scheme with graphic animation.

Starting the simulation with nonzero initial state, the action of the controller will be apparent on MATLAB 'Figure 1'.

We now add constant disturbances. Working with SIMULINK we are forced to fix the evolution of the disturbances *a priori*, but with a graphic simulation it is possible to insert disturbances asynchronously *during run time*. We consider two **check boxes**, one for each disturbance, and type the following initialization commands:

```
p2=uicontrol(gcf,'style','check','string','Disturbance 1',...
'units','normalized','pos',[0.1 0 .2 .07],'back','w');
p3=uicontrol(gcf,'style','check','string','Disturbance 2',...
'units','normalized','pos',[0.4 0 .2 .07],'back','w');
set(gcf,'userdata',[p2 p3]);
```

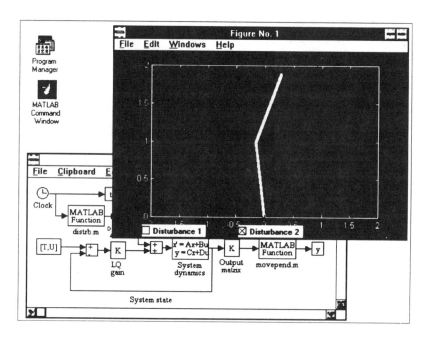

Figure B.4 Complete scheme.

The disturbances enter the sum block through the function DISTRB.M:

```
function y=distrb(x);
p=get(gcf,'userdata');
y(1)=get(p(1),'value');
y(2)=get(p(2),'value');
```

This function returns two variables, whose values are 1 if the corresponding **check box** is selected, 0 otherwise; then we can control the disturbances independently. Note the dummy input x, that may be for instance the simulation time (`clock`) to consider different disturbances (Figure B.4).

Appendix C

S-function

SIMULINK saves its schemes in ascii files called *S-function*; they are like the usual MATLAB M-file with a special syntax for the call.

However one may also write 'by hand' an *S-function*. This solution, as we will show in the following, allows us to overcome some problems.

However, it is important to emphasize the fact that to write an *S-function* by hand is a very 'specialized' operation and hence this procedure should only be used in the presence of very sophisticated applications.

An important example of the use of an '*S-function* written by hand' is the implementation of control algorithms. It is quite difficult to design a control algorithm in SIMULINK, i.e. a controller based on conditional jumps or, in other words, controllers which may easily be implemented via **if**, **then**, **else** constructs. This class of controllers is easily implemented in MATLAB. Although they could be implemented in the SIMULINK schemes via *MATLAB-Fcn* block, such a solution would reduce the computational efficiency.

A different solution is to write an *S-function*.

C.1 Writing an S-function

The ways of writing an *S-function* are as follows.

Graphic	SIMULINK scheme
M-file	file with MATLAB syntax
MEX-file	C or Fortran subroutine

In the following we focus our attention on writing an *S-function* with a MATLAB syntax.

As concerns MEX-files, the interested reader may refer to the manuals.

An *S-function* may be achieved via the SIMULINK scheme by using the *Inport* and *Outport* ports to identify the scheme inputs and outputs respectively.

S-functions obtained via the SIMULINK scheme have some peculiarities. Indeed, these files contain the topological description of the scheme. Information such as number of states, outputs, inputs, and so on, are stored in an implicit form.

On the other hand, *S-functions* written by hand or MEX-files must 'declare' the states, their dynamics, outputs and inputs.

A last observation: an *S-function* written by hand needs a computation time about twenty times greater than that needed by SIMULINK schemes (the MEX-file improves the numerical efficiency by reducing the computational time by about 30% with respect to the SIMULINK schemes.)

The syntax to call an *S-function* is

$$\texttt{[sys, x0]} = model(\texttt{t, x, u, flag, P1, P2 ...})$$

where t is the time instant in which the function is evaluated, x is the 'initial' state vector, the u are the inputs, and the parameters P1, P2, ... are variables that are passed directly to the function. Moreover, there is the variable flag, whose role is the following.

The output generated by the function depends on the flag variable values. In particular, the outputs of an *S-function* are related to the values of flag in the following way:

flag=0	structural information and initial conditions
flag=1	continuous time states derivative dx/dt
flag=2	discrete time states next value $x(n+1)$
flag=3	outputs values y
flag=4	discrete time next sample time instant
flag=5	function roots (singularities)
flag=9	operations to evaluate at the simulation end

All the information needed to evaluate the system behavior may be recovered by means of some calls to the *S-function* with suitable flag values.[1] During a simulation, at each sample instant, SIMULINK makes more calls to the *S-function* with different flag values. Some examples will clarify the SIMULINK way of working.

For the moment, it is useful to highlight the flag=0 case. During a simulation, SIMULINK calls *S-functions* with this value only one time, i.e. before starting the simulation itself. The output given by an *S-function* with flag=0 allows one to check the connection matching and the system syntax.

In particular, with flag=0 the *S-function* outputs are unrelated to other inputs values:

$$\texttt{[sys, x0]} = model(\texttt{[],[],[]},0)$$

---

[1] Due to numerical problems the flag variable may also assume the values $-1, -2$ with the same meaning of $1, 2$ respectively.

gives as output the initial conditions vector x0 and the vector sys, where the following structural information is collected:

sys(1)   continuous-time states number
sys(2)   discrete-time states number
sys(3)   outputs number
sys(4)   inputs number
sys(5)   discontinuous function roots number
sys(6)   direct feedthrough (algebraic loop)
sys(7)   number of sample times (optional)

It is possible to design an *S-function* which is able to match the input dimensions automatically. This is done by inserting -1 as the dimension of all the quantities which must be adapted.

The adaptable quantities are the number of continuous-time states, the discrete-time states and the outputs, but each of these quantities must be expressed as a function of the input dimension.

If the call to an *S-function* with flag=0 is done with three output variables, i.e. with a syntax such as

$$[\text{sys, x0, str}] = model([],[],[],0)$$

the output consists of the usual sys and x0 vector and the str vector. This last is a vector containing the 'name' of the state variables. For example, if the state variables of *model* are the outputs of two integrators, the str vector is:

/model/Integrator
/model/Integrator1

and this means that the first element of the state vector is the state of the first *Integrator* block, the second element is the state of the second *Integrator* (i.e. that named Integrator1), and so on. This information is very useful in all those operations, such as linearization, which extract the state-space model of a SIMULINK scheme.

When the call with flag=0 is performed with four output variables

$$[\text{sys, x0, str, ts}] = model([],[],[],0)$$

the variable ts is a two-element vector. The first entry is the sample time, the second is the offset.

In the presence of multi-rate or hybrid systems ts is a matrix whose columns have the same meaning as before, and consists of as many rows as the number of different sample times and/or offsets implemented.

In the following we explore how to write an *S-function* for a continuous-time system and for a discrete-time one; the extension to hybrid system is straightforward.

C.2 Continuous-time systems

The *S-function* which reproduces the behavior of the following system:

$$\dot{x} = Ax + Bu$$
$$y = Cx + Du$$

where

$$A = \begin{pmatrix} 1 & 2 \\ 0 & 1 \end{pmatrix} \quad B = \begin{pmatrix} 0 \\ 1 \end{pmatrix} \quad C = \begin{pmatrix} 2 & 1 \end{pmatrix} \quad D = 0.1$$

is

```
function [sys, x0, str, ts] = system(t,x,u,flag)
A=[1 2; 0 1];
b=[0 1]';
c=[2 1];
d=0.1;
if      abs(flag) == 1          If flag=1, returns the continuous-time
                                      state derivatives.
        sys=A*x+b*u[1];
elseif flag == 3                If flag=3, returns the output values.
        sys = c*x+d*u[1];
elseif flag == 0                If flag=0, returns the initial conditions
                                      and the system structure.
        [m,n]=size(d);          m is the number of outputs
                                n is the number of inputs
        sys = [length(b), 0, m, n, 0, any(d~=0)]';
                                length(b) is equal to the numbers of states.
                                The last parameter of sys is different from zero
                                if there is a direct input-output feedthrough.
        x0=[0 0]';              initial conditions
        str=['x1'; 'x2'];       state variable names
        ts=[0; 0];
else                            For any other values of flag the function
        sys=[];                 return an empty value.
end
```

The *S-function* structure is not very complex. In the presence of a dynamic system described by very complex equations, it may be useful to use this procedure to model the system. Note that the output variables str and ts are optional; however their presence is strongly recommended: the first one improves the clarity of the results during analysis operation (e.g. linearization), the second improves the computational performance.

C.3   Discrete-time systems

As mentioned before, *S-functions* are preferred to implement control algorithms in SIMULINK.

A simple discrete-time system with a saturation is analyzed in the following.

This *S-function* represents a saturated ZOH sample and hold, with upper state limit equal to 3, i.e. the state value cannot be greater than 3.

```
function [sys, x0, str] = system(t,x,u,flag)


offset = 0;
ts = 1.0;
```

`if      abs(flag) == 2`	*If* flag=2, *returns the next step state value.*
`         sys(2)=x(1);`	*Use a further state to store the actual state value.*
`         if abs(round((t-offset)/ts)-...`	
`            (t-offset)/ts)<1e-8`	*Check if it is a sample instant.*
`            if u[1]>3 sys(1)=3;`	*The control 'algorithm'.*
`            else`	
`                       sys(1)=u[1];`	
`            end`	
`         else`	
`             sys(1)=x(1);`	*It is not a sample instant.*
`         end`	
`elseif flag == 3`	*If* flag=3, *returns the output values.*
`         sys(1) = x(2);`	*Output depends on the actual state value instead of the next one.*
`elseif flag == 4`	*If* flag=4, *returns the value of the next sample instant.*
`         ns = (t - offset)/ts;`	
`         sys = offset + (1 + floor(ns + 1e-13*(1+ns)))*ts;`	
`elseif flag == 0`	*If* flag=0, *returns the initial condition and the system structure.*
`         sys = [0, 2, 1, 1, 0, 0]';`	
`         x0=[0 0]';`	
`         str=['z'];`	
`else`	*For any other* flag *values the function*
`         sys=[];`	*returns an empty value.*
`end`	

Compared to the continuous-time case, there is also the **flag=4** item which evaluates the next sample instant.

At the **flag=2** item there is a check on the sample instant: this is necessary to simulate multi-rate discrete-time systems.

However in the presence of a purely discrete-time *S-function* with a single sample time, the use of the `flag=4` item and the check on the sample time instant are redundant. Indeed, they may be avoided by using the fourth output parameter and specifying the sample time at initialization. In our example, the *S-function* may be rewritten in the following, more efficient, form:

```
function [sys, x0, str, ts] = system(t,x,u,flag)
if       abs(flag) == 2        If flag=2, returns the next step state value.
         sys(2)=x(1);                 Uses a further state
                                      to store the actual state value.

         if u[1]>3 sys(1)=3;          The control 'algorithm'.
             else
                   sys(1)=u[1];
         end
elseif flag == 3               If flag=3, returns the output values.
         sys(1) = x(2);               Output depends on the actual state
                                      value instead of the next one.

elseif flag == 0
         sys = [0, 2, 1, 1, 0, 0]';
         x0=[0 0]';
         str=['z'];
         ts=[1 0]';                   There is one sample time
                                      sample period=1; offset=0.
else
         sys=[];
end
```

Blocks with more sample times must use the `flag=4` item and check the sample instant. However, if the sample times are multiples of each other and have no offset, the efficiency may be improved by specifying that the block has a single rate equal to the fastest rate (obviously in this case a check on the sample instant must be performed).

C.4 Hybrid systems

In the presence of hybrid systems the *S-function* has both the `flag=1` item and that related to the discrete-time parts.

It is important to stress that the dimension of the state vector **x** is equal to the sum of the continuous-time states plus double the discrete-time ones, but the dimension of the output vector **sys** is either that of the continuous-time states or double the discrete-time ones according to the `flag` value.

C.5 Observations

During a simulation SIMULINK calls the *S-function* with different `flag` values. In particular, the SIMULINK calls are performed in the following order:

'Discrete' instant	'Continuous' instant
flag=0	flag=0
flag=4	
flag=3	flag=3
flag=2	
flag=1	flag=1

where 'discrete' instant means a time instant where the discrete part is also evaluated, whereas 'continuous' instant is a time instant where only the continuous part is evaluated. Obviously the `flag=0` call is performed only once before starting the simulation, whereas all the other calls are performed cyclically.

It is interesting to note that the `flag=3` call is evaluated at each instant, i.e. the system outputs are always evaluated.

Moreover, for discrete systems, the next sample instant is evaluated as the first operation. In other words, at time $\bar{t}$ SIMULINK, before updating the state, evaluates the next 'discrete'-time instant $\bar{t} + T$. This way of proceeding results in very complex handling of dynamically varying sample time systems.

As mentioned before, if the *S-function* is purely discrete with a single sample time the computational efficiency may be increased by means of the seventh element of the `flag=0` `sys` vector and by putting the sample time value as the fourth output variable of the `flag=0` call.

References

Numeric analysis and linear algebra

Ahlfors L. V., *Complex Analysis*, McGraw-Hill, New York, 1966.

Barnett S., *Matrix Methods for Engineers and Scientists*, McGraw-Hill, London, 1979.

Bellman R., *Introduction to Matrix Analysis*, McGraw-Hill, New York, 1970.

Coddington E. A., and Levison N., *Theory of Ordinary Differential Equations*, McGraw-Hill, New York, 1955.

Devaney R. L., *An Introduction to Chaotic Dynamical Systems*, Addison-Wesley, Redwood City, CA, 1989.

Dongarra J. J., J. R. Bunch, C. B. Moler, and G. W. Stewart, *LINPACK User's Guide*, SIAM, Philadelphia, 1979.

Egglestone H. G., *Convexity*, Cambridge University Press, Cambridge, 1966.

Gantmacher F. R., *The Theory of Matrices*, Chelsea, New York, 1959.

Golub G., and C. van Loan, *Matrix Computations*, Johns Hopkins University Press, Baltimore, 1983.

Horn R. A., and C. R. Johnson, *Matrix Analysis*, Cambridge University Press, Cambridge, 1985.

Lipschutz S., *Linear Algebra*, Schaum, McGraw-Hill, New York, 1968.

MacDuffee C. C., *The Theory of Matrices*, Chelsea, New York, 1946.

MacLane S., and Birkhoff G., *Algebra*, Macmillan, New York, 1967.

Marden M., *The Geometry of the Zeros of Polynomial in a Complex Variable*, American Mathematical Society, New York, 1949.

MathWorks Inc., *MATLAB Reference Guide*, MathWorks Inc., Natick, Massachusets, 1993.

Press W. H., B. P. Flannery, S. A. Teukolsky, and W. T. Vetterling, *Numerical Recipes*, Cambridge University Press, Cambridge, 1987.

Ralston A., and P. Robinowitz, *A First Course in Numerical Analysis*, McGraw-Hill, New York, 1978.

Rudin W., *Real and Complex Analysis*, McGraw-Hill, New York. 1966.

Stoer J., and R. Burlisch, *Introduction to Numerical Analysis*, Springer Verlag, New York, 1980.

Widder D. V., *Advanced Calculus*, Dover Publications, New York, 1989.

Dynamic systems and control theory

Anderson B. D. O., and J. B. Moore, *Optimal Control, Linear Quadratic Methods*, Prentice Hall, Englewood Cliffs, New Jersey, 1989.

Bailey N. T. J., *The Mathematical Theory of Epidemics*, Hafner, New York, 1957.

Cannon R. H., *Dynamics of Physical Systems*, McGraw-Hill, New York, 1967.

Chen C., *Linear System Theory and Design*, Holt, Rinehart and Winston, New York, 1984.

Desoer C. A., and M. Vidyasagar, *Feedback Systems: Input–Output Properties*, Academic Press, New York, 1975.

Franklin G. F., and J. D. Powell, *Digital Control of Dynamic Systems*, Addison-Wesley, Reading, Massachusets, 1980.

Gandolfo G., *Mathematical Methods and Models in Economic Dynamics*, North-Holland, Amsterdam, 1971.

Grace A., A. J. Laub, J. N. Little, and C. M. Thompson, *Control System Toolbox, User's Guide*, MathWorks Inc., Natick, Massachusets, 1992.

Hoppensteadt F. C., and C. S. Peskin, *Mathematics in Medicine and the Life Sciences*, Springer Verlag, New York, 1992.

Jazwinski A. H., *Stochastic Processes and Filtering Theory*, Academic Press, New York, 1970.

Kailath T., *Linear Systems*, Prentice Hall, Englewood Cliffs, New Jersey, 1980.

Khalil H. K., *Nonlinear Systems*, Macmillan, New York, 1992.

Kučera V., *Discrete Linear Control*, John Wiley, New York, 1979.

Luenberger D. G., *Introduction to Dynamic Systems*, John Wiley, New York, 1979.

Maciejowski J. M., *Multivariable Feedback Design*, Addison-Wesley, Wokingham, 1989.

MathWorks Inc., *SIMULINK User's Guide*, MathWorks Inc., Natick, Massachusets, 1993.

McKendrick A. G., *Mathematics Applied to Medical Problems*, Royal Society, Edinburgh, 1926.

Meirovitch L., *Elements of Vibration Analysis*, 2nd edn., McGraw-Hill, New York, 1986.

Sciavicco L., and B. Siciliano, *Modeling and Control of Robot Manipulators*, McGraw-Hill, New York, 1996.

Thompson J. M. T., and H. B. Stewart, *Nonlinear Dynamics and Chaos*, John Wiley, Chichester, 1986.

Vidyasagar M., *Nonlinear Systems Analysis*, Prentice Hall, Englewood Cliffs, New Jersey, 1993.

Miscellaneous

Boulez P., *Penser la Musique aujourd'hui*, Schott's Söhne, Mainz, 1963.

Prieberg F. K., *Musica ex Machina*, Verlag Ullstein, Berlin, 1960.

Quenau R., *Bâton, Chiffres et Letters*, Gallimard, Paris, 1965.

Schönberg A., *Harmonielehre*, Universal Edition, Vienna, 1922.

Zaripov R. C., *Kibernetika i Muzyka*, Nakua Akademii SSSR, Moscow, 1971.

Index

MATLAB commands are in `typewriter` font, SIMULINK blocks are in *italic* font and graphic objects attributes begin with capital letter.

|, 32
\, 26, 36, 48
&, 32, 40, 377
*, 26, 36
.*, 26
+, 26, 36
-, 26, 36
', 8, 36
.', 36
^, 26, 36
.^, 26
/, 26, 36
./, 26
..., 9
:, 17, 18
;, 8, 9
<, 31
<=, 31
==, 31
>, 31
>=, 31
[], 16
~, 32
~=, 31
2-D Look-Up Table block, 144, 225

abcdchk, 343
abs, 27
Abs block, 142, 222
acker, 314

acos, 29
acosh, 29
acot, 29
acoth, 29
acsc, 29
acsch, 29
Adams method, 158, 200
algebraic loop, 159
all, 40, 41
and, 32
angle, 27
ans, 16, 25
any, 40, 41
append, 269
are, 346
asec, 29
asech, 29
asin, 29
asinh, 29
AspectRatio, 356
atan, 29
atan2, 29
atanh, 29
augstate, 270
Auto-Scale Graph block, 140, 215
axes, 349, 351, 355
axis, 73, 349, 356

BackgroundColor, 364
Backlash block, 223

balreal, 264
Band-Limited White Noise block, 213
bar, 74
basis
    of the null space, 42
    of the span, 42
block
    build up, 180, 203
    properties, 134
bode, 290
Box, 357
break, 92
ButtonDownFcn, 380

c2d, 248, 329
c2dm, 249
CallBack, 365, 377
canon, 244
canonical form, 247, 272
cardioid, 89, 337
cart2pol, 30
cart2sph, 30
Cartesian to polar block, 230
Cartesian to spherical block, 230
caxis, 79, 356
Cayley–Hamilton, 59
CData, 360, 361
cdf2rdf, 45
ceil, 26
characters, 357
Check Box, 363
Checked, 377
Children, 350, 353, 357
Chirp Signal block, 213
chol, 43
clear, 8, 99, 103
CLim, 356
CLimMode, 356
Clipping, 353, 359
clock, 25
Clock block, 141, 209
Clock Logic block, 232
cloop, 268
close, 66

color, 67, 79, 354, 361
    attribute, 356
    in SIMULINK, 135
    maps, 79
    shading, 80
colormap, 79, 354
Colormap, attribute, 354
ColorOrder, 356
Combinatorial Logic block, 143, 191,
    227
compan, 50
computer, 25
cond, 42
conj, 27
Constant block, 212
continuing a line, 9
contour, 77
controllability, 244, 254
controllers
    lead–lag, 295, 303
    standard, 287, 311
    state feedback, 314
conv, 58
conversion
    among representations, 242
    continuous to discrete, 201, 247,
        252
    discrete to continuous, 328
cos, 29
cosh, 29
cot, 29
coth, 29
Coulombic Friction block, 226
covar, 260
cputime, 25
Cross correlator block, 232
csc, 29
csch, 29
ctrb, 255
ctrbf, 256
CurrentAxes, 355
CurrentFigure, 353
CurrentObject, 354
CurrentPoint, 354, 379

cycloid, 88
cylinder, 84

D flip-flop block, 231
d2c, 328
d2cm, 328
damp, 258
damping ratio, 258, 308, 329, 337
date, 25
dbalreal, 331
dbode, 335
dcgain, 259
dcovar, 329
ddamp, 329
ddcgain, 331
Dead Zone block, 223
debugger, 100
    dbclear, 103
    dbcont, 102
    dbdown, 103
    dbquit, 103
    dbstack, 103
    dbstatus, 103
    dbstep, 102
    dbstop, 101
    dbtype, 101
    dbup, 102
decomposition
    partial fraction, 60
    Schur, 47
    singular value, 46, 291
deconv, 58
Demux block, 171, 230
Derivative block, 220
destim, 339
det, 38
dexresp, 343
dfrqint, 344
dfrqint2, 345
dgram, 331
diag, 19
diary, 93, 353
Diary, attribute, 353
DiaryFile, 353

differential equations, 157, 158, 198, 200
Digital Clock block, 188, 212
digital control, 339, 341
dimpulse, 332
dinitial, 332
Diophantine equation, 96
Dis. State-space block, 217
Dis. Transfer Fcn block, 218
Dis. Zero-Pole block, 217
Dis. Integrator block, 219
Dis. Limited Integrator block, 219
Dis. regulator block, 233
Dis. state estimator block, 232
dlinmod, 201
dlqe, 339
dlqew, 339
dlqr, 340
dlqry, 340
dlsim, 332
dlyap, 346
dmodred, 331
dmulresp, 344
dnichols, 336
dnyquist, 336
DOS Commands, 13
dot operators, 26
dreg, 340
drmodel, 343
dsigma, 337
dsort, 331
dtimvec, 344

echo, 99, 353
Echo, attribute, 353
EdgeColor, 360
Editable Text, 364
eig, 45
eigenvalues and eigenvectors, 45
eps, 25
equilibrium points, 202
EraseMode, 359–361
esort, 260
estim, 323

Euler method, 158, 200
eval, 95
exist, 41
exit, 7
exp, 29, 35
expm, 39
expm1, 40
exresp, 344
eye, 19

FaceColor, 360
factorization
    Cholesky, 43
    LU, 43
    QR, 44
    spectral, 45
False, 32
fbode, 345
fclose, 12
Fcn block, 144, 171, 223
feedback, 268, 313, 324
feedback, 268
Feedback gain block, 232
figure, 66, 349, 351, 354
file .M, 93
Filter block, 217
find, 40, 41
findstr, 20
finite, 41
First-Order Hold block, 218
first-order hold, 251
fix, 26
fliplr, 21
flipud, 21
floor, 26
flops, 25
Font, 135
FontAngle, 357, 361
FontName, 357, 361
FontSize, 357, 361
FontStrikeThrough, 357, 361
FontUnderline, 357, 361
FontWeight, 357, 361
fopen, 12

for, 28, 92
ForegroundColor, 364
format, 10
Format, attribute, 353
FormatSpacing, 353
fplot, 74
Frame, 364
freqint, 345
freqint2, 345
freqresp, 345
From File block, 142, 171, 211
From Workspace block, 141, 161, 171,
    210
fscanf, 12
function, 94
funm, 40

Gain block, 220
gain margin, 298
Gaussian elimination, 43
gca, 350
gcd, 27
gcf, 350
Gear method, 158, 200
get, 350
global, 99
gram, 257
Graph block, 139, 215
graphic
    objects, 349
    pointers, 350
grid, 68, 357
GridLineStyle, 357
group, 180

hadamard, 50
hankel, 50
help, 13
help, creation of, 95
hess, 47
hilb, 50
hist, 74
Hit Crossing block, 159, 160, 216
hold, 71, 72, 356

HorizontalAlignment, 361, 364
Householder transformation, 44

i, 25, 28
if...else, 92
imag, 27
image, 352, 361
imaginary unity, 25, 28
impulse, 283
Inf, 25
initial, 280
inner product, 26
Inner Product block, 222
Inport block, 198, 229
input, 99
int2str, 20
integration method, 99, 158, 200
Integrator block, 221
interpolation, 61
    interp1, 62
    interp2, 62
    interpft, 62
Interruptible, 352, 366
inv, 36
InvertHardcopy, 355
invhilb, 50
isempty, 41
isieee, 41
isinf, 41
isnan, 41
isstr, 41

j, 25, 28
JK flip-flop block, 232

Kalman filter, 321, 339
Kalman form, 255, 257
keyboard, 100

Label, 376
Latch block, 231
lcm, 27
least square, 48
legend, 72
length, 16

Limited Integrator block, 228
line, 349, 352, 358
line and point type, 67
linearization, 200
LineStyle, 360
LineWidth, 357, 359, 360
linmod, 200
LinSim method, 158, 200
linspace, 17
load, 9
log, 29
log10, 29
log2, 29
Logical Operator block, 142, 226
loglog, 74
logm, 39
logspace, 17, 291
Look Up Table block, 144, 224
lqe, 321
lqe2, 323
lqed, 323
lqew, 323
lqr, 318
lqr2, 321
lqrd, 320
lqry, 320
lsim, 284
ltifr, 345
ltitr, 344
lu, 43
lyap, 345
Lyapunov equation, 345

magic, 50
margin, 298
MarkerSize, 359, 360
mask, 203
MATLAB Fcn block, 145, 173, 225
Matrix Gain block, 221
matrix,
    3D, 96
    condition number, 42
    echelon, 44
    Hessenberg, 47

max, 37
Max, attribute, 365
mean, 37
Memory block, 159, 228
MenuBar, 354
mesh, 77, 360
meshc, 78
meshgrid, 77
MeshStyle, 360
meshz, 78
mex-file, 7
min, 37
Min, attribute, 365
minreal, 265
model reduction, 261, 264, 331
modred, 263
mouse, use in SIMULINK, 131
mulresp, 344
multi-selection in SIMULINK, 134
multi-variable schemes, 167
Mux block, 171, 229

NaN, 25
nargin, 25, 95
nargout, 25, 95
natural frequency, 258, 308, 329, 337
NextPlot, 356
ngrid, 295
nichols, 294, 299
nnls, 49
norm, 38
not, 32
null, 42
num2str, 20, 96
NumberTitle, 354
nyquist, 292, 298
Nyquist theorem, 292, 302

observability, 257
observer, 315
obsv, 257, 316
obsvf, 257
ode23, 99
ode45, 99

ones, 19
optimal control, 317, 340
or, 32
ord2, 343
Orientation, 134
orth, 42
Outport block, 198, 229

pack, 14
pade, 343
PaperPosition, 355
PaperSize, 355
PaperType, 355
PaperUnits, 355
parallel, 267
Parent, 350, 353
partial fractions, 241
pascal, 50
patch, 352
pause, 99
pcolor, 361, 376
phase margin, 295, 298
pi, 25
PID Controller block, 232
PID approx. derivative block, 232
PID with anti-windup block, 232
pinv, 48
place, 314, 316
plane rotation, 44
plot, 66, 71, 349, 351, 358
plot3, 83, 358
Pointer, 354
PointerLocation, 354
PointerWindow, 353
pol2cart, 30
polar, 74
Polar to Cartesian block, 230
poly, 38, 58
polyder, 59
polyfit, 61
polynomial
    characteristic, 58
    Chebyshev, 88
    Legendre, 65

minimal, 63
representation, 58
roots, 59
polyval, 59
polyvalm, 59
pop-up menu, 364
Position, 355, 361, 364, 368, 377
pow2, 29
Power spectral density block, 232
print, 85
Product block, 142, 222
pseudo-inverse, 48
Pulse Generator block, 212
push button, 363
pzmap, 261, 306

qr, 45
Quantizer block, 226
quit, 7

radio button, 363
rand, 50, 74
randn, 50
Random Number block, 211
rank, 38, 95
rat, 26
Rate Limiter block, 143, 225
rats, 26
rcond, 42
real, 27
realmax, 25
realmin, 25
reg, 324
Relational Operator block, 142, 227
Relay block, 143, 225
rem, 26, 32
Repeating Sequence block, 190, 212
reroute line, 133
Reset Integrator block, 227
reshape, 21
residue, 60, 278, 333
Resize, 355
return, 100
Riccati equation, 317, 321, 339, 346

rlocfind, 308
rlocus, 307
rmodel, 343
root, 351, 353
root locus, 306, 310, 337
roots, 59
rose, 74
rot90, 21
Rotation, 361
round, 26
rref, 44
rsf2csf, 47
Runge-Kutta method, 158, 200

S-domain to Z-domain block, 233
S-function block, 145
sample time colors, 194
sampling, 248
Saturation block, 142, 224
save, 9
scalar expansion, 170
schur, 47
Scope block, 127, 139, 213
ScreenSize, 354
script file, 93
sec, 29
sech, 29
semilogx, 74, 358
separator, 377
series, 266
set, 351
sgrid, 308
shading, 80
shadows, 134
sigma, 291
sign, 27
Sign block, 143, 229
Signal Generator block, 210
SIMULINK schemes, 125, 131
sin, 29
Sine Wave block, 188, 210
singular values, 46, 291
sinh, 29
size, 16

Slide Gain block, 222
Slider, 364
sort, 37
Spectrum analyzer block, 232
sph2cart, 30
Spherical to Cartesian block, 230
spline, 62
sqrt, 26
sqrtm, 39
SR flip-flop block, 232
ss2ss, 243
ss2tf, 246
ss2zp, 246
ssdelete, 262
ssselect, 263
stability, 292, 310
stairs, 74
State Estimator block, 232
State-Space block, 221
step, 281
Step Input block, 211
STOP block, 162, 216
str2mat, 20
str2num, 20
strcmp, 20
string, 20
String, attribute, 364
Style, 363
subplot, 75, 349
sum, 37
Sum block, 219
surf, 77, 360
surface, 360
Surface, attribute, 352
surfc, 78
surfl, 360
svd, 46
Switch block, 144, 224

tan, 29
tanh, 29
text
    in SIMULINK, 132
    object, 357

static, 364
    text, 68, 352, 361
tf2ss, 246
tf2zp, 247
tfchk, 343
TickDir, 357
TickLength, 357
timvec, 344
title, 67
Title, attribute, 357, 361
To File block, 141, 171, 214
To Workspace block, 140, 171, 214
toeplitz, 50
trace, 38, 94
Transfer Fcn block, 220
transmission zeros, 261
Transport Delay block, 162, 228
transpose, 8, 36
triangularization, 43
tril, 50
trim, 202
triu, 50
Type, 351, 353
tzero, 261
tzreduce, 343

uicontrol, 352, 363
uimenu, 352, 376
Unit Delay block, 217
Units, 353–355, 361, 364
update diagram, 134, 194
UserData, 353, 370

Value, 365
vander, 50
Var. Transp. Delay block, 162, 228
variables,
    global, 99
    local, 94, 102
VerticalAlignment, 361
view, 76, 357
View, attribute, 357
Visible, 353

while, 92

who, 16, 102
whos, 16
wilkinson, 50
WindowButtonDownFcn, 379
WindowButtonMotionFcn, 379
WindowButtonUpFcn, 379

XColor, 358
XData, 359–361
XDir, 358
XGrid, 358
XLabel, 358
xlabel, 67
Xlabel, attribute, 361
XLim, 358
XLimMode, 358
xor, 32
XScale, 358
XTick, 358
XTickLabels, 358
XY Graph block, 140, 215

YData, 359–361
ylabel, 67
Ylabel, attribute, 361

ZData, 359, 360
zero order hold, 250
Zero Order Hold block, 187, 218
Zero-Pole block, 221
zeros, 19
zeros of a function, 113
zgrid, 337
zoom, 73
zp2ss, 247
zp2tf, 247